Trust and Artificial Intelligence

Trust and Artificial Intelligence: Development and Application of AI Technology explores the crucial role of trust in the development and application of artificial intelligence (AI) technology. The book discusses the challenges and opportunities associated with building trust in AI systems and highlights the importance of transparency, accountability, and ethics in creating trustworthy AI. Drawing on the latest research and case studies, the book provides valuable insights and practical strategies for building trust in AI that can be applied by developers, policymakers, and end-users. It is a must-read for anyone interested in the intersection of technology and society and the future of artificial intelligence.

Across its two distinct sections, the book delves deep into both theoretical frameworks and real-world applications. Section I, "Trust in Artificial Intelligence Technology," comprises 12 insightful chapters, each shedding light on different aspects of trust in AI. From ethical considerations and the credibility of AI systems to the intricacies of blockchain technology and digital therapists, the book offers a kaleidoscope of perspectives, showcasing how trust shapes and is shaped by AI advancements. Section II, "Trust in Artificial Intelligence Technology Applications," extends the discourse to practical implications and case studies. With 12 additional chapters, it scrutinizes the impact of AI on diverse sectors such as healthcare, agriculture, the labor market, and online shopping. It contemplates the trust dynamics in neural networks, public sector AI, and the burgeoning field of last-mile logistics.

The book is more than just an academic text; it is a vital conversation starter in the ever-evolving discourse of AI. It challenges us to rethink our relationship with technology, underlining the critical role of trust in harnessing the full potential of AI for a better, more efficient, and ethically sound future.

Joanna Paliszkiewicz is a Professor and the Director of the Management Institute at the Warsaw University of Life Sciences, Poland.

Jerzy Gołuchowski is a Professor and Dean of the Faculty of Informatics and Communication at the University of Economics in Katowice, Poland.

Routledge Studies in Trust Research
Series editors: Joanna Paliszkiewicz and Kuanchin Chen

Available Titles in this Series:

Trust and Digital Business
Theory and Practice
Edited by Joanna Paliszkiewicz, Kuanchin Chen and Markus Launer

Trust, Power and Public Sector Leadership
A Relational Approach
Steen Vallentin

Artificial Intelligence, Management, and Trust
Edited by Mariusz Sołtysik, Magda Gawłowska, Bartłomiej Śnieżyński and Artur Gunia

Privacy, Trust and Social Media
Edited by Joanna Paliszkiewicz, Kuanchin Chen and Jerzy Gołuchowski

Communication, Leadership and Trust in Organizations
Edited by Joanna Paliszkiewicz and Jose L. Guerrero Cusumano

Trust and Organisational Resilience
Iwona Otola, Marlena Grabowska and Zoran Krupka

Trust and Artificial Intelligence
Development and Application of AI Technology
Edited by Joanna Paliszkiewicz and Jerzy Gołuchowski

Trust in Social and Business Relations
Theory and Practice
Edited by Joanna Paliszkiewicz, Kuanchin Chen and Marta Mendel

Trust and Artificial Intelligence

Development and Application of AI Technology

Edited by
Joanna Paliszkiewicz
and Jerzy Gołuchowski

Routledge
Taylor & Francis Group
NEW YORK AND LONDON

First published 2025
by Routledge
605 Third Avenue, New York, NY 10158

and by Routledge
4 Park Square, Milton Park, Abingdon, Oxon, OX14 4RN

Routledge is an imprint of the Taylor & Francis Group, an informa business

ISBN: 978-1-032-62632-1 (hbk)
ISBN: 978-1-032-62722-9 (pbk)
ISBN: 978-1-032-62723-6 (ebk)

DOI: 10.4324/9781032627236

Typeset in Times New Roman
by codeMantra

I dedicate the book to my loved ones
JP
To my grandchildren Alicja, Jan, and Paweł
JG

Contents

Figures

Tables

Biographical notes

Editors

Joanna Paliszkiewicz works as a Full Professor at the Warsaw University of Life Sciences (WULS-SGGW). She is the Director of the Management Institute. She is also an Adjunct Professor at the University of Vaasa in Finland. She obtained academic title "Full Professor" from International School for Social and Business Studies in Slovenia. She is well recognized in Poland and abroad for her expertise in management issues: knowledge management and trust management. She has published over 220 papers/manuscripts and is the author/co-author/editor of 18 books.

She has been a part of many scholarship endeavors in the United States, Ireland, Slovakia, Taiwan, the United Kingdom, and Hungary. She has actively participated in presenting research results at various international conferences. Currently, she serves as the Deputy Editor-in-Chief of the *Management and Production Engineering Review*. She is an Associate Editor for the *Journal of Computer Information Systems, Expert System with Applications, Issues in Information Systems, Przegląd Organizacji, Intelligent Systems with Applications*. She is the Vice President of the Polish Association for Production Engineering. She also serves as chair of the International Cooperation in European Business Club. She serves as the Vice President of the International Association for Computer Information Systems in the United States. She is a board member of the Intellectual Capital Accreditation Association. She has successfully supervised many PhD students leading them to completion of their degrees. She has also served as an external reviewer for several PhD students in Poland, India, and Finland. She is actively involved in participating in the scientific committees of many international conferences. She was named the 2013 Computer Educator of the Year by IACIS in the United States.

Warsaw University of Life Sciences webpage: www.sggw.edu.pl

Joanna Paliszkiewicz website: https://paliszkiewicz.pl

Jerzy Stanisław Gołuchowski is a Full Professor and is associated with the University of Economics in Katowice. He served as Vice-Rector (2005–2009) and Dean of the Faculty of Information Technology and Communications (2009–2019), and Head of the Department of Communication Design and Analysis. His scientific interests are concentrated around communication issues in decision-making processes, artificial intelligence, and management of organizations, particularly the management of trust, knowledge, and communication in organizations. Selected publications: B. Filipczyk: J Gołuchowski: Cyfrowa komunikacja organizacji (Wydawnictwo UE Katowice 2020), Jay Liebowitz, Joanna Paliszkiewicz, Jerzy Gołuchowski (eds): *Intuition, Trust, and Analytics*. Boca Raton: CRC Press Taylor & Francis Group, (2018). He has participated in international projects, e.g., "European cooperation project on disinformation and fact-checking training" (2019–2022); Language Communication on eCommerce websites (2004–2007); European Master in e-Commerce and Cyber communication (2000–2003). He is a member of the Society for Social Communication and Scientific Society for Business Informatics. He has also served as President of the Scientific Society for Business Informatics and as a member of the Council of Scientific Societies of the Polish Academy of Sciences. From 2012 to 2016, he was also a member of the Programme Council of TVP o/Katowice and, from 2014 to 2016, a member of the Supervisory Board of the Polish Radio Katowice.

University of Economics in Katowice website: https://www.ue.katowice.pl/en.html

Contributors

Caner Asbaş is founder and CEO of Hamle RF Elektrik ve Elektronik A.Ş. which is a company operating in defense, telecommunication, automotive, and avionics areas. He received his undergraduate and MSc degrees from İhsan Doğramacı Bilkent University in Electrical and Electronics Engineering. He also has undergraduate degrees in International Relations from Anadolu University, Sociology from Anadolu University, and Public Finance from Anadolu University. He is proficient in C, C++, C#, Java, MSSQL, and Python programming languages. Asbaş continues his PhD education in Management at Atılım University. His research focuses on social networks and aimplementation of artificial intelligence methods to social networks.

Tetiana Balanovska is PhD in Economics, Professor, is Head of the Department of Management named after Prof. J.S. Zavadskyi, National University of Life and Environmental Sciences of Ukraine, Kyiv, Ukraine. Her scientific and research activity focuses on such issues as problems of theory and practice of management and marketing, topical issues of anti-crisis management of enterprises and human resource management, and entrepreneurship. She is the author of over 200 publications, including the co-author of eight monographs and 21 textbooks.

Marta Barna (Dr Hab, Full Professor) holds PhD and habilitation in Economics. She is a Scientific Researcher and Lecturer in tour operating, Vice-Rector of the Lviv University of Trade and Economics – LUTE, Ukraine. Her research focuses on various factors of sustainable tourism development. Marta participated in three research projects and coordinated a teaching project funded by the European Union. She published more than 180 scientific and educational works. She is a member of the editorial board of the foreign peer-reviewed scientific publication *Acta Scientiarum Polonorum. Oeconomia* (SGGW, Warsaw, Republic of Poland). She is the Head of the scientific topic: "Formation of theoretical and methodological support for the development and functioning of the tourism industry of Ukraine in terms of European integration." State registration number 0117U002092 Associate member of the NGO "National Tourist Organization of Ukraine". She has received the Ivan Vyhovsky Award from the Center for Eastern European Studies at the University of Warsaw and the College of Eastern Europe Foundation under the Honorary Patronage of the President of the Republic of Poland and in 2018 completed a four-month internship in Poland. She is an expert (member) of the focus group "Tourism" on the development of the Development Strategy of Lviv region for the period up to 2027 and the Action Plan for its implementation for 2021–2023.

Agata Berdowska is a Researcher and Lecturer at the University of Economics in Katowice. She started her work in the Department of Knowledge Engineering. Since 2017, she has been a member of the Department of Communication Design and Analysis. Her subject of scientific interest is concentrated around new information technologies: technologies supporting fact checking, conversation systems (chatbots) and electronic training (e-learning), and the labor market, new ICT professions. She has participated in two international projects: *European Cooperation Project on Disinformation and Fact-Checking Training* (2019–2022), the *CALYPSO- Collaborative Fact-Checking* (2021–2022).

Michał Borowy (PhD) is an economist and expert in industry innovation and experienced project manager with proven track record of working in the research sector. He is skilled in R&D business relationship management, change management, planning, and innovation management. Within 15 years of professional experience, Michal's key achievements relate to technology transfer in life sciences, Spin-off Academy coordination at the University of Warsaw, and mentoring in acceleration programs, like Digital Health, EIT Manufacturing, or EIT Digital. He acts as an Innovation Coach for industry sector within program of Polish Ministry of Funds and Regional Development. He is a member of Board at Polish Business and Innovation Centres Association. He developed, supervised, and managed projects related to: business & science cooperation, technology audits, researches, business offers, and implementation works. He has managed small and EUR 1MM+ projects.

Marcin Choiński is an Assistant Professor at the Warsaw University of Life Sciences. He holds a Master's degree in Mathematics, an Engineer's degree in Computer Science, a Bachelor's degree in Medical Physics, and a Doctoral degree in Mathematics. His interests lie in mathematical modeling of tuberculosis dynamics and combined therapy for HIV.

Fatih Çetin is a Professor of Management in the Department of Business Administration at Baskent University in Turkey. His research focuses on the field of positive organizational behavior and knowledge management studies. He has authored over 100 publications including journal articles, book chapters, and proceedings. His recent research works have appeared in the *Journal of Computer Information Systems*, *International Journal of Intercultural Relations*, *The Journal of General Psychology*, *Management Research Review*, *Journal for East European Management Studies*, and the 79th Annual Meeting of the Academy of Management.

Sorin Dan is an Assistant Professor of Public Management in the School of Management at the University of Vaasa. His expertise includes public management reform and innovation and digital talent management. He is the co-author of *Digital Talent Management: Insights from the Information Technology and Communication Industry* (Palgrave Macmillan, 2021) and his research has appeared in journals such as *Public Management Review* and *Public Performance & Management Review*.

Teresa Gądek-Hawlena earned her PhD from the Department of Logistics, Warsaw University of Life Sciences (SGGW). Earlier, she completed her PhD in the Department of Transport, University of Economics, Katowice. Her main research interests include security issues in transport and logistics. She has authored and co-authored numerous articles and papers published in journals and national and international conference proceedings.

Urszula Grzybowska received her PhD in Mathematics in 2000. She has been an Assistant Professor in the Faculty of Applied Informatics and Mathematics at the Warsaw University of Life Sciences since November 2000. Her key qualifications concern application of mathematical methods in economics, data analysis, and data mining. She is an author and co-author of more than 20 scientific publications, including two university textbooks.

Nataliya Horin is an Associate Professor in the Department of International Economic Relations, Faculty of International Relations, Ivan Franko National University of Lviv (IFNUL), Ukraine. Her research interests focus on international economics issues, including the international economic integration, foreign economic policy, eco-innovative development of the global economy, and competitiveness in global markets. She lectures on Trade Policy, Strategic Management, Organization of Foreign Trade Operations, International Logistics, and World Economy. She is the author and co-author of more than 70 scientific works, including five manuals and four monographs. During her career, she was a member of the two specialized academic council on thesis defense in specialty World Economy and International Economic Relations, and a scientific supervisor of many bachelor's and master's theses, and defended PhD dissertation in the field of the World Economy and International Economic Relations. Participant in several international postdoctoral fellowships: Lane Kirkland Scholarship Program, National Scholarship Program of the Government of the Slovak Republic, Erasmus Mundus Program, Erasmus+ Program. Coordinator for the Cooperation with the Department of Career Development and Business Cooperation at IFNUL. She is the reviewer of the *Scientific Journal Bulletin of the Lviv University*, Series in International Relations (Ukraine), the *Scientific Journal of the Warsaw University of Life Sciences* (SGGW), *Problems of World Agriculture* (Poland), the *Academic Scientific Journal of Geography, Politics and Society* (Poland).

Szymon Jarosz earned his Master's degree in Accounting and Controlling from the University of Economics in Krakow. Currently, he contributes to academia as a Research and Teaching Assistant at the Cracow University of Economics, specifically in the Department of Technology and Ecology of Goods. His academic journey was notably enriched by participating in the "Szkoła Orłów" Tutoring Programme. Demonstrating his leadership skills, Szymon served as the President of the Students' Scientific Association of Data Analysis in 2021. He has been recognized for his academic excellence, receiving the Best Student of the Institute of Management at the Cracow University of Economics award in 2020, and the prestigious Students' Nobel Prize in Economics in 2021. His main research interests focused on innovation, sustainable development, sustainable transformation of enterprises, and environmental management.

Anna Jasiulewicz is an Assistant Professor at the Institute of Management, Warsaw University of Life Sciences, Head of Postgraduate Public Procurement Studies, and a member of the Polish Scientific Society of Marketing. She is a researcher in international and national projects and author of several dozen scientific publications, organizer of international conferences and reviewer of scientific articles in national and international journals. She specializes in issues related to consumer behavior, digital marketing, and sports marketing.

Tomasz Jastrząb received his MSc (2011) and PhD (2019) in Computer Science from the Silesian University of Technology, Gliwice, Poland. His research interests focus on parallel computing, natural language processing, and finite automata. He is also a Java Developer and Team Leader at Technicenter Sp. z o.o., where he works on interdisciplinary projects related to economics, neuropsychology, and computer science.

Irena Jędrzejczyk is a Professor of Economic Science and has many years of experience in teaching, research, and administration. Her field of interest has been economy, business management, finances and investment. Irena Jędrzejczyk earned her MA and PhD from the University of Economics in Cracow and completed her Habilitation and her Professorship at the University of Economics in Katowice – Silesia. She spent many years as a Research Scholar at the University in Trier, Germany and at the University of Zurich, Switzerland. She was first appointed as the Head of the Educational Centre of Insurance and Risk Management, Coordinator and Project Manager of JEP No 11015 TEMPUS PHARE, Head of the Department of Insurances Market Studies at the University of Economics in Katowice – Silesia, Head of Branch of Finance Markets Studies at the University of Life Sciences, Warsaw, Coordinator and Project Manager of JEP No: DKS/DEF-III/POPT/04/185/08, and since 2017 Professor Irena Jędrzejczyk is working as a researcher and teacher in the Management Department, Faculty of Management and Transport, University of Bielsko-Biała in Silesia, Poland.

Jakub Kraciuk is a Professor at the Warsaw University of Life Sciences (SGGW) employed at the Institute of Economics and Finance of this university. His research interests focus on the issues of international economics, with particular emphasis on economic globalization and its effects, such as concentration processes taking place in the world economy, as well as the processes of fragmentation and relocation of production. In his research work, he also dealt with the activities of transnational corporations, financial crises and their impact on the world economy and the economies of individual countries, as well as the flows of goods, services and factors of production on an international scale. He is the author of many publications in this field, as well as the promoter of three defended doctoral theses and many bachelor's and master's theses in the field of Economics and Finance and Accounting. He is a member of the Editorial Committee of Scientific Journals of the Warsaw University of Life Sciences, Problems of World Agriculture. He was the Deputy Dean for Science and the Dean at the Faculty of Economics of the Warsaw University of Life Sciences. Currently, he is the Head of the Department of International Economics and Agribusiness. He was also the coordinator of the research of the members of the rural elite in Poland, carried out under the 6th Framework Program of the European Union, carried out by the Institute of Agricultural Economics at the University of Kiel.

Elżbieta Kacperska is an Adjunct at the Institute of Economics and Finance of the Warsaw University of Life Sciences. Her research interests focuses on international economics issues, including foreign direct investment, international migration, international management, European integration, and foreign trade. She is the author of many publications in this field, as well as the supervisor of many bachelor's and master's theses in the field of Economics as well as Finance and Accounting. She is a member of the Editorial Board of the *Scientific Journal of the Warsaw University of Life Sciences* (SGGW) Problems of World Agriculture. Coordinator of the Academy of Entrepreneurship at the Faculty of Economics at SGGW. She is the coordinator for the Promotion and Cooperation with High Schools of the Faculty of Economics at SGGW.

Johanna Kalliokoski is a doctoral student in Regional Studies in the School of Management at the University of Vaasa. She is conducting her doctoral research on inclusive innovation policies and urban development.

Krzysztof Karpio (PhD) is currently an Assistant Professor at the Institute of Information Technology, Warsaw University of Life Sciences. He is also the Head of Postgraduate Studies in Databases and postgraduate studies in Administration of Systems and Applications. He earned his Master of Science degree at Warsaw University in 1994. His PhD was awarded in High Energy Nuclear Physics from the Institute for Nuclear Studies (currently: National Centre for Nuclear Research). The main subject of his thesis was the study of correlations in high energy heavy ion collisions at SPS accelerator at CERN. The work utilized the preparation, cleaning, and analysis of huge amount of experimental data. After completing his PhD, Karpio accepted the position of Assistant Professor at Warsaw University of Life Sciences. He is the author of works in the fields of Physics, Information Technologies, and Economics. Currently, his main scientific interests are: modern database systems, data mining techniques, and economic data modeling. Karpio has 49 publications listed in the Web of Science, 2583 citing articles and an H-index of 20.

Marek Karwański is an Assistant Professor in the Faculty of Applied Informatics and Mathematics at the Warsaw University of Life Sciences. He conducts research in the field of data analysis. He specializes in solutions using both statistical and algorithmic models of artificial intelligence. His interests concern the methods and technologies of creating and using information systems in enterprise management. Many publications confirm his scientific activity.

Damian Kedziora (PhD) works as an Associate Professor in Software Engineering Department of LUT University in Lahti, Finland. He is also Docent of Business Process Automation Software at LUT University. His research concentrates on the low-code software, process automation, and commercial aspects of software design.

Aleksandra Konopka (MSc) is a Research and Teaching Assistant at the Institute of Information Technology, Warsaw University of Life Sciences – SGGW. In 2021, she earned a Master of Science degree in Computer Engineering with distinction. Her primary research focuses on the theoretical exploration and practical application of classical artificial intelligence methods, advanced neural networks, and digital image processing within biological context. Her research projects involve the analysis of soil bacteria classification using machine learning and image processing techniques, aimed at enhancing data analysis, with potential applications in various fields such as agriculture and horticulture.

Prof. Ryszard Kozera is currently a Director of The Institute of Information Technology at Warsaw University of Life Sciences – SGGW in Poland. He obtained a Habilitation Degree (Dr Sc) from Silesian University of Technology (Gliwice, Poland) in 2006, a PhD from Flinders University of South Australia (Adelaide, Australia) in 1991 and a MSc degree from Warsaw University (Warsaw, Poland) in 1985. Selected previous research positions include Associate Professor at The University of Western Australia, Perth, Australia 1991–2008 (current adjunct Associate Professor), and Professor at Warsaw University of Technology, Poland 2009–2012. In addition, he was awarded three times Alexander von Humboldt Research Fellowships (Technical University of Berlin 1996–1997; Christian Albrechts University of Kiel, 2000, 2004) in Germany. His research interests are artificial intelligence, optimization, interpolation and approximation, computer vision and image analysis, numerical analysis, neural computation and networks, partial differential equations, and in general applied mathematics and modeling processes in physics, engineering, biology, medical sciences, and agriculture. He published over 140 scientific papers in prestigious international journals, research monographs, and conference proceedings including two single authored papers exceeding 120 pages. He organized many international conferences, workshops, and sessions among all in Australia, Germany, Greece, Mexico, Poland and presented over 70 invited talks and conference presentations worldwide. Prof. R. Kozera served also as coeditor of conference proceedings, research monographs, journal, conference, and book reviewer as well as he has also won international grants and rewards. He supervised over 70 BSc, Honors, or MSc students and two PhD conferred students in Australia and Poland. He delivered lectures on over 30 different topics at the undergraduate, postgraduate, and PhD levels in Computer Science and Applied Mathematics.

Paulina Kramarz holds a PhD in Economics. She received her PhD at the Faculty of Economic Sciences of the Warsaw University of Life Sciences. Her specialization is focused on issues such as the economics of sustainable development, rural development, and health economics. Her scientific interests are related to the study of regions with specific conditions for economic development, the use of statistical and econometric methods in the study of regional differentiation of socio-economic development, the relationship between the level of the population's health resource and the directions of economic development. Currently, she is employed in the Department of Economics and Organization of Enterprises – Institute of Economics and Finance, Warsaw University of Life Sciences.

Marzena Lemanowicz is an Assistant Professor at the Institute of Economics and Finance, Warsaw University of Life Sciences, member of Polish Scientific Society of Marketing, Polish Association of Agricultural and Agribusiness Economists, member of the Board of Economics and Finance Discipline (WULS), former head of the Division of Regional and European Studies. She is a researcher in international and national projects. She is the author and reviewer of many scientific publications. She is active participant in international scientific conferences. She is a member of the editorial committee and scientific council in international journals. Her research interests focus on marketing, consumer behavior in the food and tourism market, consumer trends, and integration processes in agribusiness.

Anna Losa (PhD in Management) is an Assistant Professor in the Department of Communication Design and Analysis of the University of Economics in Katowice. Her research interests focus on the issues of communication management, trust and risk in communication, as well as ESG and sustainability disclosures. She is the author of scientific publications, articles, and book chapters. She also has been a reviewer for international journals and publications. At the university, Anna Losa conducts courses in the field of social responsibility of organizations, communication management in crisis situations, and creating the image of companies and institutions. She also serves as a certified academic tutor for students. In 2015–2020, she worked as a member of the Sustainable Development and Corporate Social Responsibility Advisory Board at the Polish Ministry of Development Funds and Regional Policy. Anna Losa is also a member of the European Risk Research Network Committee and a member of The Research Team of Centre for Economics of Climate Change.

Katarzyna Łukasiewicz (PhD) work as Assistant Professor at the Institute of Management at the Warsaw University of Life Sciences. Her research interests focus on tourism, management, quality management, trust, and customer relationship management. She is author of many publications in this field, as well as the supervisor of many bachelor's and master's theses in the field of Management and Tourism and Recreation (two of them awarded with the prize of the Minister of Sport and Tourism in 2009 and 2013). She is secretary of the Scientific Discipline Board for Management and Quality Sciences.

Piotr Łukasiewicz, is currently an assistant professor at the Institute of Information Technology, Warsaw University of Life Sciences. He earned his Master of Science degree in Mathematics at Warsaw University. He awarded PhD in Economics at the Warsaw University of Life Sciences in 2006. His thesis concerned the modeling of households' income distributions. After completing his PhD, Łukasiewicz accepted the position of Assistant Professor at the Warsaw University of Life Sciences. The main scientific interests are concentrated on (1) modeling of distributions of incomes and wages, and (2) studying of time series by utilizing modern database technologies: business intelligence and data mining techniques. Łukasiewicz is the author of tens of publications including 15 papers listed in the Web of Science, an H-index of 5.

Dorota Marquardt (PhD) is an Assistant Professor in the Department of Communication Design and Analysis at the University of Economic in Katowice. She conducts research in the field of computational linguistic, natural language processing, media discourse analysis, and new media. She is the author of many publications connected with linguistic factors in argumentative dialogs, new media, and identity discourse.

Konrad Michalski earned his PhD from the Department of Logistics (earlier at Management Institute), Warsaw University of Life Sciences (SGGW). He is a graduate of the University of Szczecin (PhD thesis in Economics on Logistics). He also completed postgraduate studies in social psychology and HRM. His main scientific interests include logistics, changes in processes of adjustment of organizational structures, and processes within an organization to requirements of innovative technologies. He has many years of experience in the business.

Artur Modliński (PhD) is the Head of Center for Artificial Intelligence and Cybercommunication Research, working also in Research Labs in Portugal, Italy, and Latvia. Artur is interested in using such technologies as artificial intelligence, robotic process automation, and metaverse for business development.

Marco Papa (PhD in Business Administration) is an Associate Professor in Accounting at the University of Aldo Moro Bari. He was visiting researcher at the University of Glasgow. His research interests lie in the areas of corporate disclosure, business valuation, and international accounting. He has published books and articles on risk reporting and target price run-ups in M&A. Marco is currently a board member of the BAFA Financial Reporting and Accounting Special Interest Group. He is also member of the European Risk Research Network Committee.

Piotr Pietrzak (PhD) is an Adjunct Professor at the Management Institute, Warsaw University of Life Sciences (SGGW), Poland. His research interests include several topics like public management, efficiency measurement, strategic management, process management, sustainable development, and digital management. He is the author of over 90 articles and four books. He has published work in journals such as *International Journal of Innovation and Learning*, *Human Systems Management*, *Foundations of Management*, and *Energies*. Dr Pietrzak is currently the Head of the Rector's Rankings Committee.

Wojciech Piotrowicz (PhD, Brunel, UK; MA Gdańsk, PL; PGDipLATHE Oxon, UK) is an Associate Professor in Supply Chain Management and Social Responsibility at the Hanken School of Economics. His current research is related to supply chain management, including humanitarian logistics, resilience disaster preparedness, and security.

Wojciech Pizło is an Associate Professor at the Institute of Management of the Warsaw University of Life Sciences (WULS). Previously, he was a Professor at the Institute of Management at the Jan Kochanowski University in Kielce (2012–2018) and the Institute of Economics and Finance (2014–2021) of the Warsaw University of Life Sciences. He completed one-year studies in Germany in the field of management. His main interests relate to management in the era of information technology revolution, in particular the theory of marketing, institutional economics, leadership, and creativity. Professor Pizło is the author of, among others, the monographs *Enterprise in the Information Society in the Light of the Theory of Institutional Economics* (2009), *Marketing. Economic Knowledge and Market Activity* (2017), *International Marketing – European Point of View* (2019), as well as the author and co-author of the work on virtual organizations in international research teams: Pizło W., Parzonko A., *Virtual Organizations and Trust in: Trust, Organizations and the Digital Economy. Theory and Practice*. Paliszkiewicz J., Chen K., (eds.), 2022, Taylor & Francis Group, pp. 61–78, and Pizło W., *The Role of Trust and Cooperation in Virtual Teams in: Trust and Digital Business: Theory and Practice*. Paliszkiewicz J., Chen Kuanchin, Launer Markus (eds.), 2023, Routledge, pp. 47–64. Author of 140 articles and chapters in monographs, scientific, and thematic editor in scientific journals.

Aleksandra Popiel is a PhD student in Management and Quality Studies at the University of Economics in Katowice, where she teaches copywriting, and

AI in communication. She conducts research in the field of trust in chatbots, AI, and social media, focusing on the generational cohort theory. She is the author of several publications on trust, such as *What Should We Learn About Trust in Chatbots?*, *Trust Factors in Chatbots Used in Organization's Communication with a Client*, *User's Trust, and Attachment to Social Media after Crisis: The Case of Facebook*.

Joanna Rakowska (Dr Hab, Associate Professor) holds PhD and habilitation in Economics. She is a Scientific Researcher and Lecturer in Local and Regional Development at the Institute of Economics and Finance, Warsaw University of Life Sciences – SGGW, Poland. Her research focuses on different factors of local and regional sustainable development and their interconnectedness. Joanna participated in eight research projects and coordinated a teaching project funded by the European Union. She published four books and over 100 papers in international peer-reviewed journals and conference proceedings. She has served as a visiting scholar at, among others, the Scotland's Rural College, the United Kingdom; the University of East London, the United Kingdom; and Agroscope, the Federal Office for Agriculture, Switzerland. She is the Head of the European and Regional Studies Division at WULS-SGGW.

Henryk Runowski (Dr Hab) is a Professor of Economics. He specializes in the issues of agricultural economics, in particular in the field of organization and economics of enterprises and farms. He is the Head of the Department of Economics and Organization of Enterprises at the Institute of Economics and Finance of the Warsaw University of Life Sciences. His main scientific interests include adjustment processes in farms, issues related to the measurement, and assessment of agricultural income and the dissemination of progress and innovation in agriculture. He conducted dozens of research and training projects for farmers, financed by the European Union and the state budget. He was an expert of the Senate of the Republic of Poland, president of the Association of Agricultural and Agribusiness Economists. Currently, he is the President of the Scientific Council of the Institute of Rural Development and Agriculture of the Polish Academy of Sciences, a member of the Scientific Council of the Agricultural and Food Economics and a member of the Economic Sciences Committee of the Polish Academy of Sciences. He is the author or co-author of about 400 publications, including monographs. He has implemented several dozen projects of enterprise restructuring, recovery programs, and business plans. The main hobby is the restoration of rural monuments.

Khuram Shahzad is an Assistant Professor of Digital Innovation Management in the School of Technology and Innovations and the Innovation and Entrepreneurship InnoLab research platform at the University of Vaasa. His research is focused on digital innovation management and blockchain technology and has been published in several outlets including *Industrial Marketing Management* and *International Business Review*.

Edyta Skarzyńska is a PhD student in the Doctoral School of the Warsaw University of Life Sciences. She graduated in management, writing her master's thesis on knowledge management. Her areas of interest are intellectual capital of individuals, trust, and social media.

Kari Smolander (PhD) is a Professor of Software Engineering in the School of Engineering Science, LUT University. He has published more than 250 refereed research papers in international journals and conferences, especially on software and systems development practices.

Mariusz Sołtysik is a University Professor at UEK, College of Management and Quality Sciences at the University of Economics in Krakow. He is also the Vice-Chairman of the Małopolska Regional Group of the International Project Management Association.

H. Cenk Sözen (PhD) is a Professor of Management at Baskent University, Faculty of Economics and Administrative Sciences, Department of Management. Dr. Sozen has also been working as a Part-time Professor in Turkish Military Academy. His interests and research areas include organization theory, organizational change, social networks, social capital, complex systems, strategic HRM, and knowledge management. He lectures on organizational behavior, organization theory, human resources management, and strategic management. He has worked as a reviewer and published articles in journals such as *Journal of Management Inquiry*, *Personnel Review*, *European Management Review*, *Property Management*, and *International Journal of Human Resource Management*. He received books and book chapters on social network theory, organizational behavior, and organization theory. He is a member of EURAM (European Academy of Management) and AOM (Academy of Management).

Karol Struniawski (MSc) is a Research and Teaching Assistant at the Institute of Information Technology, Warsaw University of Life Sciences – SGGW. He received his Master of Science in Computer Engineering, earned with distinction, at the Warsaw University of Life Sciences – SGGW in 2021. His primary research focus centers around the application of the highly efficient machine learning method known as Extreme Learning Machine (ELM). This innovative approach holds significant potential for various fields, and in the current research, an emphasis is placed on leveraging ELM, metaheuristic algorithms, and image processing. The ELM is utilized in research to precise classify and identify soil bacteria species as well as other classical machine learning approaches and convolutional neural networks.

Rehan Syed (PhD) is a Lecturer in the School of Information Systems at Queensland University of Technology. He has published research works in top tier information system conferences and journals. Dr. Syed's interests are in the areas of socio-technical aspects of process automation, digital leadership, and business process management.

Agnieszka Tul-Krzyszczuk (PhD) is an Assistant Professor at the Institute of Management at the Warsaw University of Life Sciences, Poland. She is a consultant, trainer, and educator specializing in the organization and management of production and service enterprises in the food industry, particularly in the hotel and foodservice sectors. She conducts scientific research in the fields of management, marketing, innovation, and confidence in digitization. She actively participates in national and international grants and projects. Furthermore, she is the author of numerous scientific and popular science publications in both national and international conferences.

Şule Erdem Tuzlukaya is a Professor in the Faculty of Management, Department of Management, Atılım University. She completed her PhD in Management and Organization from Başkent University. She teaches classes in organization theory, organizational behavior and strategic management. Her research interests include social networks and cyber space-related topics.

Zygmunt Waśkowski is a Professor of Poznań University of Economics and Business, Director of Marketing Institute and Head of Marketing Strategies Department, Chairman of the Board of The Polish Scientific Society of Marketing. His scientific areas include Value for Customer, Experience Marketing, Marketing of the Non-Profit Organizations Marketing Strategies, Sport Marketing, Management of Sport Organisations. He is the author of over 120 publications on the above-mentioned issues.

Monika Wieczorek-Kosmala has a post-PhD in Finance. She is currently an Associate Professor at the Faculty of Finance, University of Economics in Katowice (Poland). Her expertise merges risk management and financial management, with a focus on the buffering role of financial slack, financial outcomes of business interruptions, and environmental risks. Her current research interests cover climate change risk mitigation and adaptation, climate risk reporting and challenges of green transition. She is a member of the Committee of European Risk Research Network. She gained professional experience in risk management and property insurance as an insurance broker.

Barbara Wyrzykowska (PhD in Economics) is a Research Fellow at the Institute of Management and a Lecturer at the Warsaw University of Life Sciences. Her scientific and research activities focus on issues related to human capital, sustainable development, social responsibility of organizations, and new business models in organizations. She is the author and co-author of over 100 scientific and popular papers. She actively participates in numerous national and international conferences where she presents her research results. She participates in various teams carrying out research projects. She is a reviewer of scientific articles in national and international journals.

Samuel Yaw Kusi is an Assistant Professor of Marketing in the School of Management, Faculty of Management, Law and Social Science, University of Bradford, UK. His research is focused on international marketing and entrepreneurial branding and has appeared in journals such as the *Journal of World Business* and *International Business Review*.

Andrew J. Zaliwski holds an MSc in Mathematics from M.C. Sklodowska University (1986) and a PhD in Computer Science from the University of Mining and Metallurgy (1991), both in Poland. He served in various academic roles in Poland and Ireland from 1984 to 2001, including at UMCS Lublin and the University of Economics in Krakow. From 2001 to 2006, he was an Information Systems Lecturer at the City University of New York, followed by a stint as a Business and IT Lecturer at Whitireia Polytechnic in Auckland, New Zealand (2008–2020). In 2020, he was one of the founders of Freeos DAO, and in 2023, he founded and is CEO of LearnLayers Ltd., a startup aimed at incorporating blockchain and AI into educational systems."

Katarzyna Zdanowicz-Cyganiak has a PhD in Humanities Sciences and is an Assistant Professor in the Department of Communication Design and Analysis at the University of Economics in Katowice, Poland. She is the author of two monographs: *Aliens. Reactivation – Sketches* (Katowice 2013) and *Who Is Afraid of Maria K.? Art and Exclusion* (Katowice 2005), and co-author of the monograph *In the World of Axiology, Politics and Culture. Students towards Media – Based on Empirical Research* (Katowice 2018). She was also the co-editor of the monograph *(CO)medy – Contexts of Medial Discourse* (Warsaw 2014). Her research interests include digital communication, media narratives, virtual communities, cyberpsychology, and storytelling in digital media. She is working on a monograph on online confessions and their function in digital communication.

Foreword

There are six years now since ***Routledge Studies in Trust Research*** book series was successfully launched, and part of this success should be granted to their industrious series co-editors: professors Joanna Paliszkiewicz and Kuanchin Chen. Counting ten titles to date, this book series is the result of editing work of 16 series authors as well as collective endeavor on trust-related knowledge of over 100 chapter authors across the world. And it is counting...

This youngest member of the ***Trust Research*** family (edited by professors Joanna Paliszkiewicz and Jerzy Gołuchowski) is the second dedicated to artificial intelligence (AI), specifically to trust-related applications of AI technology. The volume is well-balanced as theory-practice ratio as well as number of chapters: 12 chapters dealing with trust-related conceptual background of AI technology, and 12 chapters reporting a variety of AI applications in diverse sectors (agriculture to regional development, healthcare services to insurance industry), from micro- to macro-level, robots to logistics, online shopping to market forecasting.

Currently, the topic of AI-driven technology progress is so ardent and so organically linked to trust that governments and international bodies are deeply concerned: to develop or not a legal framework for AI regulation and watching, for both security reasons and protection against fake news; and, if so, *how comprehensive* should it be? Moreover, the rapid development of generative AI (GenAI) toward hypothetical AGI (artificial general intelligence) brings new questions and challenges. In this context, Elon Musk stated that we just witnessed the birth of an entity more intelligent than any human on this planet.

Last December, over 13,000 people attended in New Orleans at the 37th Conference on Neural Information Processing Systems (NeurIPS 2023). A large number of scholars and experts from advanced technology firms (Intel and Microsoft included) have discussed the future of machine learning and AI models. *The Economist* (January 20, 2024) reports that 17 (out of 23 participants asked) considered that *AI-generated media would eventually become undetectable.*

The significant body of knowledge coalesced in this novel book, genuinely recommended to the world top-professionals active in a multitude of AI-related applications, can be a positive attempt to answer all these questions and address the new challenges associated to AI.

It also has a distinct print: the influx of Polish experts in AI (since more than half of the papers are authored by AI professionals originating from this country) possibly is a weak signal anticipating an emerging *Polish pole* of research on AI.

Dr. Cezar Scarlat
Professor Emeritus, Doctoral School
National University of Science & Technology POLITEHNICA
Bucharest, Romania

Forewords, cont.

AI has transitioned from being a technical domain accessible only to computer scientists and those with technical training, to user-friendly tools in recent decades. Traditionally, the division between those who had access to AI and those who did not was based solely on technical expertise. However, the consumerization of AI in recent decades has expanded this divide to include factors such as usage, users, and specific use case scenarios. Consequently, AI is no longer solely a technical subject. Psychological and behavioral factors now play an equally significant role in determining usage patterns. Among these factors, trust has been identified as a key driver of satisfaction, intention to use, and ongoing engagement with technology.

The book's clear distinction between trust in technology itself and trust in its application is a clever way to illustrate the fundamental difference of how AI is used. On one hand, there are concerns regarding trust in the functionality of technology such as virtual assistants, digital therapy, AI's learning capabilities, and data storage. Since most users don't have access to the inner workings of AI applications, their assessment of the technology capabilities usually relies on interacting with its features. This experience-based assessments further shape user expectations and satisfaction.

The second section delves into various use case scenarios, including healthcare, non-financial disclosures, market forecasting, agriculture, online shopping, and logistics. Here, the focus extends beyond evaluating the core functionality of the tool to examining how this functionality is applied across different use cases and business domains. This perspective broadens the audience's understanding of AI not only in terms of its predefined capabilities but also through innovative applications. Viewing this section through the lens of generalizability reveals insights that transcend disciplinary, national, and technical-behavioral boundaries.

The research presented in this book is both timely and relevant to a wide range of audiences. It takes readers from foundational knowledge to practical technology applications. The carefully selected articles and the structured presentation vividly showcase current research in a thematic manner, which will capture the interest of both academics and practitioners alike.

Kuanchin Chen, John W. Snyder Fellow
Director of Center for Business Analytics
Professor of Computer Information Systems and Business Analytics
Western Michigan University, U.S.A.

Preface

In an era where artificial intelligence (AI) is not just a distant possibility but a present reality, the dialogue around trust in AI has become increasingly vital. The development and application of AI technology pose unique challenges and opportunities for individuals, organizations, and societies at large. This book, "Trust and Artificial Intelligence: Development and Application of AI Technology," aims to delve deep into this conversation, exploring the multifaceted relationship between trust and AI. The book discusses the challenges and opportunities associated with building trust in AI systems and highlights the importance of transparency, accountability, and ethics in creating trustworthy AI. Drawing on the latest research and case studies, the book provides valuable insights and practical strategies for building trust in AI that can be applied by developers, policymakers, and end-users.

The purpose of this book is to bring together the theory and practice of trust in AI technology and its applications in different areas.

This book is structured in two parts: the first focuses on the conceptual underpinnings of trust in AI technology, while the second part explores practical applications and case studies. Through this structure, we aim to provide a comprehensive overview that is both theoretically rich and pragmatically relevant.

Section I is entitled "Trust in Artificial Intelligence Technology" and consists of 12 chapters. The aim of Chapter 1 is to describe the future directions of trust in AI. Chapter 2 focuses on blockchain technology, leadership competencies, and organizational change. Chapter 3 explores the role of ethical aspects in using AI in organizations. The purpose of Chapter 4 is to analyze the understanding of online trust in the digital era. Chapter 5 examines trust in modern digital technologies. Chapter 6 aims to determine trust in computer simulations used for predicting epidemic scenarios in heterogeneous populations. The objective of the research presented in Chapter 7 is to explore consumers' digital trust and their engagement in social media, focusing on potential predictors such as knowledge-based familiarity, perceived benefits, security, and privacy protection. Chapter 8 covers the topic of trust in the performance of modern database systems, particularly the issue of multithreading. Chapter 9 is focused on testing the credibility of randomness in Extreme Learning Machine performance. Chapter 10 presents the results of a qualitative study on the reasons for distrust in AI among Polish Generation Z students. Chapter 11 explains the role of digital therapists, detailing the potential and limitations of AI. Chapter 12 seeks answers on how to shape trust and cooperation with virtual assistants.

Section II is entitled "Trust in Artificial Intelligence Technology Applications" and also consists of 12 chapters. Chapter 13 analyzes the impact of AI on the labor market and various economic sectors. Chapter 14 focuses on a study aimed at measuring trust in chatbots used in healthcare. Chapter 15 presents trust in machine learning applied to the analysis of non-financial disclosures. Chapter 16 discusses trust in the application of neural networks for long-run market forecasting. Chapter 17 explores the role of AI in online shopping and trust. Chapter 18 identifies factors affecting Generation Z's trust in modern last-mile logistics solutions, including those in the implementation stage such as bots and drones. Chapter 19 presents trust in AI in agriculture. Chapter 20 conducts a comprehensive review of trust in public sector AI. Chapter 21 analyzes the application of trust in regional development. Chapter 22 examines the impact of AI on the information asymmetry in the insurance market. Chapter 23 describes a collaborative framework for building trust in software robots. Chapter 24 explores the intersection of trust and machine learning.

This book is intended for a wide audience, including scholars, practitioners, policymakers, and anyone interested in the intersection of AI and trust. We hope that it serves as a valuable resource for understanding the current state of AI technology, the importance of trust in its adoption and development, and the future directions this field is likely to take.

We are grateful to all the contributors whose insights and research make this book a rich repository of knowledge. Their diverse perspectives and in-depth analysis provide a nuanced understanding of the challenges and opportunities that lie ahead in the journey of integrating AI into our daily lives.

As editors, our goal has been to curate a collection of writings that not only informs but also inspires critical thinking and dialogue. We hope that this book contributes meaningfully to the ongoing discourse around AI and trust, encouraging further research and exploration in this dynamic field.

Joanna Paliszkiewicz & Jerzy Gołuchowski

Part I

Trust in artificial intelligence technology

1 Trust in artificial intelligence – future directions

Joanna Paliszkiewicz and Jerzy Gołuchowski

Introduction

Artificial intelligence (AI) is widely recognized as one of the pivotal technologies of both today and the future (Atabekov, 2023; Gerlich, 2023; Heinlein and Huchler, 2023). Within the dynamic field of AI, the concept of trust emerges as a crucial construct and predictor of the willingness to adopt AI systems, garnering increasing attention from researchers, developers, and policymakers (Riedl, 2022; Kreps et al., 2023). The notion of trust in AI spans a broad spectrum of considerations – it extends beyond the accuracy and efficiency of algorithms, delving primarily into ethical implications, transparency, accountability, and the alignment of AI systems with human values, privacy concerns, and societal norms. There is a notable gap in research regarding the comprehensive understanding of the complex nature of trust in AI. This gap presents not just academic challenges but also real-world implications for the acceptance and integration of AI by individuals, businesses, and governments. Considering the critical role of trust in the successful development and deployment of AI, it becomes essential to review and synthesize existing research on this subject.

This chapter aims to outline future research directions, policy considerations, and practical applications to foster trust in AI. The structure of the chapter is organized as follows: initially, it delves into the definitions of trust and AI and then examines the interrelationships between these two components. It also shows the threats and challenges associated with AI and identifies crucial future research and development directions aimed at enhancing trust in AI. The chapter concludes with a summary of its findings, discussing the broader implications and identifying the limitations of the study.

The concept of trust and artificial intelligence

Trust, as a multifaceted concept, can be defined from various perspectives, including psychological, sociological, and economical exchange. Generally, trust involves a willingness to be vulnerable based on positive expectations of the intentions or behavior of another entity, whether an individual, organization, or system.

DOI: 10.4324/9781032627236-2

Viewed from the individual level, trust is best understood by looking at the psychology of the person. Wrightsman (1966) and Rotter (1967) suggest that trust is a characteristic inherent in an individual's personality, shaped by one's general belief in the reliability of others. This belief is influenced by socialization processes and past personal experiences. Further, Gibb (1978) argues that trust is innate and, as an emotion, is closely related to love. Various scholars have provided definitions of trust in the context of social and economic exchanges. Hosmer (1995) views trust as the dependence of one individual, group, or firm on a duty voluntarily undertaken by another to acknowledge and safeguard the rights and interests of all involved in a cooperative venture or economic transaction. Sztompka (1999) perceives trust as the expectation that other people, groups, or institutions with whom we interact will act in ways conducive to our well-being. James (2002), from an economic perspective, regards trust as the anticipation that one will not be exploited by others, which prevails in the absence of strong incentives for opportunistic behavior. Trust has also been shown to influence technology adoption, online transaction, cloud adoption, social media and website usage decisions (e.g., Bahmanziari et al., 2003; Gefen et al., 2003; Habib et al., 2012; Paliszkiewicz and Koohang, 2016).

Building and sustaining trust is a challenging task, yet it can be rapidly lost. Regaining trust, once eroded, often requires a prolonged effort (Gill and Crane, 2017). Trust seems to function as a unique factor that, when mutually shared, fortifies relationships over time without diminishing, unlike other forms of capital.

AI is the capability of machines to mimic human skills such as reasoning, learning, planning, and creativity. AI is an important aspect of modern life that has had a transformative impact on how we live and work (Grace et al., 2018). AI enables technical systems (robots) to perceive their environment, manage what they perceive, and solve problems while working toward a specific goal. Computers equipped with AI receive data (prepared or gathered through their sensors, e.g., cameras), process it, and respond accordingly. AI systems are capable of adapting their behavior to some extent by analyzing the effects of previous actions and operating autonomously. Some AI technologies have been around for over 50 years, but advancements in computational power, the availability of vast amounts of data, and new algorithms have led to significant breakthroughs in this field.

Numerous examples demonstrate the benefits of utilizing AI in various areas. For instance, in business, AI makes many jobs safer for humans, as robots are used for performing dangerous tasks. AI supports the development of a new generation of products and services, including in sectors such as the green economy, circular economy, and machinery production. Undoubtedly, it also impacts sales increase, thanks to the use of chatbots in customer service. AI assists people by improving healthcare, making cars and other transport systems safer, and ensuring products and services are more user-friendly, cheaper, and more durable. AI facilitates access to information, education, and training, which has become particularly important in the era of the coronavirus pandemic when the need for remote education increased.

Nowadays, in the world of AI, trust remains a cornerstone for widespread acceptance and integration into various sectors. As AI systems become more complex and autonomous, comprehending, and developing trust in these systems is crucial.

Threats and challenges associated with artificial intelligence

The increasing application of AI in various areas of social and economic life, leading to a dependence on AI systems, also brings potential threats. For instance, it's unlikely that computers will become conscious in the near future in a manner similar to human consciousness.

The outcomes delivered by AI depend on its design and the data used. Both the design and data can be biased, either intentionally or unintentionally. For example, crucial aspects of an issue may not be programmed into the algorithm, or they might be programmed to reflect and reproduce structural errors. Furthermore, using numbers to represent complex social realities might give AI an unwarranted impression of being fact-based and precise. AI can significantly impact privacy and data protection rights, for example, being used in facial recognition technology or for tracking and profiling individuals online. AI also enables the combination of provided information into new data, leading to unexpected outcomes.

One area particularly threatened by changes in the digital world and the culture of post-truth and algorithms is democracy. AI applications pose threats to fundamental rights and democracy. It seems possible to both limit and strengthen democracy by applying data-based control, preventing and combating misinformation and cyberattacks, and ensuring access to high-quality information, enabling voters to make rational decisions. On the other hand, excessive control threatens individual freedoms and the potential misuse of collected data.

Democracy might also be threatened by the use of AI in creating online "information bubbles." These bubbles arise from user recommendations based on previous online behavior. By personalizing content presentation, they display only content that the algorithm believes will appeal to the individual, thereby, instead of supporting a pluralistic and open public debate, AI-based recommendations keep us within our existing circle of knowledge. Improperly used AI also threatens the freedom of assembly, as it can be used to track and profile individuals associated with certain beliefs or actions. Moreover, AI is used to create fake news and incredibly realistic fake videos, audio recordings, and images using techniques such as deepfake. This phenomenon poses financial risks, damages reputations, and complicates decision-making, potentially leading to increased divisions and polarization in the public sphere and manipulation of elections.

The collection and analysis of information with AI systems also contribute to a form of digital exclusion, leading to further restrictions on competition. This arises because companies with more information could gain an advantage over competitors and effectively eliminate them. Richer companies and countries have greater capabilities. Missed opportunities in using AI are considered a significant threat, potentially leading to poor implementation of important programs by companies, countries, and even entire regions, loss of competitive advantage, economic stagnation, and worse opportunities for citizens. Insufficient use may result from poor infrastructure, lack of initiative, low investment, or, as machine learning depends on data, fragmented digital markets and a lack of trust in AI by society and business.

It is anticipated that the use of AI will lead to the elimination of many jobs. Although AI will likely also create better employment opportunities. Education and training will play a key role in preventing long-term unemployment and ensuring a skilled workforce. The vision of a world where people lose jobs entirely due to AI is mistaken. Combines used in agriculture did not eliminate farmers, and Excel did not make accountants redundant. Similarly, AI will not replace highly skilled programmers. Indeed, it will perform tedious tasks for humans in every mentioned field, but thanks to digital transformation aided by AI, new professions and specializations are emerging (and are already emerging).

Another area of risk concerns safety. AI creates risks for the safety of individuals and institutions when AI applications have physical contact with people or are integrated with the human body. They can pose a security threat because they can be poorly designed, misused, or hacked. Poorly regulated use of AI in weapons could lead to a loss of human control over them. In military matters, AI could be used for defense and attack strategies in hacking and phishing or for attacking important systems in cyberwars.

An important challenge for AI development is legal solutions that require overcoming difficulties in determining who is responsible for damages caused by a device or service controlled by AI: for example, in the case of an accident involving a self-driving car, should the damage be covered by the owner, the car manufacturer, or the programmer? This raises many controversies. Note that if the manufacturer were completely exempt from liability, it might not stimulate the delivery of a good product or service, consequently harming people's trust in technology. On the other hand, regulations can also be too restrictive, consequently hindering innovation.

Inequalities in access to information can be used for unethical actions. For example, based on a person's behavior on the internet or other data and without their knowledge, an online seller might use AI to predict how much that person is willing to pay. Similarly, political campaigns can tailor their message to them. This results from the opacity of actions. Another issue related to transparency is that it can sometimes be unclear to people whether they are interacting with AI or a person.

Future directions in research on trust to AI

Trust is often viewed as a psychological mechanism for reducing uncertainty and increasing the likelihood of a successful interaction with entities in the environment. Trust in AI is a vital aspect for its acceptance and evolution. As AI becomes more prevalent in our daily lives and sophisticated systems, establishing trust is incredibly important (Lukyanenko, Maass and Storey, 2022; Venger and Dozortsev, 2023). We expect AI to continue revolutionizing various sectors like journalism, accounting, sales, and programming, but its exact impact remains unpredictable. Issues of trust in AI and its applications are rising due to increasing skepticism among private users and media critics. AI creates an array of qualitatively different trust challenges compared to more traditional information technologies (Makridakis, 2017).

Literature reviews highlight the importance of continued research in this area, suggesting a focus on understanding tools like Chat GPT-4 and developing key mechanisms for AI based on openness and transparency. Many AI systems, especially those based on complex algorithms like deep learning, are often considered "black boxes" due to their lack of transparency. It makes it difficult for users to understand how decisions are made, leading to mistrust and skepticism, especially in high-stakes situations. Establishing rigorous testing and validation frameworks for AI systems, especially in critical areas like healthcare and finance, can enhance their reliability and accuracy, fostering greater trust. Research should also address potential knowledge monopolization threats, the economic power of data-rich chatbots, and AI technology's concentration in a few corporations' hands. AI systems are vulnerable to various security threats, including data breaches, cyber-attacks, and misuse of personal data. The increasing sophistication of AI also raises concerns about privacy, as AI can be used to analyze and predict human behavior in invasive ways. Developing operational rules for these systems, legal regulations, and new ethical codes for AI use are urgent needs. AI should comply with privacy protection laws like the GDPR, with examples such as Italy's regulatory actions on ChatGPT highlighting potential implementation challenges. Ethical guidelines and standards need to be integrated into AI development processes to ensure that AI systems adhere to societal values and norms. Work on technologies that enhance personal data protection and privacy, like Blockchain and digital product copyright, is also essential.

Future researcher's work should be centered around deepening and broadening the understanding and applications of AI, for example, interdisciplinary research is needed to encourage collaborations across fields like ethics, psychology, sociology, and computer science to approach AI from multiple perspectives. Concentrating on developing AI systems that are ethical, fair, and unbiased. Working on making AI decision-making processes more transparent. Prioritizing research on protecting user data and enhancing the security of AI systems. Exploring the use of AI in newer fields such as environmental conservation, mental health, and more. indicated directions can help to advance AI research in a way that is ethically responsible, socially beneficial, and technologically innovative.

Conclusions

Trust in AI is a key aspect of its adoption and development. As AI becomes more integrated into our daily lives, both in routine applications and advanced systems, building trust is increasingly crucial. The potential for AI research and its applications is vast, and this technology can significantly alter our lifestyle and work in the future. However, this comes with challenges like ethics, security, and regulations, requiring attention and effort from researchers and society. The need for research on trust levels in AI and factors enhancing it is essential. Building trust in AI is a collaborative process involving researchers, companies, governments, and society, with surmountable challenges leading to increased AI acceptance and usage that benefits people and society as a whole.

This chapter outlines future research directions, policy considerations, and practical steps to cultivate trust in AI. The study contributes significantly to our understanding of trust's role in AI and suggests numerous research avenues for developing greater support for AI's trustworthy advancement.

However, the study has limitations. By focusing on specific databases like ProQuest, Google Scholar, EBSCO, and Google Books, it might not encompass all empirical studies in diverse fields. Despite this, it manages to capture major research trends.

References

Atabekov, A. (2023). Artificial intelligence in contemporary societies: Legal status and definition, implementation in public sector across various countries. *Social Sciences, 12*(3), 178.

Bahmanziari, T., Pearson, J. M., & Crosby, L. (2003). Is trust important in technology adoption? A policy capturing approach. *The Journal of Computer Information Systems, 43*(4), 46–54.

Gefen, D., Karahanna, E., & Straub, D. W. (2003). Trust and TAM in online shopping: An integrated model. *MIS Quarterly, 27*(1), 51–90.

Gerlich, M. (2023). Perceptions and acceptance of artificial intelligence: A multi-dimensional study. *Social Sciences, 12*(9), 502.

Gibb, J. R. (1978). *Trust, a New View of Personal and Organizational Development*. Los Angeles: Guild of Tutors Press, International College.

Gill, M., & Crane, S. (2017). The role and importance of trust: A study of the conditions that generate and undermine sensitive information sharing. *Security Journal, 30*(3), 734–748.

Grace, K., Salvatier, J., Dafoe, A., Zhang, B., & Evans, O. (2018). When will AI exceed human performance? Evidence from AI experts. *Journal of Artificial Intelligence Research, 62*, 729–754.

Habib, S. M., Hauke, S., Ries, S., & Mühlhäuser, M. (2012). Trust as a facilitator in cloud computing: A survey. *Journal of Cloud Computing, 1*(1), 1–18.

Heinlein, M., & Huchler, N. (2023). Artificial intelligence in the practice of work: A new way of standardising or a means to maintain complexity? *Work Organisation, Labour & Globalisation, 17*(1), 34–60.

Hosmer, L. T. (1995). Trust: The connecting link between organizational theory and philosophical ethics. *Academy of Management Review, 20*(2), 379–403.

James, H. S. (2002). The trust paradox: A survey of economic inquiries into the nature of trust and trustworthiness. *Journal of Economic Behavior & Organization, 47,* 291–307

Kreps, S., George, J., Lushenko, P., & Rao, A. (2023). Exploring the artificial intelligence "Trust paradox": Evidence from a survey experiment in the United States. *PLoS One, 18*(7).https://journals.plos.org/plosone/article?id=10.1371/journal.pone.0288109

Lukyanenko, R., Maass, W., & Storey, V. C. (2022). Trust in artificial intelligence: From a foundational trust framework to emerging research opportunities. *Electronic Markets, 32*(4), 1993–2020. https://doi.org/10.1007/s12525-022-00605-4

Makridakis, S. (2017). The forthcoming Artificial Intelligence (AI) revolution: Its impact on society and firms. *Futures, 90*, 46–60.

Paliszkiewicz, J., & Koohang, A. (2016). *Social Media and Trust: A Multinational Study of University Students*. Santa Rosa, California: Informing Science Press.

Riedl, R. (2022). Is trust in artificial intelligence systems related to user personality? Review of empirical evidence and future research directions. *Electron Markets*, *32*, 2021–2051.

Rotter, J. B. (1967). A new scale for the measurement of interpersonal trust. *Journal of Personality, 35*, 651–665.

Sztompka, P. (1999). *Trust: A Sociological Theory.* Cambridge: Cambridge University Press.

Venger, A. L., & Dozortsev, V. M. (2023). Trust in artificial intelligence: Modeling the decision making of human operators in highly dangerous situations. *Mathematics, 11*(24), 4956. https://doi.org/10.3390/math11244956

Wrightsman, L. S. (1966). Personality and attitudinal correlates of trusting and trustworthy behaviors in a two-person game. *Journal of Personality and Social Psychology, 4*, 328–332.

2 Blockchain technology, leadership competencies, and organizational change

A conceptual exploration and future research agenda

Samuel Yaw Kusi, Khuram Shahzad and Sorin Dan

Introduction

This research conceptually explores the interplay between the adoption of blockchain as a new disruptive technology and organizational change as a process and argues that a new kind of leadership (Clegg et al., 2021) and competencies are required as organizations engage in the implementation of blockchain. By emphasizing competencies for digital transformation, this research studies primarily the peripheral, yet important, elements of leadership theory (Rost, 1993). Digital transformation is being witnessed in both public and private organizations. It contributes to the creation and development of existing and new products and services and facilitates customer experience and a reconsideration of existing organizational arrangements and processes (Vial, 2021). The understanding of how humans and technology interact in digital transformation has been developed by the socio-technical systems (STS) theory and applied to organizational change (Appelbaum, 1997; Cherns, 1976; Pasmore et al., 2019; Sony & Naik, 2020).

The STS theory has been considered in the context of several technologies, including the ones in Industry 4.0. However, there is an increasing understanding that "even the successful tech firms... have not found ways to keep their social systems advancing as quickly as their technology" (Pasmore et al., 2019, p. 72). Therefore, STS design in contemporary organizations should shift from linear to iterative, from project-based to ongoing, and from a single organization to an ecosystem-driven process (Pasmore et al., 2019). This is particularly true when designing organizational change based on emerging, open-source technologies and decentralized ledgers (Kouhizadeh, Saberi & Sarkis, 2021). Blockchain removes intermediaries and can establish trustworthy transactions without external supervision (Woodside et al., 2017). Therefore, new thinking about leadership and its capacities is required to organize work and retain and enhance organizational functionality and relevance.

DOI: 10.4324/9781032627236-3

Although the STS theory offers a detailed overview of change resulting from the adoption of new technologies (Sony & Naik, 2020), it has not yet considered the conditions under which leadership might be seen as redundant. We argue that understanding what leadership competencies and capacities are necessary when adopting a novel and insufficiently understood technology warrants a separate inquiry, in which we engage in this chapter. This chapter's central research question is: *What kind of leadership and competencies are needed to change organizations and implement blockchain technology?*

Building on the specific features of blockchain and the theoretical underpinnings of the STS theory, we design a framework to shed light on leadership competencies and capacities that leaders require to implement blockchain within their organizations. By doing so, we aim to contribute to rethinking leadership and organizational change in view of the growing digitalization of our lives, societies, and economies (By, 2021; Clegg et al., 2021). We aim to extend the theorization of STS and reframe and adapt its theoretical explanations to the newest technologies in Industry 4.0. We posit that blockchain technology, in particular, given its unique features, has been considered in both digital transformation (Imran et al., 2021; Noorit et al., 2020) and STS theories (Pasmore et al., 2019; Sony & Naik, 2020). This raises questions about the relevance of the STS theory in a blockchain context, which has the potential to act as a driver in changing our current understanding of leadership and its contribution to stimulating and sustaining organizational change (Ford et al., 2021).

Socio-technical systems theory and digital transformation: connecting the dots

Different theories have been employed to study the technological phenomenon. In socio-material theory, technologies are viewed as a group's experiences gained from the local environment. Thus, environmental actions, business, and human interactions among others occur. According to Orlikowski (2007, p. 1437), the material and social are inseparable – "there is no social that is not material and no material that is not also social". Structuration theory (Giddens, 1979) has also been utilized to examine the technological phenomenon. Nonetheless, it has suffered from criticism such as not being consistent in terms of structural details of technology (Jones & Karsten, 2008).

The most widely adopted theory is STS (Imran et al., 2021), which considers the interplay between social and technological factors during change processes (Pasmore et al., 2019). It also emphasizes the necessary skills for navigating technological change (Sony & Naik, 2020). STS underscores the importance of addressing both technical and social aspects to drive change within an organization (Cherns, 1976). At its core, this theory views organizations as intricately connected within a complex work framework, acknowledging the dynamic interactions between individuals and technology, as well as the complexities inherent in the interplay between societal infrastructure and human behavior (Cherns, 1976). STS advocates for joint optimization, where social and technical systems are concurrently designed to facilitate their collaborative functioning (Appelbaum, 1997; Cherns, 1976). The STS philosophy emerged from the UK Tavistock Institute's research into introducing new mining machinery (Davis et al., 2014). It emphasizes the interconnectedness of technology and the social aspects of work, stressing the need to consider the human factor in technological change (van Eijnatten, 2013). This theory provides a comprehensive understanding of how organizational transformation results from technological changes and the necessary competencies for managing them (Sony & Naik, 2020). STS aims to integrate humans and technical systems to create new opportunities for work and pave the way for technological change (Trist, 1981).

Studies have applied STS theory to explore digital transformation and leadership competencies in organizations (e.g., Imran et al., 2021). However, these studies often lack a detailed examination of the specific individual capabilities and competencies required for driving change related to particular technologies. Furthermore, Pasmore et al. (2019) suggest that STS principles should evolve alongside Industry 4.0, viewing digital transformation as a continuous, iterative process that considers the ecosystem, organization, and individuals, especially the self-directed management crucial for the emerging Next Gen workforce. Therefore, there's a need for a deeper understanding of these actors and the new competencies required to facilitate change in the implementation of advanced technologies like blockchain, emphasizing workforce independence, ongoing change, and ecosystem thinking (Kouhizadeh et al., 2021).

The distinctive features of blockchain technology

Blockchain is a decentralized ledger that records transactions in data blocks linked by cryptographic pointers (Kouhizadeh et al., 2021). It promotes transparency and trust, granting all parties access to information through nodes and chain blocks (Downie & Parker, 2019). Longer chains enhance trust by making modifications difficult (Cole et al., 2019; Morkunas et al., 2019). It leverages the internet as infrastructure for transaction records and consensus rules (Smith, 2017), making it a versatile technology for digital asset exchange in distributed networks (Ølnes, Ubacht & Janssen, 2017). Blockchain gained popularity in 2008 with Nakamoto's article "Bitcoin – A Peer–to–Peer Electronic Cash System," revolutionizing information transactions and business operations. Scholars, including Morkunas et al. (2019), advocate for more research on how blockchain impacts organizational innovation models and value creation. Numerous organizations and countries recognize its potential. For instance, El Salvador embraced Bitcoin as an official legal tender, Lithuania issued the world's first state-backed digital collectible coin, and China introduced its digital

Yuan, akin to Bitcoin (Areddy, 2021). It's no surprise that blockchain is projected to contribute 10% to global GDP by 2027 (World Economic Forum, 2015).

Blockchain technology has been empirically researched in various industries, such as its impact on supply chain management (Cole et al., 2019), its role in disrupting real estate development in the Netherlands and international markets (Veuger, 2018), its relevance to addressing barriers in sustainable fashion retailing (Caldarelli et al., 2021), and the hotel-hospitality industry (Chan, Okumus & Chan, 2015). It has also been examined in the context of talent retention during change management (Vineyard et al., 2020; Woodside et al., 2017). Several distinctive blockchain features, including open disruptive innovation, flexibility, transparency and trust, and networking, necessitate specific leadership competencies when driving organizational change toward blockchain adoption. We will further explore these features and their significance as catalysts for change in blockchain adoption within organizations.

Open disruptive innovation – disruptive technologies have become a prominent concern in boardrooms (Evans, 2017). Blockchain's decentralized nature is causing disruption across various industries by authenticating traded goods, disintermediating processes, and reducing transaction costs. This compels practitioners to adapt their business models urgently to maintain competitiveness (Nowiński & Kozma, 2017). Veuger's (2018) research on blockchain's impact on the real estate industry in the Netherlands and internationally highlights how real estate is innovating in response to blockchain-induced disruptions, capitalizing on new opportunities. Woodside et al. (2017), using a mixed-method approach to investigate blockchain technology acceptance from a managerial perspective, also confirm blockchain's status as a disruptive innovation affecting diverse sectors, including finance, accounting, and various marketplaces. Blockchain is fundamentally reshaping business models and operations across industries. In the insurance sector, blockchain has revolutionized claim settlement, enhancing speed, transparency, and trust (Grima et al., 2020). Claimants are now active participants in the process, accessing real-time information, unlike the previous practice of organizations controlling information dissemination. Ordinary individuals are increasingly collaborating with organizations, resulting in significant improvements in operations and supply chain management. These improvements encompass inventory management, reduced intermediaries, and the stimulation of new product development, all facilitated by blockchain (Grima et al., 2020). Customer order management processes have become more efficient and streamlined with blockchain implementation (Martinez et al., 2019). On a philosophical note, blockchain has been found to actively shape financial transactions, solidifying them gradually, replacing human interaction with automated code, and exerting a substantial impact on social aspects like cryptocurrency (Reijers & Coeckelbergh, 2018).

Flexibility – blockchain technology has been hailed as bringing flexibility to business operations and transactions. For example, blockchain implementation and its absorption by industries have resulted in a change in orientation by users and owners of real estate which has contributed to bringing about flexibility in property management (Veuger, 2018). The financial industry is experiencing the features of blockchain in the delivery of services and the entirety of business operations. Peterson's (2018) study highlights this line of discussion that blockchain advances flexibility in different capacities in the financial industry. For instance, sleeve-level performance can be computed with ease, and the prospect of after-tax performance computation, simplifies total sleeve-level bases – thus, easing financial reporting and enhancing unified managed accounts (Peterson, 2018).

Transparency and trust – blockchain is seen as a technological advancement, particularly notable for its inherent transparency, which empowers all stakeholders and creates trust rather than concentrating power on top management. This transparency and trust address the growing demand among practitioners and the public for comprehensive insights into transactions and business operations, making blockchain a promising solution (Caldarelli et al., 2021). For instance, in an exploratory study, Caldarelli et al. (2021) found that blockchain records the entire production process, from raw materials to the final product, allowing anyone to check and verify, instilling confidence in consumers and other stakeholders regarding production and sustainability. This inclusive approach makes previously overlooked individuals critical partners in the production process. Similarly, in the context of supply chain management, blockchain offers transparency that enhances quality management by providing visibility and easy access to batch information, aiding in recalls, advancing services, automating contracts, and reducing the need to establish trust within the supply chain (Cole et al., 2019). Martinez et al. (2019) also highlight the transparency and trust brought about by blockchain in customer orders and services. In the banking sector, blockchain adoption streamlines lengthy verification processes and supports the automation of accounting services through AI. Major global accounting firms such as KPMG, PwC, Deloitte, and Ernst & Young utilize blockchain in diverse contexts to enhance their financial service provision and delivery (Downie & Parker, 2019). The insurance industry utilizes blockchain to improve transparency, addressing issues like manual procedures (Grima et al., 2020). Peterson's (2018) research recognizes blockchain's potential to simplify reporting in the financial sector, enabling real-time access to ledger records for authentic transaction monitoring. In healthcare, blockchain enhances transparency in managing patients' data (Shahzad et al., 2024).

Networking – blockchain relies on ledgers involving multiple stakeholders within a network. Beyond financial applications, blockchain has demonstrated its capacity to foster collaboration among diverse groups, including scientists, businesses, policymakers, and research funding bodies, through platforms like IdeaChain (Burmaoglu et al., 2020). IdeaChain functions by initiating ideas from scientists and researchers, translating them into products and services by businesses and industry, and finally, guiding the process through policymaking to address legal, social, and ethical aspects (Burmaoglu et al., 2020). In the medical and healthcare sector, blockchain facilitates networking, enabling secure two-way communication and information exchange among doctors, nurses, patients, and other relevant stakeholders (Shahzad et al., 2024). Blockchain yields both economic and noneconomic outcomes.

In *economic terms*, numerous studies (Caldarelli et al., 2021; Grima et al., 2020; Kouhizadeh et al., 2021; Martinez et al., 2019; Michelman, 2017; Peterson, 2018; Woodside et al., 2017) emphasize cost savings. For instance, Martinez et al. (2019) highlight reduced customer order processing times, Grima et al. (2020) note decreased insurance claims costs, and McKinsey & Company (2017) projects significant cost savings of $50–$60 billion in cross-border B2B payments. Additionally, blockchain helps combat illegal counterfeiting by providing information about authentic fashion products (Cole et al., 2019). Noneconomic benefits manifest as enhanced efficiency (e.g., Burmaglou et al., 2020; Kouhizadeh et al., 2021; Martinez et al., 2019; Nowiński & Kozma, 2017) and improved transparency and trust (e.g., Cole et al., 2019; Peterson, 2018; Veuger, 2018). Lumineau et al. (2021) explored how the unique features of blockchain change governance models noting that one of the key questions for future research is "*What kinds of actors can be the most effective in using blockchains to govern collaborations?*" (p. 514). As organizations keep experimenting with blockchain adoption and there are few use cases that would allow empirical investigation and generalized conclusions about the leadership competencies needed to lead the change toward blockchain implementation, we engage in a conceptual investigation and build our suggestions bridging insights from STS, change management, leadership, and blockchain research.

Conceptual framework: leadership and organizational change for blockchain Implementation

Leadership is a pivotal factor in effective change management, involving the initiation of change awareness within the organization, planning, explaining implementation strategies, organizing training, and assigning responsibilities. According to Dzwigol et al. (2019), leadership is essential across various subsystems at the strategic management level of a firm to drive change, advance human resources, and foster innovation within the organization's lifecycle. The absence of strong leadership, as seen in supply chains, can hinder the successful implementation of blockchain and ethical sustainability practices (e.g., Hughes et al., 2019; Luthra et al., 2016). Leadership competencies encompass the knowledge, skills, abilities, and attributes required for effective leadership (Das et al., 2011). These competencies are instrumental in driving digital transformation within organizations (Horlacher et al., 2016). While existing literature has predominantly focused on developing profiles like Chief Digital Officers or

Chief Technical Officers responsible for leading digital innovation initiatives, there is less explicit guidance on the competency requirements for leaders tasked not only with operational digital innovation projects but also with preparing their organizations for broader organizational change.

Complexity leadership theory, as described by Uhl-Bien et al. (2007), identifies adaptive, administrative, and enabling leadership as crucial qualities in driving technological transformation. Adaptive leadership involves creative, learning actions emerging from interactions within complex adaptive systems to address tension or constraints. Administrative leadership pertains to formal managerial roles planning and coordinating activities efficiently to achieve organizational goals. Enabling leadership aims to create conditions for adaptive leadership to thrive and manage the interplay between bureaucratic and emergent functions in the organization (Uhl-Bien et al., 2007). From an adaptive leadership perspective, Kane et al. (2019) conducted a global survey and interviews, concluding that effective digital transformation hinges on leaders possessing qualities like adaptability, creativity, and decisive action. Administrative leadership qualities within groups and individuals also play a vital role in successful technology adoption (Dzwigol et al., 2019; Noorit et al., 2020; Philpot & Monahan, 2017). Scholars, including Imran et al. (2021), have emphasized the importance of leadership competencies encompassing adaptability, a positive attitude, effective communication, open-mindedness, technical expertise, and a willingness to experiment. These competencies act as catalysts for achieving agility, customer-centricity, and collaboration. We, therefore, move to offer the following proposition:

Proposition 1 (P1): For any blockchain implementation, leadership flexibility is required to undergo technological adoption by the organization.

The preceding discussion underscores that the adoption and implementation of blockchain are not isolated endeavors. Organizations must embrace new technology and disruptive innovations to ensure continuity and gain a competitive edge. This transition inherently brings about change, necessitating appropriate leadership competencies to steer management policies forward. The introduction of blockchain into an organization necessitates adjustments that drive digital transformation, a process inherently characterized by change, as evident in the literature. Several studies shed light on the prerequisites for embracing new technologies like blockchain. For instance, Vineyard et al. (2020), drawing on the Learning, Enhancement, Advancement, and Development model, elucidate that organizations looking to successfully adopt blockchain must first embrace change, cultivate a customer-centric perspective, establish career development paths, and devise retention strategies to achieve their objectives. The authors emphasize adhering to ADKAR guidelines: creating awareness about the necessity for change, fostering a desire for collaboration, ensuring understanding of how to implement change, and providing the necessary resources and training to facilitate the transition (Vineyard et al., 2020).

The allure of blockchain's benefits prompts organizations to transition to its use. Studies have delved into how blockchain adoption unfolds within organizations. Employing a mixed-method approach grounded in the diffusion of innovation theory, Woodside et al. (2017) highlight factors like transparency, trust, user control, and quality (particularly in safeguarding against fraudulent activities) that motivate organizations to embrace blockchain technologies. Michelman (2017) also illustrates that organizations are adapting to new technological advancements through blockchain due to its simplicity and the benefits it offers, which outweigh existing business models. This adaptation process entails a thorough evaluation of current systems in terms of cost benefits and resource commitments that will benefit the organization following the integration of the new technology (Michelman, 2017). However, it's essential to acknowledge that not all changes encounter smooth acceptance. Some organizations face resistance to adopting new technology due to the comfort associated with existing systems and uncertainties regarding the benefits of transitioning to new technology, which can impede change. Consequently, it is apparent that adopting blockchain, driven by its associated benefits, necessitates organizational changes, and leadership qualities are indispensable in facilitating this transformation. In light of these considerations, we propose the following:

Proposition 2 (P2): Blockchain adoption by the organization brings about more inclusiveness of stakeholders; thus, leaders should have the knowledge and the ability to embrace these stakeholders when taking organizations over the transformation.

Discussion and implications

The current study addresses a conceptual examination of the following research question concerning blockchain implementation, global leadership competencies, and change management underpinned by STS theory. *What leadership competencies are needed to transform organizations to adopt and implement blockchain technologies?* The literature on digital transformations broadens our knowledge about technological systems and the social aspects that need to work in harmony to progress with change that occurs from technologies such as blockchain. The social aspects of STS theory exhibit the importance of the human element in the transformation of technology in an organization (Singh & Hess, 2020). Nonetheless, our understanding of these individual actors who play a meaningful role in the implementation of new technologies has been limited. Through our conceptual model development, we have shown that other individual actors also play a meaningful role in the transformation of digitalization such as blockchain. In that respect, we extend the STS theory and its application in understanding the adoption of new technologies in organizations.

Leadership competencies are crucial for digital transformation, particularly for those spearheading the shift to new technologies. Top-level management consistently emerges as a key driver of technological change in organizations (e.g., Kane et al., 2019; van Hoek, 2019). Leadership attributes, encompassing knowledge, skills, and abilities, are intrinsic qualities that assume paramount importance during change processes (Das et al., 2011), especially in the context of digital transformation (Horlacher et al., 2016). However, possessing these attributes alone is insufficient. Effective leadership roles extend to various facets, including raising awareness of digital transformation topics, fostering collaboration, driving technological change, cultivating digital transformation culture, setting exemplary standards, promoting mentoring and coaching, exhibiting strategic followership, enhancing transparency and trust throughout actions and the organizational value chain, and aligning with core values (Imram et al., 2020). This underscores the fact that digital transformation is not solely the domain of top-level management; the leadership skills of individuals and groups involved in administrative tasks within the organization also wield significant influence (Dzwigol et al., 2019; Noorit et al., 2020; Philpot & Monahan, 2017). These individual leadership skills contribute effectively to support managers and the lifecycle of firms within organizational and innovation subsystems (Dzwigol et al., 2019).

Theoretical contribution

An important gap in the conceptualization of blockchain has been especially the neglect of the kind of leadership and capacities that are required for its implementation (Downie & Parker, 2019). These are critical for organizational success in the digital era (Kane et al., 2019). Existing research has found that driving digital transformation requires leadership qualities and capacities, such as the ability to spread awareness of digital technology, promote collaboration, and sustain efforts over time (Janssen et al., 2020). Besides, little is known about the process of digital transformation and the role that blockchain actors play in enacting it. We argue that these actors play a meaningful role in the transformation process to ensure that the new technology is integrated smoothly into the organization. We propose that blockchain technology implementation should incorporate leadership competencies and organizational change (Figure 2.1).

Figure 2.1 illustrates that blockchain implementation is a complex process involving multiple considerations. Embracing the social perspective of technology is pivotal for organizational transformation through the adoption of new technologies like blockchain. In this context, the successful implementation of blockchain in organizations relies on well-defined processes executed by human elements within the organization. Effective supervision, typically provided by managers, is essential to ensure a seamless transition. Our framework underscores the importance of various attributes working harmoniously to facilitate this transition, recognizing the interplay between technological systems, represented by blockchain, and social systems, exemplified by human actors including top-level

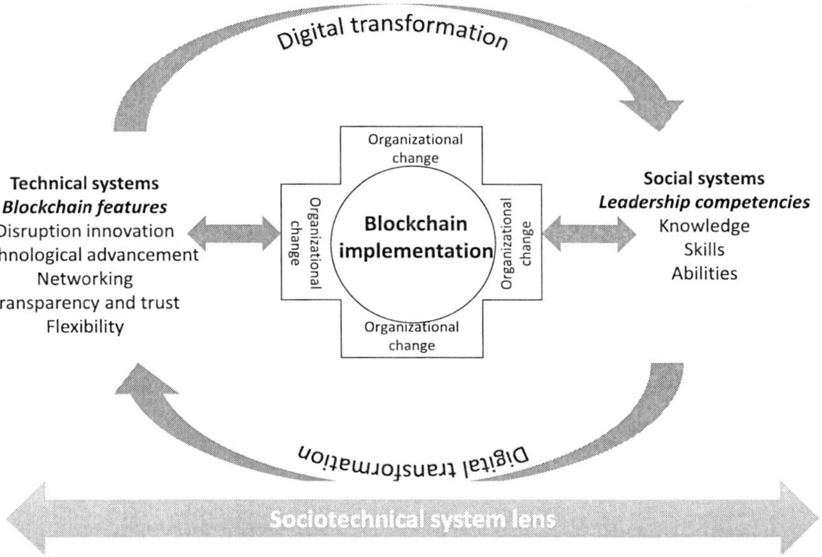

Figure 2.1 Conceptual framework for blockchain implementation.

Source: Authors' own interpretation.

management and individuals. Given the dynamic nature of the digital era and the disruptive potential of blockchain in revolutionizing existing technological capabilities, there arises a pertinent question regarding the suitability of current theoretical perspectives in the literature. Consequently, scholars in the emerging fields of digital transformation and disruptive innovation are encouraged to explore these issues, thus advancing our collective knowledge in this domain.

Conclusion and future research agenda

We have aimed to emphasize the integration of blockchain implementation, digital transformation, and leadership competencies. Our expectation is that leadership competencies, particularly those exhibited by individual actors involved in blockchain implementation in both private and public organizations, have been underscored as a crucial and timely focus within the nascent field of blockchain studies. We hope to contribute an integrative lens for studying blockchain, recognizing the existing omission of leadership competencies essential for driving change in blockchain implementation. Furthermore, leadership attributes have been established as imperative for digital transformation (Reck & Fliaster, 2019). Our conceptual framework strives to address this gap in the literature by bridging studies on blockchain, digital transformation, and leadership competencies. Building on this foundation, we contend that a comprehensive understanding of blockchain necessitates the integration of change management and leadership competencies into blockchain technology implementation. Additionally, we posit that the change process, management interventions, and requisite leadership competencies present an opportunity to explore an empirical gap, shedding light on the managerial procedures undertaken during the transition to blockchain and their alignment with established knowledge regarding radical organizational change and digital transformation. Our proposed conceptual framework in Figure 2.1 is expected to offer practical implications for practitioners. Change is not a straightforward exercise whether for a small or large organization. Therefore, to implement blockchain in an organization our framework can guide managers to understand the causes of, mediums, resources, and capabilities to effect the change.

Although there are academics, practitioners, and policymakers interested in blockchain, knowledge gained about the concept is somehow fragmented and nevertheless not incorporated into a higher level of theorization within the confines of change management and leadership competencies literature. This important displacement existing among theory and practice is our contention here (Shepherd & Suddaby, 2017) and we argue that needs to re-examine how critical leadership competencies and change management are to blockchain adoption and implementation. This study provides a foundation for future research directions. Firstly, it is important to transition from conceptual exploration to empirical testing, validating the ideas presented here. Additionally, while we've focused on social contexts in this study, exploring the cultural implications of blockchain adoption remains an intriguing avenue for further research. Cultural differences may impact blockchain's application and implementation across borders, warranting investigation. To realize blockchain's potential for various stakeholders, future studies should delve into its implementation and outcomes, incorporating perspectives from change management and leadership competencies. By examining how these factors intersect with blockchain adoption, we can identify the necessary leadership competencies at each stage of the transformation process and understand their resulting impact. This knowledge is particularly valuable for large organizations, including government institutions like tax and social benefit authorities, seeking to adopt blockchain effectively.

Management and organizational theory have historically overlooked the significance of context (Matthews et al., 2018). However, blockchain technology has the potential not only to disrupt policies and services, but also to introduce a new institutional governance technology that can enhance existing organizational processes. Consequently, a comprehensive conceptual framework for blockchain adoption and implementation, incorporating change management and leadership competencies, represents a crucial advancement for scholars investigating the antecedents, barriers, and causal mechanisms of blockchain technology. This is particularly relevant for theory-building efforts that underscore the pivotal role of context in understanding blockchain technology.

Some authors have already begun employing theories such as STS to examine digital transformation and the requisite leadership competencies within specific contexts (e.g., Imran et al., 2021). Given the relevance of context in the business landscape and the growing interest in blockchain research among academics and practitioners alike, we argue that exploring approaches to gain insights into this disruptive innovation, which has already demonstrated its revolutionary potential, is essential.

Studies have enlightened us about the critical role of leadership in digital transformation, including the domain of blockchain. However, digital transformation is a multifaceted challenge in management and organizations. We require insights into the mechanics of technological transformation, encompassing the change process itself. This could be achieved through real-time observations (Pettigrew, 1992), longitudinal studies detailing the duration and evolution of the change process (Hassett & Paavilainen-Mäntymäki, 2013; Van de Ven, 1992), or retrospective analyses (Sonenshein, 2010) examining past change efforts. While a few studies (e.g., Kane et al., 2019; Michelman, 2017) have touched upon leadership qualities essential for digital transformation, including blockchain, we align with Imran et al. (2021) in advocating for expanded research on leadership, organizational structure, culture, and their performance implications within the context of blockchain. Furthermore, longitudinal research is needed to explore interconnected issues that may pose challenges to the seamless implementation of blockchain technology within organizations, and to evaluate the resulting impacts on efficiency, transparency, and traceability (Kouhizadeh et al., 2021).

Disclosure statement

No potential conflict of interest was reported by the author(s).

Funding

This study was supported by the Foundation for Economic Education in Finland under Grant agreement 200264, 210257 project Brothers from a different mother? How businesses and public sector organizations develop leadership competencies and change processes for blockchain implementation (BizPub).

References

Appelbaum, S. H. (1997). Socio-technical systems theory: an intervention strategy for organizational development. *Management Decision*, *35*(6), 452–463.

Areddy, J. T. (2021). China creates its digital currency, a first for major economy. *The World Street Journal*, Available at: https://www.wsj.com/articles/china-creates-its-own-digital-currency-a-first-for-major-economy-11617634118

Burmaoglu, S., Saritas O., & Sesen, H. (2020). IdeaChain: a conceptual proposal for blockchain-based STI policy development. *Foresight*, *22*(2), 189–204.

By, R. T. (2021) Leadership: in pursuit of purpose. *Journal of Change Management*, *21*(1), 30–44.

Caldarelli, G., Zardini, A., & Rossignoli, C. (2021). Blockchain adoption in the fashion sustainable supply chain: Pragmatically addressing barriers. *Journal of Organizational Change Management, 34*, 507–524.

Chan, E. S., Okumus, F., & Chan, W. (2015). Barriers to environmental technology adoption in hotels. *Journal of Hospitality and Tourism Research*, *38*, 361–387.

Cherns, A. (1976). The principles of sociotechnical design. *Human Relations, 29*(8), 783–792.

Clegg, S., Crevani, L., Uhl-Bien, M., & By, R. T. (2021). Changing leadership in changing times. *Journal of Change Management*, *21*(1), 1–13.

Cole, R., Stevenson, M., & Aitken. J. (2019). Blockchain technology: implications for operations and supply chain management. *Supply Chain Management: An International Journal, 24*(4), 469–483.

Das, A., Kumar, V., & Kumar, U. (2011). The role of leadership competencies for implementing TQM: an empirical study in Thai manufacturing industry. *International Journal of Quality & Reliability Management*, *28*(2), 195–219.

Davis, M. C., Challenger, R., Jayewardene, D. N., & Clegg, C. W. (2014). Advancing socio-technical systems thinking: a call for bravery. *Applied Ergonomics*, *45*(2), 171–180.

Downie, G., & Parker, D. (2019). Blockchain technology: the missing link in services management? Part I. *Management Services*, *Winter 2019*, 32–35.

Dzwigol, H., Shcherbak, S., Semikina, M., Vinichenko, O., & Vasiuta, V. (2019). Formation of strategic change management system at an enterprise. *Academy of Strategic Management Journal*, *18*, 1–8.

Evans, G. L. (2017). Disruptive technology and the board: the tip of the iceberg. *Economics and Business Review*, *3*(1), 205–233.

Ford, J., Ford, L. & Polin, B. (2021) Leadership in the implementation of change: functions, sources, and requisite variety. *Journal of Change Management*, *21*(1), 87–119.

Giddens, A. (1979). *Central Problems in Social Theory: Action, Structure, and Contradiction in Social Analysis* (Vol. 241). Univ of California Press.

Grima, S., Spiteri, J., & Romānova, I. (2020). A STEEP framework analysis of the key factors impacting the use of blockchain technology in the insurance industry. *The Geneva Papers on Risk and Insurance-Issues and Practice*, *45*(3), 398–425.

Hassett, M. E., & Paavilainen-Mäntymäki, E. (Eds.) (2013). *Handbook of longitudinal research methods in organisation and business studies*. Edward Elgar Publishing.

Horlacher, A., Klarner, P., & Hess, T. (2016). Crossing boundaries: organization design parameters surrounding CDOs and their digital transformation activities. In *AMCIS 2016: Surfing the IT Innovation Wave -22nd Americas Conference on Information Systems*, San Diego.

Hughes, L., Dwivedi, Y. K., Misra, S. K., Rana, N. P., Raghavan, V., & Akella, V. (2019). Blockchain research, practice and policy: applications, benefits, limitations, emerging research themes and research agenda. *International Journal of Information Management*, *49*, 114–129.

Imran, F., Shahzad, K., Butt, A., & Kantola, J. (2020). Leadership competencies for digital transformation: evidence from multiple cases. In *Advances in Human Factors,*

Business Management and Leadership: Proceedings of the AHFE 2020 Virtual Conferences on Human Factors, Business Management and Society, and Human Factors in Management and Leadership, July 16-20, 2020, USA (pp. 81–87). Springer International Publishing.

Imran, F., Shahzad, K., Butt, A., & Kantola, J. (2021). Digital transformation of industrial organizations: toward an integrated framework. *Journal of Change Management, 21*(4), 451–479.

Janssen, M., Weerakkody, V., Ismagilova, E., Sivarajah, U., & Irani, Z. (2020). A framework for analysing blockchain technology adoption: Integrating institutional, market and technical factors. *International Journal of Information Management, 50*, 302–309.

Jones, M. R., & Karsten, H. (2008). Giddens's structuration theory and information systems research. *MIS Quarterly, 32*(1) 127–157.

Kane, G. C., Phillips, A. N., Copulsky, J., & Andrus, G. (2019). How digital leadership is (n't) different. *MIT Sloan Management Review, 60*(3), 34–39.

Kouhizadeh, M., Saberi, S., & Sarkis, J. (2021). Blockchain technology and the sustainable supply chain: theoretically exploring adoption barriers. *International Journal of Production Economics, 231*, 107831.

Lumineau, F., Wang, W., & Schilke, O. (2021). Blockchain governance—A new way of organizing collaborations? *Organization Science, 32*(2), 500–521.

Luthra, S., Mangla, S. K., Xu, L., & Diabat, A. (2016). Using AHP to evaluate barriers in adopting sustainable consumption and production initiatives in a supply chain. *International Journal of Production Economics, 181*, 342–349.

Matthews, R. S., Chalmers, D. M., & Fraser, S. S. (2018). The intersection of entrepreneurship and selling: An interdisciplinary review, framework, and future research agenda. *Journal of Business Venturing, 33*(6), 691–719.

Martinez, V., Zhao, M., Blujdea, C., Han, X., Neely, A., & Albores, P. (2019). Blockchain-driven Customer Order Management. *International Journal of Operations & Production Management, 39*(6/7/8), 993–1022.

McKinsey & Company. (2017). Blockchain technology in the insurance sector. In *Proceedings of the Quarterly Meeting of the Federal Advisory Committee on Insurance (FACI)*. New York: McKinsey & Company.

Michelman, P. (2017). Seeing beyond the blockchain hype. *MIT Sloan Management Review, 58*(4), 17–19.

Morkunas, V. J., Paschen, J., & Boon, E. (2019). How blockchain technologies impact your business model. *Business Horizons, 62*(3), 295–306.

Nakamoto, S. (2008). Bitcoin: A peer-to-peer electronic cash system. *Decentralized Business Review*, 21260. Accessed 20 September 2023. https://bitcoin.org/bitcoin.pdf

Noorit, N., Thapayom, A., & Pornpundejwittaya, P. (2020). Guidelines for adaptation of the Thai industrial business to support the digital economy. *Academy of Strategic Management Journal, 19*(6), 1–15.

Nowiński, W., & Kozma, M. (2017). How can blockchain technology disrupt the existing business models? *Entrepreneurial Business and Economics Review, 5*(3), 173–188.

Ølnes, S., Ubacht, J., & Janssen, M. (2017). Blockchain in government: benefits and implications of distributed ledger technology for information sharing. *Government Information Quarterly, 34*(3), 355–364.

Orlikowski, W. J. (2007). Sociomaterial practices: exploring technology at work. *Organization Studies, 28*(9), 1435–1448.

Pasmore, W., Winby, S., Mohrman, S. A., & Vanasse, R. (2019). Reflections: sociotechnical systems design and organization change. *Journal of Change Management*, *19*(2), 67–85.

Peterson, M. (2018). Blockchain and the future of financial services. *The Journal of Wealth Management, 21*(1), 124–131.

Pettigrew, A. M. (1992). The character and significance of strategy process research. *Strategic Management Journal*, *13*(2), 5–16.

Philpot, S., & Monahan, K. (2017). A data-driven approach to identifying future leaders. *MIT Sloan Management Review, 58*(4), 19–21.

Reck, F., & Fliaster, A. (2019). Four profiles of successful digital executives. *MIT Sloan Management Review*, *60*(3), 1–7.

Reijers, W., & Coeckelbergh, M. (2018). The blockchain as a narrative technology: investigating the social ontology and normative configurations of cryptocurrencies. *Philosophy & Technology, 31*, 103–130.

Rost, J. C. (1991). *Leadership for the Twenty-First Century*. Bloomsbury Publishing.

Shahzad, K., Helo, P., Ranta, M., & Nousiainen, E. (2024). Blockchain technology for operational excellence and supply chain resilience: a framework based on use cases and an architecture demonstration. *Technology Analysis & Strategic Management*, 1–18.

Shepherd, D. A., & Suddaby, R. (2017). Theory building: a review and integration. *Journal of Management*, *43*(1), 59–86.

Singh, A., & Hess, T. (2020). How chief digital officers promote the digital transformation of their companies. In Robert D. Galliers, Dorothy E. Leidner, Boyka Simeonova (Eds.), *Strategic Information Management* (pp. 202–220). New York: Routledge. https://doi.org/10.4324/9780429286797

Smith, T. D. (2017). The blockchain litmus test. In Robert D. Galliers, Dorothy E. Leidner, Boyka Simeonova (Eds.), *2017 IEEE International Conference on Big Data (Big Data)* 5th Ed. (pp. 2299–2308). IEEE.

Sonenshein, S. (2010). We're changing—Or are we? Untangling the role of progressive, regressive, and stability narratives during strategic change implementation. *Academy of Management Journal*, *53*(3), 477–512.

Sony, M., & Naik, S. (2020). Industry 4.0 integration with socio-technical systems theory: a systematic review and proposed theoretical model. *Technology in Society*, *61*, 101248.

Sun, B., Lv, Z., & Li, Q. (2021). Obstetrics nursing and medical health system based on Blockchain technology. *Journal of Healthcare Engineering*, *2021*.

Trist, E. L. (1981). *The Evolution of Socio-Technical Systems* (Vol. 2). Toronto: Ontario Quality of Working Life Centre.

Uhl-Bien, M., Marion, R., & McKelvey, B. (2007). Complexity leadership theory: shifting leadership from the industrial age to the knowledge era. *The Leadership Quarterly*, *18*(4), 298–318.

Van de Ven, A. H. (1992). Suggestions for studying strategy process: a research note. *Strategic Management Journal*, *13*(Special Issue 1), 169–191.

van Eijnatten, F. M. (2013). Developments in socio-technical systems design (STSD). In Drenth, P.J.D., Thierry, H., & Wolff, Ch.J. de (Eds.), *A Handbook of Work and Organizational Psychology* (pp. 61–88). Psychology Press.

van Hoek, R. (2019). Exploring blockchain implementation in the supply chain. *International Journal of Operations and Production Management, 39(*6/7/8), 829–859.

Veuger, J. (2018). Trust in a viable real estate economy with disruption and blockchain. *Facilities 36(*1/2), 103–120.

Vial, G. (2021). Understanding digital transformation: a review and a research agenda. *Managing Digital Transformation*, *28*(2), 13–66.

Vineyard, S., Fitz, D., Larkins, L., & Parker, A. (2020). Reimagine the financial management workforce for success in an age of innovation. *The Journal of Government Financial Management, 68*(4), 44–49.

Woodside, J. M., Augustine, F. K. Jr., & Giberson, W. (2017). Blockchain technology adoption status and strategies. *Journal of International Technology and Information Management, 26*(2), 65–93.

World Economic Forum (2015). Deep Shift: Technology Tipping Points and Societal Impact. Retrieved from: www3.weforum.org/docs/WEF_GAC15_Technological_Tipping_Points_report_2015.pdf (Accessed: 30 July 2021).

3 Trust in artificial intelligence

Ethical aspects of using artificial intelligence in organizations

Barbara Wyrzykowska, Agnieszka Tul-Krzyszczuk and Tetiana Balanovska

Introduction

Artificial intelligence (AI) is one of the major technologies of the 21st century. Recently, it has been growing in importance in business, where it has an increasingly wide range of applications, from relatively simple "chatbots" used for customer service to more complex analytical solutions based on deep learning (Adetayo, 2023; Tlili et al., 2023). Artificial intelligence is a subdiscipline of computer science that deals with the creation of information systems that can learn and act autonomously. One area where artificial intelligence is gaining prominence is ChatGPT (Costello, 2023; Ekanayake & Saputhanthri, 2020).

Companies implementing AI technology are realizing tangible benefits, and new applications are attracting a lot of attention from both theorists and practitioners. Major industries using specialized AI solutions include telecommunications (Terashima, 2002), financial services (Longbing, 2022), internet companies, healthcare and medicine (Bohr & Memarzadeh, 2020; Buch et al., 2018; Imran & Jawaid, 2020), media (Chintalapati & Pandey, 2021), the entertainment industry (Casero-Ripolles et al., 2022), and industrial manufacturing (Siarry, 2022). We are also increasingly noticing decisions to implement AI in the area of Smart City (Bielińska-Dusza et al., 2021; Luckey et al., 2020) or education (Caballé et al., 2021). Experts predict that AI technology will continue to improve and will significantly impact the operation of businesses (Calp, 2019). It will contribute to changes in terms of their business models.

Despite the benefits associated with the development of artificial intelligence, it is also necessary to recognize serious risks for legal entities, such as discrimination, loss of autonomy, restrictions on access to certain services, and manipulation of information (Bérubé et al., 2021; Calp, 2019). The scale of these problems depends on the sector in which AI is used (automatons performing medical operations, determining drug dosages for patients, military drones, autonomous vehicles, etc.). The prospect of the spread of "intelligent" machines, thinking, and making decisions independently, is a source of concern not only for users and manufacturers (Bielińska-Dusza, 2022). It is necessary to set and

DOI: 10.4324/9781032627236-4

disseminate certain ethical frameworks that guarantee respect for human values and interests by those developing technologies using artificial intelligence (Agostini & Filippini, 2019; Dalenogare et al., 2018; Müller et al., 2021). The use of new technologies is also associated with a number of new legal challenges (Amariles, & Baquero, 2023).

The issue of the impact of AI technology on the operations of businesses is quite widely reported in the literature (Brem et al., 2021; Casero-Ripolles et al., 2022; Johnson et al., 2021; Lu, 2019; Raisch & Krogh, 2018; Wagner et al., 2021). In contrast, there is a lack of studies diagnosing the level of customer trust in artificial intelligence and its ethical aspects in organizations. This provided the rationale for undertaking research in this area and allowed us to identify the research questions: (1) What should organizations do to move toward artificial intelligence systems that are ethical? (2) What benefits they can derive from this, and what will be the consequences if they do not implement such systems? (3) What actions/practices are organizations taking in relation to ethical artificial intelligence?

The purpose of the chapter is to present the results of a report on customer confidence in artificial intelligence and the ethical aspects of using artificial intelligence in organizations.

The chapter begins with a literature review on ethics in artificial intelligence. The next section presents the methodological approach and presents results from the Capgemini Research Institute report on trust and ethics followed by a discussion. Finally, the conclusions, implications, and limitations of this research are presented.

Literature review – ethical aspects of using artificial intelligence in organizations

The term "artificial intelligence" was first used in 1955 by John McCarthy who, formulated the concept of "machines whose actions are similar to human intelligence" (Fountaine, McCarthy & Saleh, 2019). Despite the passage of time, it is difficult to point to a single universally accepted definition. The concept of artificial intelligence is broad and capacious, encompassing various technologies and solutions based on diverse and often complex algorithms. The problem of definition is a challenge for many scientists and lawmakers (Calp, 2019; Nowakowski, 2022).

Recent studies by international organizations and institutions, including those of the EU, are moving away from trying to define artificial intelligence itself to so-called "artificial intelligence systems. "Artificial intelligence system" means software that uses available and known self-learning techniques [by that software], which has the functionality to – for a given set of cells defined by a human – generate (human-determined) results, such as content, predictions, recommendations, or decisions affecting external environments, which such results can affect both positively and negatively, including by causing harm (European Commission, 2019).

In addition, we are dealing with the evolution of solutions in this area. Thus, we should not so much focus on defining artificial intelligence and trying to "humanize" it, but rather put the emphasis on such creation of legal, regulatory and social (and infrastructure) solutions that will ensure that AI is created in an ethical manner. The discussion of the definition of AI includes such issues as the explainability and auditability of models, algorithmic bias and the issue of discrimination, as well as the creation of ethical (trustworthy) artificial intelligence (Nowakowski, 2022).

According to the European Commission (2019), AI ethics is a subfield of applied ethics and technology that focuses on ethical issues related to the design, development, implementation, and use of artificial intelligence. According to UNESCO (2021) recommendations, AI ethics is understood as a dynamic basis for evaluating and guiding AI technologies from the perspective of normative standards, aiming to respect human dignity and human well-being and prevent harm, and based on the ethics of science and technology.

Given the impact of artificial intelligence technologies on humans and their relations in society, many international organizations and countries have taken steps to prepare ethical and legal standards for the development and operation of artificial intelligence. Among those most involved in the discourse in this area are: European Union, Council of Europe, UNESCO, UNICRI, and OECD.

A special contribution to the rule-making process in the field in question was made by the OECD, which, in 2018, established the Artificial Intelligence Expert Group (AIGO). The result of the group's work was the draft recommendations on the ethical aspects of the use of artificial intelligence, which were subsequently adopted by the OECD Council at the ministerial level on May 22, 2019. This is the first act of this stature, created and accepted in intergovernmental cooperation, setting standards for the development of artificial intelligence (Kubiak-Cyrul 2020).

The OECD's recommendations have no binding force, but are taken into account when setting international standards and designing laws in this area. They form a solid foundation for future legislation. As a result, it is possible to speak of the global reach of this standard because the OECD recommendations accept countries that are considered leaders in the implementation of artificial intelligence in the economy. The main recommendation of the OECD is the principle of respect for human autonomy and dignity, human rights, and democratic values. They promote artificial intelligence systems that can be considered trustworthy. In this aspect, the proposed principles complement the demands of the European Commission, which published a document called Ethics Guidelines for Trustworthy AI by an independent high-level expert group on artificial intelligence on April 8, 2019.

Trustworthy artificial intelligence has three characteristics that must characterize artificial intelligence throughout its life cycle (European Commission, 2019; Hsu, 2022).

1 it should be lawful, that is, ensuring compliance with all applicable laws and regulations;
2 it should be ethical, ensuring compliance with ethical principles and values;
3 it should be robust from both a technical and social perspective, as AI systems can cause unintended harm even when used in good faith.

In addition, the European Commission on February 19, 2020 (2020) published a White Paper on artificial intelligence. Its contents cover two main topics: "a policy framework setting out measures to combine efforts at the European, national and regional levels" to help create an "ecosystem of excellence," and "key elements of a future regulatory framework for AI in Europe" to help create an "ecosystem of trust." The entity responsible for the effects of AI should provide an effective path to redress for all victims. Knowledge of how to seek redress should be publicly available, as this is crucial to building trust in AI (Hsu, 2022).

In 2021, under the resolution of the 41st session of the General Conference, UNESCO adopted a recommendation on the ethics of artificial intelligence. The UNESCO Recommendation to Member States, also known as the UNESCO Recommendation, is a normative instrument of a nonbinding nature. It is a set of indications that member states implement in a manner tailored to their own circumstances (Amariles & Baquero, 2023). Countries and international organizations are developing sets of ethical rules for developers of artificial intelligence algorithms. In them, they specify what rules they must follow in the operation of artificial intelligence systems, in addition to achieving business goals. According to the World Economic Forum, we already have about 200 separate codes of ethics for artificial intelligence (Androniceanu, Georgescu & Sabie, 2022).

Methodology

A Capgemini Research Institute report titled Artificial Intelligence and the Ethics Conundrum: How Organizations Can Build Ethically Sound Artificial Intelligence Systems and Earn Trust was used to get a picture of the state of ethics in artificial intelligence. Capgemini Research Institute publishes research on the impact of digital technologies on businesses. The team draws on Capgemini's global network of experts and works closely with academic and technology partners. The institute has dedicated research centers in India, the United Kingdom, the United States, and Singapore. It ranks first in the world in terms of the quality of research conducted by independent analysts (Capgemini Research Institute Report, 2020).

The survey was conducted in April–May 2020. The survey was global in scope was conducted among both customers and executives. The survey to customers included 2,900 consumers from six countries, while the survey to executives included 884 managers from ten countries (the United States – 15%, the United Kingdom – 12%, Germany – 11%, France – 11%, China – 10%, India – 10%, Italy – 8%, Sweden – 8%, Spain – 8%, and Netherlands – 7%). The results were compared with 2019 data from the same countries and sectors (retail –24%, banking – 23%, insurance – 22%, consumer products – 17%, and public/government – 14%) and functions (sales, marketing, and customer support – 70%, and IT – 30%). In addition, in-depth interviews were conducted with executives, data scientists, and experts in the field of ethics in artificial intelligence in August–September 2020. The research sample size was 884 executives in 2020 and 722 in 2019.

In order to determine the progress in implementing ethical AI, several key questions were posed, among others (Capgemini Research Institute Report, 2020):

1 How can ethical AI interactions benefit organizations?
2 What progress have organizations made in the various dimensions of ethical AI (explainability, transparency, integrity, and auditability)?
3 How have organizations evolved in terms of AI ethics awareness and management?
4 Do they have: an ethics leader, a code of conduct, defined policies and practices, and a deeper awareness of ethical issues?
5 How can organizations transition to AI systems that are ethical?

Analysis of report results and discussion

Trust in technology manifests itself in a person's willingness to be influenced by technology based on its usefulness, the predictability of its effects, and the reliability of its providers (Paliszkiewicz & Chen, 2023). The concept of trust, therefore, refers to the belief that the other party in the relationship, in this case technology, will act in a predictable and reliable manner that ensures positive results (Ejdys, 2017; Jasiulewicz, Pietrzak & Wyrzykowska, 2022). A huge challenge in the context of identifying and measuring trust is becoming AI as the most advanced form of technology development (Korzyński et al., 2023; Rzeźnik, 2023; Sala & Tańska, 2022). Based on a survey of more than 800 organizations and 2,900 customers, this section of the chapter will present the key findings of the Capgemini Research Institute Report: AI and the Ethical Conundrum. How organizations can build ethically robust AI systems and gain trust (2020).

In 2020, nearly half of customers (49%) said they trust interactions with AI-based organizations. Meanwhile, while in 2019 as many as 73%, organizations informed users about how AI-related decisions could affect them, that percentage dropped to 59% in 2020. The survey found that 65% of managers were aware that some AI systems they used discriminated against different social groups. The report shows that the biggest social and cultural concerns remain about the data-collection process, especially for healthcare and the public sector. The percentage of customers judging that the organization's actions are fully transparent when it comes to the use of their personal data has dropped by more than 10% (from 76% in 2019 to 62% in 2020).

As the use of AI, such as facial recognition, increases, the ethics of AI are coming under increasing legal and social scrutiny (Floridi & Cowls, 2019). Nearly 60% of organizations have come under legal scrutiny for this reason, and 22% of organizations have faced negative customer feedback due to decisions made by AI systems. Negative customer experiences translate into losses for organizations, with 45%. of customers admitting that they will share their negative feedback with family and friends, 39% will voice their concerns and demand explanations, and 39% will reject AI in favor of more costly human interactions.

The survey also shows that the vast majority of customers (70%) – expect organizations to ensure transparency and fairness in interactions with artificial intelligence. To understand the progress of ethical AI implementation compared to 2019, four key dimensions of AI were focused on: (1) explain ability (AI systems can explain how they work in a language that humans can understand), (2) transparency (AI systems operate in a clear and understandable way), (3) fairness (AI systems ensure fair treatment of all customer groups), and (4) auditability (AI systems can be audited so that the results can be trusted). Although managers are more aware of ethical AI, progress in its implementation has been uneven. An example of this is that almost as many managers (65%) in 2020 said they had taken steps to build "integrity" compared to 2019. (66%) and "auditability" (45% in 2020 compared to 46% in 2019) of their AI systems. The rating for transparency of operations dropped: from 73% in 2019 to 59% in 2020. This is despite the fact that 58% of companies say they are building awareness among employees of the problems that can arise from the use of artificial intelligence.

Organizations are also lagging behind in developing the internal practices needed to ethically implement artificial intelligence. The research shows that there are significant differences in understanding of ethical issues between artificial intelligence developers and users (Table 3.1).

As many as 51% of AI developers (programmers) believe that their systems have made decisions that are inconsistent with the organization's values. But only 40% of users feel the same way, suggesting that problems are not reaching back to those who develop AI systems. This may also be due to differences in understanding of ethical parameters especially in the areas of knowledge, design and management. For example, 40% of AI developers, such as data specialists, have knowledge of why their AI systems produce the results they do and not others, while only 27% of AI users, such as sales and marketing executives.

Table 3.1 Share of executives agreeing with each statement (%)

Statements	*IT/AI and data professionals (AI developers)*	*Sales and marketing executives (AI users)*
(1) Knowledge and awareness		
We have detailed knowledge of how and why our systems produce the output that they do	40	27
(2) Design and implementation		
We focus on 'decision-making process' not the 'decision outcome' to evaluate the performance of AI	65	56
(3) Governance		
We have a dedicated team to monitor the use and implementation of AI from ethics perspective	49	58
We have a confidential hotline/ombudsman to raise ethical issues in our AI systems	62	43
We provide clear options for our users to report issues with automated decisions	62	49
We have a leader who is responsible and accountable for the ethical issues in AI	63	49
(4) Outcomes		
Realized that our AI systems sometimes make decisions which are incompatible with our corporate values.	40	51
Attracted legal scrutiny due to AI system and data handling procedures	53	62

Source: own elaboration based on: Capgemini Research Institute Ethics in AI executive survey, March–May 2020, N = 884 executives.

The number of organizations that have a defined AI ethics charter on how AI systems are developed and used is growing. In 2019, only 5% of organizations had such a charter, and in 2020, 45%. In order for ethics issues to be prioritized in organizations, leadership needs to be established at the highest level. However, according to the report, only 53% of organizations have a person responsible for AI systems ethics, for example, in the role of Chief Ethics Officer. Half of those surveyed admitted that they have an appropriate cell for reporting ethical issues related to AI systems, such as a confidential hotline or ombudsman.

The report also indicates the direction organizations should take to meet consumer expectations. The proposed set of seven key principles is based on leadership, management, and internal best practices (Capgemini Report):

1 Clearly define the purpose of using AI and assess the overall and potential impact on the environment.
2 Implementation of AI for the benefit of society and the environment.

3 Consideration of the principle of diversity and integration throughout the life cycle of AI systems.
4 Ensuring transparency of the rules of system use through technological tools.
5 Humanizing the AI experience and ensuring human supervision of AI systems.
6 Ensuring the technological robustness of AI systems.
7 Protecting people's privacy and giving them agency and responsibility for their interactions with AI.

The seven key dimensions of ethical artificial intelligence are derived from the seven principles of ethical artificial intelligence identified by the European Commission. Organizations should develop and implement ethically sound artificial intelligence by taking a series of actions addressing each dimension (Bielińska-Dusza, 2022).

Conclusion

Ethics – in addition to regulations and proper governance – is essential to properly guide AI development. There is a need for further discussion, modification of laws, and adaptation of existing rules to the specific requirements of AI. Recommended solutions in this regard are (Nowakowski, 2022):

1 Regulating the concept of artificial intelligence systems and linking this definition (or excluding it from the scope) to the concept of software, as well as introducing categories of risk that such systems may generate;
2 Regulating the scope of duties and responsibilities of the various entities in the chain of operation of AI;
3 To establish a certain catalog of ethical standards (codes) on which IS systems should be based;
4 Consideration of the introduction of requirements for the participation of an ethics (artificial intelligence) specialist in the process of design, development and implementation of artificial intelligence (high-risk) systems, the so-called ethics by default and design;
5 Solidifying the nature of ethics solutions with self-governance and the consequences of not adhering to internal (or sectoral) standards.

The study undertaken by the authors contributes to further discussion and research on trust and the ethical, legal, and social aspects of artificial intelligence in management and economics. The issue of AI and its applications and ethical implications for business is multifaceted and requires a comprehensive approach. This study has some limitations, namely, being a kind of research report, it can only be considered exploratory. As a result, subsequent studies should take into account a broader research perspective and a more methodological approach.

The presented research results, as well as numerous publications, indicate the need for further research. The rapid development of artificial intelligence and its exponentially growing impact on individual and collective life make its ethical and various social aspects central issues from the point of view of: sustainable development, security, social equality, peacekeeping, and the future of human rights, democracy and personal freedom, as well as today's situation in these areas. In the course of further research, it is also worth answering the question: in addition to a horizontal legal and ethical framework, a sectoral approach that takes into account the context-specificity of AI systems would also be useful.

References

Adetayo, A. J. (2023). Artificial intelligence chatbots in academic libraries: The rise of ChatGPT. *Library Hi Tech News*, 40 (3), 18–21.

Agostini, L., & Filippini, R. (2019). Organizational and managerial challenges in the path toward Industry 4.0. *European Journal of Innovation Management,* 22 (3): 406–421. https://doi.org/10.1108/EJIM-02–2018–0030/FULL/XML.

Amariles, D. R., & Baquero, P. M. (2023). Promises and limits of law for a human-centric artificial intelligence. *Computer Law & Security Review*, 48: 1–9.

Androniceanu, A., Georgescu, I., & Sabie, O. -M. (2022). The impact of digitalization on public administration, economic development, and well-being in the EU countries. *Central European Public Administration Review*, 20 (1): 7–29.

Bérubé, M., Giannelia, T., & Vial, G. (2021). *Barriers to the Implementation of AI in Organizations: Findings from a Delphi Study*. Proceedings of the 54th Hawaii International Conference on System Sciences, 10. https://hdl.handle.net/10125/71425.

Bielińska-Dusza, E. (2022). Transformacja technologiczna przedsiębiorstw jako skutek zastosowania sztucznej inteligencji. [Transformation of technology in enterprises as a result of the artificial intelligence application] *Organization and Management,* 191 (2): 189–208. Retrieved from https://www.proquest.com/scholarly-journals/transformacja-technologiczna-przedsiębiorstw-jako/docview/2826827153/se-2.

Bielińska-Dusza, E., Hamerska, M., & Zak, A. (2021). *Sustainable Mobility and the Smart City: A Vision of the City of the Future.* Poland: The Case Study of Cracow. https://doi. org/10.3390/en14237936.

Bohr, A., & Memarzadeh, K. (2020). *Artificial Intelligence in Healthcare*. Academic Press, Elsevier. https://shop.elsevier.com/books/artificial-intelligence-in-healthcare/bohr/978-0-12-818438-7

Brem, A., Giones, F., & Werle, M. (2021). The AI digital revolution in innovation: A conceptual framework of artificial intelligence technologies for the management of innovation. *IEEE Transactions on Engineering Management*, 70 (2), 1–7. https://doi. org/10.1109/TEM.2021.3109983.

Buch, V. H., Ahmed, I., & Maruthappu, M. (2018). Artificial intelligence in medicine: Current trends and future possibilities. *British Journal of General Practice,* 68 (668): 143–144. https://doi.org/10.3399/BJGP18X695213.

Caballé, S., Demetriadis, S., Gómez-Sánchez, E., Papadopoulos, P., & Weinberger, A. (2021). *Intelligent Systems and Learning Data Analytics in Online Education*. Academic Press,

Elsevier Inc. https://shop.elsevier.com/books/intelligent-systems-and-learning-data-analytics-in-online-education/caballe/978-0-12-823410-5

Calp, M. H. (2019). The role of artificial intelligence within the scope of digital transformation in enterprises. In: Ekren, G., Erkollar, A. & Oberer, B. (ed.), *Advanced MIS and Digital Transformation for Increased Creativity and Innovation in Business* (pp. 122–146), IGI Global, Uniwersytet Techniczny Karadeniz/ [Karadeniz Technical University]. https://doi.org/10.4018/978-1-5225-9550-2.CH006.

Capgemini Research Institute Report. (2020). *AI and the Ethical Conundrum. How Organizations Can Build Ethically Robust AI Systems and Gain Trust?* https://www.capgemini.com/wp-content/uploads/2020/10/AI-and-the-Ethical-Conundrum-Report.pdf.

Casero-Ripolles, A., De-Lima-Santos, M.-F., & Ceron, W. (2022). Artificial intelligence in news media: Current perceptions and future outlook. *Journal and Media,* 3 (1): 13–26. https://doi.org/10.3390/JOURNALMEDIA3010002.

Chintalapati, S., & Pandey, S. K. (2021). Artificial intelligence in marketing: A systematic literature review. *International Journal of Market Research,* 64 (1): 38–68. https://doi.org/10.1177/14707853211018428.

Costello, E. (2023). ChatGPT and the educational AI Chatter: Full of bullshit or trying to tell us something? *Postdigital Science and Education.* https://doi.org/10.1007/s42438-023-00398-5.

Dalenogare, L. S., Benitez, G. B., Ayala, N. F., & Frank, A. G. (2018). The expected contribution of Industry 4.0 technologies for industrial performance. *International Journal of Production Economics,* 204: 383–394. https://doi.org/10.1016/J. IJPE.2018.08.019.

Ejdys, J. (2017). Determinanty zaufania do technologii [Determinants of trust in technology]. *Przegląd Organizacji* [Organization Overview], 12: 20–27.

Ekanayake, J., & Saputhanthri, L. (2020). E-Agro: Intelligent chat-bot. IoT and artificial intelligence to enhance farming industry. *Agris on-Line Papers in Economics and Informatics*, 12 (1): 15–21. https://doi.org/10.7160/aol.2020.120102

European Commission. (2019). *Wytyczne w zakresie etyki dotyczące godnej zaufania sztucznej inteligencji. Grupa ekspertów wysokiego szczebla ds. sztucznej inteligencji* [Ethics guidelines for trustworthy artificial intelligence. High-level expert group on artificial intelligence]. Bruksela.

European Commission. (2020). *Biała Księga. W sprawie sztucznej inteligencji Europejskie podejście do doskonałości i zaufania*, Komisja Europejska [*White Paper. On Artificial Intelligence A European Approach to Excellence and Trust*, European Commission]: 1–31. https://ec.europa.eu/info/sites/default/files/commission-white-paper-artificial-intelligence-feb2020_pl.pdf.

Floridi, L., & Cowls, J. (2019). A unified framework of five principles for AI in society. *Harvard Data Science Review*, 1 (1), 1–5. https://doi.org/10.1162/99608f92.8cd550d1

Fountaine, T., McCarthy, B., & Saleh, T. (2019). Building the AI-powered organization. *Harvard Business Review.* https://wuyuansheng.com/doc/Databricks-AI-Powered-Org__Article-Licensing-July21-1.pdf

Hsu, J. (2022). Europe's AI regulations could lead the way for the world. *New Scientist,* 256 (3419). https://www.newscientist.com/article/mg25634192-300-eus-artificial-intelligence-act-will-lead-the-world-on-regulating-ai/

Imran, N., & Jawaid, M. (2020). Artificial intelligence in medical education are we ready for it? *Pakistan Journal of Medical Sciences,* 36 (5): 857–859. https://doi.org/10.12669/PJMS.36.5.3042.

Jasiulewicz, A., Pietrzak, P., & Wyrzykowska, B. (2022). Trust and the digital economy: A framework for analysis. In: Paliszkiewicz, J. & Chen, K. (ed.), *Trust, Organizations and the Digital Economy. Theory and Practice* (pp. 96–107). New York: Taylor & Francis Group.

Johnson, C. D., Bauer, B. C., & Niederman, F. (2021). The automation of management and business science. *Academy of Management Perspectives,* 35 (2): 292–309. https:// doi. org/10.5465/AMP.2017.0159.

Korzyński, P., Mazurek, G., Altmann, A., Ejdys, J., Kazlauskaite, R., Paliszkiewicz, J., Wach, K., & Ziemba, E. (2023). Generative artificial intelligence as a new context for management theories: analysis of ChatGPT. *Central European Management Journal,* 31 (1), 3–13. https://doi.org/10.1108/CEMJ-02-2023-0091.

Krogh, G. von. (2018. Artificial intelligence in organizations: New opportunities for phenomenon-based theorizing. *Academy of Management,* 4 (4): 404–409. https://doi. org/10.5465/AMD.2018.0084.

Kubiak-Cyrul, A. (2020). Rekomendacje OECD jako ramy etyczne i prawne dla rozwoju sztucznej inteligencji. [OECD recommendations as an ethical and legal framework for the development of artificial intelligence]. *Studia Prawnicze. Rozprawy i Materiały. [Studies in Law: Research Papers*], 1 (26), 103–116.

Longbing, C. (2022). AI in finance: Challenges, techniques, and opportunities. *ACM Computing Surveys (CSUR),* 55 (3): 1–38. https://doi.org/10.1145/3502289.

Lu, Y. (2019). Artificial intelligence: A survey on evolution, models, applications and future trends. *Journal of Management Analytics,* 6 (1): 1–29. https://doi.org/10.1080/ 23270012. 2019.1570365.

Luckey, D., Fritz, H., Legatiuk, D., Dragos, K., & Smarsly, K. (2020). Artificial intelligence techniques for smart city applications. In: Toledo, S. E. & Scheer, S. (ed.), *Proceedings of the 18th International Conference on Computing in Civil and Building Engineering.* ICCCBE, Lecture Notes in Civil Engineering (Vol. 98, pp. 3–15). Cham: Springer. https://doi.org/10.1007/978-3-030-51295-8_1.

Müller, J. M., Buliga, O., & Voigt, K. I. (2021). The role of absorptive capacity and innovation strategy in the design of industry 4.0 business Models-A comparison between SMEs and large enterprises. *European Management Journal,* 39 (3): 333–343. https:// doi. org/10.1016/j.emj.2020.01.002.

Nowakowski, M. (2022). O moralnej odpowiedzialności HAL-a 9000, czyli etyka sztucznej inteligencji w praktyce. Czy potrzebujemy definicji sztucznej inteligencji? [On the moral responsibility of HAL 9000, or the ethics of artificial intelligence in practice. Do we need a definition of artificial intelligence?] *PME,* 1, 4–10. https://repozytorium. uni.wroc.pl/

Paliszkiewicz, J., & Chen, K. (2023). Building digital trust in business. In: Paliszkiewicz, J., Chen, K. & Launer, M. (ed.), *Trust and Digital Business, Theory and Practice* (pp. 4–6). New York and London: Routledge Taylor & Francis.

Rzeźnik, G. (ed.) (2023). Zastosowanie sztucznej inteligencji w gospodarce. Przegląd wybranych inicjatyw i technologii z rekomendacjami dla przedsiębiorców [Application of Artificial Intelligence in the Economy. Review of selected initiatives and technologies with recommendations for entrepreneurs]. Raport tematyczny numer 33 [Topical Report Number 33]. Warsaw. Krajowa inteligentna specjalizacja [National smart specialization]. PARP.

Sala, J., & Tańska, H. (2022). Instrumentalizacja sztucznej inteligencji w polityce aktywności naukowej w Polsce [Instrumentalization of artificial intelligence in science activity policy in Poland]. *Nierówności Społeczne a Wzrost Gospodarczy*, [*Social Inequalities and Economic Growth]*, 71 (3/2022): 99–102. https://doi. org/10.15584/nsawg.2022.3.6 s.

Siarry, P. (2022). *Engineering Applications of Artificial Intelligence* 110. https://www.sciencedirect.com/journal/engineering-

Sztuczna inteligencja dla Europy [Artificial intelligence for Europe]. (2018). Komunikat Komisji do Parlamentu Europejskiego, Rady Europejskiej, Europejskiego Komitetu Ekonomiczno-Społecznego i Komitetu Regionów, [Communication from the Commission to the European Parliament, the European Council, the European Economic and Social Committee and the Committee of the Regions].COM, 237 final, 25.04.2018, Bruksela.

Terashima, N. (2002). *Intelligent Communication Systems*. Waseda University Tokyo: Academic Press.

Tlili, A., Shehata, B., Adarkwah, M. A., Bozkurt, A., Hickey, D. T., Huang, R., & Agyemang, B. (2023). What if the devil is my guardian angel: ChatGPT as a case study of using chatbots in education. *Smart Learning Environments, 10* (1), 1–24.

UNESCO. (2021). Zalecenia UNESO w sprawie sztucznej inteligencji przyjęte 23 listopada 2021 roku [UNESO recommendations on artificial intelligence adopted on November 23, 2021]. https://www.unesco.pl/fileadmin/user_upload/pdf/Sztuczna_inteligencja-4.pdf

Wagner, G., Lukyanenko, R., & Paré, G. (2021). Artificial intelligence and the conduct of literature reviews, *Journal of Information Technology, 37* (2), 209–226. https://doi. org/10.1177/02683962211048201.

4 Understanding online trust in the digital era

Joanna Rakowska and Marta Barna

Introduction

Trust has been long acknowledged to be a key factor of development in different social and economic contexts, in different institutional and organizational backgrounds, as well as in different human-to-human, business-to-business and human-to-business relations. However, trust itself is not an independent phenomenon – it is influenced by certain factors, which nowadays are significantly shaped up by the conditions of the digital era – also called 'the digital era of plenty'.

In recent years, digital transformation has accelerated at an unprecedented pace. The advent of online platforms, big data, artificial intelligence, virtual reality, virtual assistants, and blockchain technology (Bandara et al., 2020) has caused dramatic technological changes resulted in the digitization of human offline activities, enhancing, and often even replacing face-to-face interactions via Internet- or network-connected devices. The communication moved to, for example, Twitter, Facebook, blogospheres, Instagram, while services became available from fast-developing service platforms (Uber and Airbnb), accompanied by search engines (Google), crowdsourcing intermediaries, such as Upwork and ZBJ, real estate sites, telecommunication sites, sports and entertainment sites, travel sites (Lin, 2021) and many other.

The digitalization process was intensified by COVID-19 pandemic, also called 'infodemic' (Bunker, 2020). It made irreversible changes in numerous aspects of our daily routines and businesses, shifting them during lockdowns to digital platforms (Kumar & Malik, 2021) to an even larger extent than before. As people have got more and more tools and easier access to doing business and their private activities online, the researchers began to distinguish the offline and the online trust of digital era.

Paliszkiewicz and co-authors (2021) highlight that online trust is a basis for collaboration and further advancement; however, it is a very fragile and intangible asset, easy to lose, and challenging to rebuild. That is why studies on online trust are of high importance to many scientific disciplines and many practical aspects of sustainable development in the digital era.

DOI: 10.4324/9781032627236-5

Online trust has been studied extensively in the literature. Researchers have investigated its various aspects in different social, cultural, economic, and demographic contexts, providing valuable insights. However, online trust as a complex and multi-faceted concept remains understudied, as the literature lacks a meta synthesis of antecedents of online trust. It is the research gap which this chapter aims to fill in.

The following parts of this chapter present methodology of the research, literature review, findings on antecedents increasing online trust, on antecedents limiting online trust, as well as conclusions, limitations, and recommendations for further studies on online trust in digital era.

Literature review

Trust and trust relationships have been a subject of extensive research across various disciplines since the 1950s (McKnight & Chervany, 2001; Kramer, 2009). The literature on trust is vast in many fields, just to mention philosophy, sociology, psychology, management, marketing, ergonomics and economics, and industrial psychology. And though each of them has produced its own concepts, definitions, and findings, researchers acknowledge the value of trust and the fact that it enables people to live in risky and uncertain situations.

Blöbaum (2016) argues that digitalization opens digital possibilities for the development of equivalents of trust. That is why trust in the digital era has been categorized as offline and online phenomenon. Still, many authors suggest that the research on online trust should build on the findings on offline trust (Corritore et al., 2003). Connolly (2020) explains that 'trust is a psychological state of mind that is multi-dimensional and comprised of factors such as expectancy, faith, consistency, and risk'. Further, trust can be modeled either as a state or as a relation between trustee and trustor (Blöbaum, 2016). A trustor is the one who places trust, and the trustee, who is being trusted, which should lead to gains for the trustor (Tsfati & Cappella, 2003). Taking the trustor-trustee relation into consideration, trust can be defined as the willingness to depend upon and make oneself vulnerable to the trustee (Mayer et al., 1995). The ability to build trust is 'the disposition to engage in social exchanges that involve uncertainty and vulnerability, but are also potentially rewarding' (Bicchieri et al., 2004, p. 286).

However, despite the common application of the understanding of offline trust for online trust research, some authors point out that the latter one is conditioned by different factors than the offline trust and thus requires a different definition. Additionally, digitalization poses new challenges to both online trustees and online trustors, different from those existing offline. For instance, one of the biggest challenges for the online trustees is to find new forms of presenting their trustworthiness to the online trustors. In this approach, online trust is also called 'e-trust' or 'digital trust'.

Investigating e-trust, Peštek, Resić and Nožica (2011) define it as: 'the consumer's commitment to trust that an offeror will successfully fulfil their online deal, i.e. in accordance to established terms'. While, due to the complexity of the phenomenon of trust in e-commerce, Grabner-Kräuter and Kaluscha (2003) propose a set of trust constructs that include system trust as institutional phenomenon and personal and interpersonal forms of trust as an expression of dispositional trust, trusting beliefs and intentions, as well as trust-related behaviors.

Despite the lack of agreement on the definition of online trust, it has been generally acknowledged that it is a complex and multifaceted concept, requiring recognition of its antecedents in different context of digital activities.

Methodology

The aim of the chapter is to enhance our understanding of online trust in the digital era by answering the following research questions [RQ]

RQ 1: What contexts of online trust were studied so far?
RQ 2: What antecedents increasing online trust have been diagnosed?
RQ 3: What antecedents limiting online distrust have been diagnosed?
RQ 4: What diagnosed antecedents build up online trust in different digital contexts?

Answer to research question 4 is intended as a concluding section.

To retrieve the data, we searched Scopus and Web of Science for peer-reviewed papers presenting findings on online trust. We used the term and keyword 'online' OR 'e-trust' OR 'digital' AND 'trust' in both cases, with the range publication from 2018 until June 2023. Next manual search in Google Scholar was conducted to complete the set of papers. Furthermore, the articles were checked according to the eligible criteria of inclusion and exclusion.

Consequently, we obtained 60 papers, which enabled an integrative review of the empirical literature on online trust in digital era and provided qualitative data for content analysis and meta-synthesis (Levitt, 2018; Paterson et al., 2001).

Results and discussion

As so many aspects of our lives shifted to the digital world, more opportunities and at the same time more risks have appeared, conditioning online trust in various ways. And so, researchers have been investigating more and more aspects and contexts of online trust, often referring to different fields of online activities. As new forms of online business, information exchange and entertainment emerge all the time, findings on contexts of so far studies in online trust presented in this chapter focus on the main and most often discussed ones.

Since millions of people do business online every day, online trust in business-to-business (B2B), business-to-consumer (B2C), and consumer-to consumer (C2C) has been investigated most often (e.g., Bandara & Syed, 2023; Hill & de Zubielqui, 2023; Zhang & Li, 2019). In the business context, special attention has been paid to trust in e-commerce, m-commerce (Sarkar et al., 2020), and digital banking, enabling online sale and purchase of products and services (Al Halbusi et al., 2022; Bandara et al., 2020; Connolly, 2013; Fernández-Bonilla et al., 2022; Lim et al., 2006; Soleimani, 2022; Van Dyke et al., 2007; Yunusa et al., 2022). So far studies in online trust concerning online business also discuss trust in digital entrepreneurship, for example, online service markets as Crowdsourcing intermediaries, Upwork, and ZBJ (Du & Mao, 2018), while online trust in digital innovation provided by start-ups (Konya-Baumbach et al., 2019) is another crucial context explaining the high failure rate in this kind of online businesses. Trust in online recycling platforms (Huang et al., 2023) and trust in sharing economy platform providers (Ko et al., 2022; Ko et al., 2022) are other examples of business contexts for research in online trust.

Online trust or e-trust has also been studied in the context of customer's trust in sellers or service providers and seller's or provider's trust in customers, which refers to changing positions of customers and sellers as either trustors or trustees. Moriuchi and Takahashi (2022) claim that depending on the complexity of the purchase-sale process, online trust can be multidimensional, representing two distinct aspects: trust in the technology/platform and trust in third-party sellers.

Online and offline businesses are supported by e-marketing (also: digital marketing) and social media advertising, which together with trust in online reviews (Dong et al., 2019) make other broadly surveyed contexts for studies in online trust (Carlson et al., 2022; Popova et al., 2019, Sarkar et al., 2020).

Social media and social networks have become overwhelmingly popular among their users, thus providing an extensive context for studies on e-trust in social media/social networks (Briones et al., 2011; Flanagin & Metzger, 2000; Grabner-Kräuter & Bitter, 2015; Laifa et al., 2019; Torres et al., 2018) and at the same time in online communication platforms (Tong, & Chan, 2022) and search engines, for example, trust in Google (Shaker, 2006).

Internet has become a very popular and impactful medium, attracting the attention of researchers by giving them a new context for analyzing trust in different types of information provided online (Fleming, 2009), trust in online journalism (Grosser, 2016; Hayes et al., 2007), as well as trust in online news, in message and in the source (Russmann & Hess, 2020).

Antecedents increasing online trust

Online trust is critical for all e-activities and influenced by various factors. Numerous studies provide findings on antecedents of online trust in different contexts.

Hopmann et al. (2015) suggest that people need to trust a particular media outlet or type to select it in the first place, and the repeated positive experiences with (or lack of disappointment in) a media outlet or type can increase trust in it. Ye et al. (2020) also argue that consumer develops trust through the number of interactions with a retail Web site over time and that the more a consumer interacts with it, the more they are likely to trust it.

The positive social perceptions of a website can also play a role in building trust. MIP (2015) found that recommendations from peers can be an important factor in building trust in online content as young adults actively follow the word-of-mouth. This finding was confirmed also by Pomianek and Muça (2023), who found that the most important factor of choice is the popularity of a given platform among friends. Besides, people most often choose free platforms.

Credibility, ease of use, risk (Corritore et al., 2003), website accessibility, privacy, quality, post-sale treatment (Peštek et al., 2011), positive referrals, and posts (Awad & Ragowsky, 2008) can positively impact e-trust. Moriuchi and Takahashi (2022) indicated that communication decreases uncertainty and increases confidence and trust in consumers' purchases, leading to higher engagement on the C2C platform (eBay or Mercari).

Mistrust or skepticisms toward the mainstream media are other antecedents of online trust since they make people go online to seek alternative information sources (Tsfati & Cappella, 2003). Russman and Hess (2020) found that user trust in online and social media if they assess the published content as reliable and objective.

Konya-Baumbach et al. (2019) indicate trustworthiness as the antecedent of online trust, while Carlson and so-authors (2022) argue that trust in social media advertising is higher when users use the Internet to pass time socializing with friends (social interaction motivation) or engaging in enjoyable activities (entertainment motivation).

Antecedents decreasing online trust

Along with the antecedents stimulating online trust, there are many factors limiting it. Hill and de Zubielqui (2023) found that online fraud, identity theft, security and privacy breaches, and online scams have a negative effect on online trust. Similarly, McKnight, Choudhury, and Kacmar (2002) identified hackers stealing personal information or unknown sellers as determinants of consumers' decisions not to buy online. Corritore, Kracher, and Wiedenbeck (2003) concluded that the security of a website determines trust in e-commerce.

Privacy concerns (Bandara et al., 2020; Konya-Baumbach et al., 2019; Van Dyke et al., 2007), together with fear of personal data misuse, doubts regarding the performance of online platforms, and consumers' unfamiliarity with the launching company (Konya-Baumbach et al., 2019), are indicated as one of the strongest factors limiting online trust. Privacy concerns and information privacy previously indicated as antecedents limiting online trust in C2C and B2C negatively impact trust in B2B and e-marketing as well Bandara and Syed (2023).

Grabner-Kräuter and Kaluscha (2003) noted that online transactions are characterized by uncertainty, anonymity, lack of control, and potential opportunism due to the lack of simultaneous exchange of goods and money which altogether limits online trust. Lack of established reputation of online stores decreases consumers' trust (Lim et al., 2006; Škerháková et al., 2022), similarly to service failure (Ye et al., 2020).

In social media and social networks context, poor privacy management also decreases online trust (Lin, 2021), as the users wish to control the data about themselves (Bélanger & Crossler, 2011). Online offense is another factor strongly diminishing online trust in the context of social media (Laifa et al., 2019). Trust in social media advertising is lower when people use the Internet to learn about current events or to do research (so-called informational motivation) (Carlson et al., 2022).

Inconsistent findings on antecedents increasing and limiting online trust

Russman and Hess (2020) conclude that they found no differences in online trust of young adults dealing with the media in their studies or work and those with no direct connections to this field, although the authors assumed that the first group would be more critical with the online media. Findings concerning the role played by age and education as predictors of trust in media are also inconsistent (Gronke & Cook, 2007), as in the paper by Tsfati and Ariely (2014), education was a negative predictor of trust, whereas in other studies education appeared a positive predictor of trust.

Although so many authors indicated identity theft, fear of personal data misuse and privacy concerns as crucial antecedents limiting online trust, other studies revealed 'privacy paradox' (e.g., Barnes, 2006) or 'privacy dilemma' (Brandtzaeg et al., 2010). The paradox refers to the situation where users are aware of the risks and threats connected with sharing their thoughts, experiences, images, files, videos etc. online, but they do share them anyway (Grabner-Kräuter & Bitter, 2015). The paradox presents inconsistency between the declared fears and doubts that should presumably diminish online trust and actual online actions of the users of the Internet.

Conclusions, limitations, and further research

The contribution of this chapter is consolidating and synthesizing findings on online trust antecedents in different digital contexts, which enhances the theoretical discussion, gives grounds for further research, and provides knowledge to the users and providers of digital products and services.

Online trust in **e-business context** can emerge and develop thanks to repeated positive experiences or lack of disappointment of customers, the number of interactions with the retail website, positive social perception, website accessibility, privacy, quality, postsale treatment, positive referrals, and seller-customer communication. Online trust in e-business context can be limited by unknown sellers, fear of personal data misuse, doubts concerning online platforms performance, consumers' unfamiliarity with the launching company.

Online trust in **e-marketing context** can increase if the user finds adverts when socializing with friends online or seeking entertainment. Privacy and information privacy concerns limit online trust in this context.

Online trust in **social media or Internet information context** is … by recommendations from peers, popularity of the platform among friends, lack of payment for the use of the services, mistrust, or skepticisms toward mainstream media. Poor privacy management and online offense significantly limit online trust in this context.

Findings show that credibility, ease of use, perceived low risk are universal antecedents increasing online trust in **different contexts**. On the other hand, online fraud, identity theft, security and privacy breaches and online scams decrease online trust in different e-contexts. The results also revealed 'privacy paradox' or 'privacy dilemma' concerning the difference between consumers' online distrust and their actual online behavior.

Limitations of this study, which presents meta synthesis, result from the limitations of the available existing qualitative data on online trust. These are constraints in terms of demographic features on online trustors and trustees, location of most studies in North American, European, and East Asian countries, while skipping other national and cultural contexts, as well as concentration of the research on several main online contexts.

It is recommended that further research in online trust should focus on extending the scope of investigation to dimmish the above listed limitations.

With our findings, we hope to provide a comprehensive and solid basis for the ongoing discussions on online trust, indicate directions for future research and provide stakeholders of the Internet with guidance on developing online trust and successful trust-based online relations.

References

Al Halbusi, H., Al-Sulaiti, K., Abbas, J., & Al-Sulaiti, I. (2022). Assessing factors influencing technology adoption for online purchasing amid COVID-19 in Qatar: Moderating role of word of mouth. *Frontiers in Environmental Science, 10*, 942527. https://doi.org/10.3389/fenvs.2022.942527

Awad, N. F., & Ragowsky, A. (2008). Establishing trust in electronic commerce through online word of mouth: An examination across genders. *Journal of Management Information Systems, 24*(4), 101–121. https://doi.org/10.2753/MIS0742-1222240404

Bandara, R., Fernando, M., & Akter, S. (2020). Privacy concerns in E-commerce: A taxonomy and a future research agenda. *Electronic Markets, 30*, 629–647. https://doi.org/10.1007/s12525-019-00375-6

Bandara, W., & Syed, R. (2023). The role of a protocol in a systematic literature review. *Journal of Decision Systems, 0*(0), 1–18.

Barnes, S. (2006). A privacy paradox: Social networking in the United States. *First Monday, 11*. https://doi.org/10.5210/fm.v11i9.1394
Bélanger, F., & Crossler, R. E. (2011). Privacy in the digital age: A review of information privacy research in information systems. *MIS Quarterly, 35*(4), 1017–1041. https://doi.org/10.2307/23044002
Bicchieri, C., Duffy, J., & Tolle†, G. (2004). Trust among strangers*. *Philosophy of Science, 71*(3), 286–319. https://doi.org/10.1086/381411
Blöbaum, B. (2016). Key factors in the process of trust: On the analysis of trust under digital conditions. In B. Blöbaum (Ed.), *Trust and Communication in a Digitized World* (pp. 1–16). Berlin: Springer.
Brandtzæg, P. B., Lüders, M., & Skjetne, J. H. (2010). Too many Facebook "friends"? Content sharing and sociability versus the need for privacy in social network sites. *Journal of Human-Computer Interaction, 26*(11–12), 1006–1030.
Briones, R. L., Kuch, B., Liu, B. F., & Jin, Y. (2011). Keeping up with the digital age: How the American Red Cross uses social media to build relationships. *Public Relations Review, 37*(1), 37–43.
Bunker, D. (2020). Who do you trust? The digital destruction of shared situational awareness and the COVID-19 infodemic. *International Journal of Information Management, 55*, 102201. https://doi.org/10.1016/j.ijinfomgt.2020.102201
Carlson, J. R., Hanson, S., Pancras, J., Ross, W. T., & Rousseau-Anderson, J. (2022). Social media advertising: How online motivations and congruency influence perceptions of trust. *Journal of Consumer Behaviour, 21*(2), 197–213. https://doi.org/10.1002/cb.1989
Connolly, B. (2020). *Digital Trust: Social Media Strategies to Increase Trust and Engage*. London: Bloomsbury Business.
Connolly, R. (2013). Trust in commercial and personal transactions in the digital age. In W. H. Dutton (Ed.), *The Oxford Handbook of Internet Studies* (online ed., pp. 1–22). Oxford: Oxford Academic. https://doi.org/10.1093/oxfordhb/9780199589074.013.0013
Corritore, C. L., Kracher, B., & Wiedenbeck, S. (2003). On-line trust: Concepts, evolving themes, a model. *International Journal of Human-Computer Studies, 58*(6), 737–758. https://doi.org/10.1016/S1071-5819(03)00041-7
Dong, B., Li, M., & Sivakumar, K. (2019). Online review characteristics and trust: A cross-country examination. *Decision Sciences, 50*, 537–566. https://doi.org/10.1111/deci.12339
Du, W. (Derek), & Mao, J.-Y. (2018). Developing and maintaining clients' trust through institutional mechanisms in online service markets for digital entrepreneurs: A process model. *The Journal of Strategic Information Systems, 27*(4), 296–310. https://doi.org/10.1016/j.jsis.2018.07.001
Fernández-Bonilla, F., Gijón, C., & De la Vega, B. (2022). E-commerce in Spain: Determining factors and the importance of the e-trust. *Telecommunications Policy, 46*(1), 102280. https://doi.org/10.1016/j.telpol.2021.102280
Flanagin, A. J., & Metzger, M. J. (2000). Perceptions of Internet information credibility. *Journalism Mass Communication Quarterly, 77*(3), 515–540.
Fleming, J. (2009). "Truthiness" and trust: News media literacy strategies in the digital age. In K. Tyner (Ed.), *Media Literacy: New Agendas in Communication* (1st ed., pp. 1–20). New York: Routledge. https://doi.org/10.4324/9780203867273
Grabner-Kräuter, S., & Bitter, S. (2015). Trust in online social networks: A multifaceted perspective. *Forum for Social Economics, 44*(1), 48–68.

Grabner-Kräuter, S., & Kaluscha, E. A. (2003). Empirical research in on-line trust: A review and critical assessment. *International Journal of Human-Computer Studies, 58*(6), 783–812. https://doi.org/10.1016/S1071-5819(03)00043-0

Gronke, P., & Cook, T. E. (2007). Disdaining the media: The American public's changing attitudes toward the media. *Political Communication*, *24*(3), 259–281.

Grosser, K. M. (2016). Trust in online journalism. *Digital Journalism, 4*(8), 1036–1057. https://doi.org/10.1080/21670811.2015.1127174

Hayes, A. S., Singer, J. B., & Ceppos, J. (2007). Shifting roles, enduring values: The credible journalist in a digital age. *Journal of Mass Media Ethics, 22*(4), 262–279.

Hill, S. R., & Corral de Zubielqui, G. (2023). Managing privacy in B2B marketing: A systematic literature review. *Journal of Computer Information Systems, 63*(3), 574–591. https://doi.org/10.1080/08874417.2022.2087792

Hopmann, D. N., Shehata, A., & Strömbäck, J. (2015). Contagious media effects: How media use and exposure to game-framed news influence media trust. *Mass Communication and Society*, *18*(6), 776–798.

Huang, Y., Liang, Y., Gong, Y., & Yuan, Z. (2023). Role of trust-building in online recycling platforms. *International Journal of Production Economics, 263*, 108962. https://doi.org/10.1016/j.ijpe.2023.108962

Ko, G., Amankwah-Amoah, J., Appiah, G., & Larimo, J. (2022). Non-market strategies and building digital trust in sharing economy platforms. *Journal of International Management, 28*(1), 100909. https://doi.org/10.1016/j.intman.2021.100909

Konya-Baumbach, E., Schuhmacher, M. C., Kuester, S., & Kuharev, V. (2019). Making a first impression as a start-up: Strategies to overcome low initial trust perceptions in digital innovation adoption. *International Journal of Research in Marketing, 36*(3), 385–399. https://doi.org/10.1016/j.ijresmar.2019.01.008

Kramer, R. M. (2009). Rethinking trust. *Harvard Business Review, 87*(6), 68–77.

Kumar, A., & Malik, A. (2021). A comparative analysis of various models for assessment of trust in digital age accelerated by Covid-19. In *2021 International Conference on Computing Sciences (ICCS)* (pp. 156–160). IEEE.

Laifa, M., Giglou, R. I., Samir, A., & Maamri, R. (2019). Forgiveness predictors and trust in a digital age. *International Journal of Technology and Human Interaction, 14*(4), 1–15.

Levitt, H. M. (2018). How to conduct a qualitative meta-analysis: Tailoring methods to enhance methodological integrity. *Psychotherapy Research*, *28*(3), 367–378. https://doi.org/10.1080/10503307.2018.1447708

Lim, K. H., Sia, C. L., Lee, M. K., & Benbasat, I. (2006). Do I trust you online, and if so, will I buy? An empirical study of two trust-building strategies. *Journal of Management Information Systems, 23*(2), 233–266.

Lin, Y. J. (2021). An exploration into trust and privacy management in a digital age. *International Journal for Applied Information Management, 2*(1), 34–43.

Mayer, R. C., Davis, J. H., & Schoorman F. D. (1995). An integrative model of organizational trust. *The Academy of Management Review*, *20*(3), 709–734, https://doi.org/10.2307/258792

McKnight, D. H., & Chervany, N. L. (2001). Trust and distrust definitions: One bite at a time. In N. L. Chervany & D. J. Galletta (Eds.), *Trust in Cyber-Societies: Integrating the Human and Artificial Perspectives* (pp. 27–54). Springer, Berlin Heidelberg.

McKnight, D. H., Choudhury, V., & Kacmar, C. (2002). Developing and validating trust measures for e-commerce: An integrative typology. *Information Systems Research, 13*(3), 334–359. https://doi.org/10.1287/isre.13.3.334.81

Media Insight Project. (2015). *How Millennials Get News: Inside the Habits of America's First Digital Generation.* Chicago, IL: American Press Institute and the Associated Press-NORC Centre for Public Affairs Research, University of Chicago.

Moriuchi, E., & Takahashi, I. (2022). The role of perceived value, trust and engagement in the C2C online secondary marketplace. *Journal of Business Research, 148*, 76–88. https://doi.org/10.1016/j.jbusres.2022.04.029

Paliszkiewicz, J., Gołuchowski, J., & Chen, K. (2021). The meaning and interdisciplinary nature of trust in the digital economy – Future directions. In Paliszkiewicz, J., Gołuchowski, J., & Chen, K. (Eds.) *Trust, Organizations and the Digital Economy* (pp. 12–31). New York: Routledge.

Paterson, B. L., Thorne, S. E., Canam, C., & Jillings, C. (2001). *Meta-Study of Qualitative Health Research: A Practical Guide to Meta-Analysis and Meta-Synthesis.* Thousand Oaks, CA: Sage.

Peštek, A., Resić, E., & Nožica, M. (2011). Model of trust in E-transactions. *Economic Research-Ekonomska Istraživanja, 24*(3), 131–146. https://doi.org/10.1080/1331677X.2011.11517472

Pomianek, I., & Muça, E. (2023). Expectations and trust of users of digital platforms in Albania and Poland. In J. Paliszkiewicz, J. L. G. Cusumano, & J. Gołuchowski (Eds.), T*rust, Digital Business and Technology. Issues and Challenges* (pp. 61–72). Routledge, New York.

Popova, N., Kataiev, A., Skrynkovskyy, R., & Nevertii, A. (2019). Development of trust marketing in the digital society. *Economic Annals-XXI, 176*(3–4), 13–25. https://doi.org/10.21003/ea.V176-02

Russmann, U., & Hess, A. (2020). News consumption and trust in online and social media: An in-depth qualitative study of young adults in Austria. *International Journal of Communication, 14*, 3184–3201.

Sarkar, S., Chauhan, S., & Khare, A. (2020). A meta-analysis of antecedents and consequences of trust in mobile commerce. *International Journal of Information Management, 50*, 286–301. https://doi.org/10.1016/j.ijinfomgt.2019.08.008

Shaker, L. (2006). In Google we trust: Information integrity in the digital age. *First Monday, 11*(4). https://doi.org/10.5210/fm.v11i4.1320

Škerháková, V., Ali Taha, V., Tirpák, D., & Král, Š. (2022). Perception of corporate reputation in the era of digitization: Case study of online shopping behavior on young consumers. *Sustainability, 14*(21), 14302. https://doi.org/10.3390/su142114302

Soleimani, M. (2022). Buyers' trust and mistrust in e-commerce platforms: A synthesizing literature review. *Information Systems and e-Business Management, 20*(1), 57–78. https://doi.org/10.1007/s10257-021-00545-0

Tong, S. C., & Chan, F. F. Y. (2022). Revisiting trust in the digital era: The interplay of online trust and online dialogic communication from the practitioners' perspective. *Journal of Communication Management, 26*(3), 271–293. https://doi.org/10.1108/JCOM-08-2021-0094

Torres, R. R., Gerhard, N., & Negabhan, A. (2018). Epistemology in the era of fake news: An exploration of information verification behaviors among social networking site users. *Data Base for Advances in Information Systems, 49*(3), 78–97. https://doi.org/10.1145/3242734.3242740

Tsfati, Y., & Ariely, G. (2014). Individual and contextual correlates of trust in media across 44 countries. *Communication Research, 41*(6), 760–782.

Tsfati, Y., & Cappella, J. N. (2003). Do people watch what they do not trust? Exploring the association between news media skepticism and exposure. *Communication Research, 30*(5), 504–529.

Van Dyke, T., Midha, V., & Nemati, H. (2007). The effect of consumer privacy empowerment on trust and privacy concerns in e-commerce. *Electronic Markets, 17*(1), 68–81.

Ye, C., Hofacker, C. F., Peloza, J., & Allen, A. (2020). How online trust evolves over time: The role of social perception. *Psychology Marketing, 37*, 1539–1553. https://doi.org/10.1002/mar.21400

Yunusa, M., Saputra, J., & Muhammad, Z. (2022). Digital marketing, online trust and online purchase intention of e-commerce customers: Mediating the role of customer relationship management. *International Journal of Data and Network Science, 6*(3), 935–944. https://doi.org/10.5267/j.ijdns.2022.2.003

Zhang, C.-B., & Li, Y.-N. (2019). How social media usage influences B2B customer loyalty: Roles of trust and purchase risk. *Journal of Business Industrial Marketing, 34*(7), 1420–1433. https://doi.org/10.1108/JBIM-07-2018-0211

5 Trust in modern digital technologies – an example of artificial intelligence

Michał Borowy and Krzysztof Karpio

Introduction

The mainstream of development in Knowledge Based Economy (KBE) is Industry 4.0, also known as the fourth industrial revolution (David et al., 2022). Modern companies, seek competitive advantages mainly among digital solutions designed, for example, to create new products and services, increase efficiency and process profitability, or improve relationships within the economic environment. In doing so, they make extensive use of artificial intelligence, which is the basis of solutions in the areas of IT, IoT, automatics, robotics, machine learning, etc. (Glikson & Wooley, 2020). Those processes occur so fast and on such a huge scale that it is worthy of considering an issue of trust in such solutions, which were completely unknown until recently. This is important, especially taking into account the fact that they actually concern almost every area of human life. Trust in products, services, and technology has many aspects and can manifest in many forms. It is most commonly understood as a willingness to use new solutions, which can be different across the countries (Omrani et al., 2022). This is related to cultural differences, different economic levels and general readiness for change both in a business and in a social environment. However in our study, we do not analyze the mental aspects of trust in artificial intelligence (AI). We evaluate the results of pragmatic decisions by entrepreneurs and future professionals, which are a result of their belief in the economic potential of artificial intelligence.

Companies, as well as individuals, being convinced of future benefits invest – companies in research and implementation activities, and potential staff in self-development (knowledge). Both sides hope to achieve an advantage in competition with other market players and consequently higher profits. The scale of interest in such activities confirms the degree of trust in AI and its universality in countries around the world.

The main purpose of the chapter is to evaluate the degree of trust in AI based on the involvement of business entities in the exploration of this technology and the level of interest in it in the context of future staff professional self-development.

DOI: 10.4324/9781032627236-6

This chapter is organized in the following way: the chapter's topic and a presentation of its aim were introduced in the first section. The second section is devoted to a literature review, giving an overall view of the problem studied. The methodology of the study, the procedure, and the presentation of the data being used are in the third section, along with a description of the population and the statistical significance. The presentation of the results and their description are in the fourth section, which is followed by the section containing the discussion of the results. The summary and conclusions are included in the last section.

Review of literature

Trust is an issue undertaken by different social sciences and interpreted in various dimensions (Paliszkiewicz, 2018). In psychology, in the greatest generalization, the issue comes down to an individual's belief in the sincere intentions of other participants in human relationships (Rotter, 1967, 1980; Watson, 2005). In management science, it is based more on competences, goodwill, and integrity in interpersonal relations (Mayer et al.,1995; Watson, 2005). Cognitive judgment of associate competence or reliability – *cognition-based trust* and emotional interpersonal bonds – *affect-based trust*, are of high importance here (McAllister, 1995; Watson, 2005, p. 8). In the case of marketing, trust is the result of proper communication (dialog) between seller and buyer to build a positive interaction in exchange relationship. It is to create consumer trust in the brand (Schurr & Ozanne, 1985; Morgan & Hunt, 1994; Delgado-Ballester et al., 2003; Watson, 2005). When analyzing trust in the economic sciences, the dimension of rational and calculation analyses is primarily indicated. In this context, trust is perceived as a complement to various types of control and coordination mechanisms, which may have a positive impact on reducing transaction costs (Kramer & Cook, 2004). The last approach is the closest to the concept of our study. This is because we analyze the rational decisions of entrepreneurs and individuals that confirm the degree of trust in AI new technologies.

AI technologies can be understood more generally as *intelligence demonstrated by machines* (Delipetrev et al., 2020, p. 5) or more specifically as *computer systems that can sense their environment, think, learn, and take action in response to what they're sensing and their objectives* (PwC, 2020, p. 2). They have been developed since the 1950s (Turing, 1950). Its development experienced its ups and downs until the 1990s, when machine learning and, consequently, deep learning were elaborated. Since then, there has been a rapid increase in interest in AI in various areas of socio-economic life (Delipetrev et al., 2020). Trust in AI technologies depends, however, on many different factors. These include cultural and socioeconomic conditions related to the geographic region, as well as the area of influence of the technology itself or the intention to use it.

AI technology has many forms, like robotics, virtual reality, embedded systems, etc. Forms of AI, as well as their capabilities, influence the users' level of trust (Glikson & Wooley, 2020). Separate studies regarded chatbots implemented by the Japanese government in order to answer people's questions were performed (Aoki, 2020). The authors concluded that, even though the technology was the same for a wide range of inquiries' topics, the trust was different. For instance, people trusted more in responses regarding waste separation than in parental support. That leads us to another aspect of trust, which is related to the users' risk during the decision-making process. That's why people trust more AI and other technologies when bad decisions are not so harmful. This behavior of people is justified by the nature of AI operations, which often go beyond human capabilities. Moreover, AI can learn, so its results and way of operation may change over time. Both aspects were investigated, with the conclusion that a level of trust is influenced by the lack of control over how AI operates (Stanton & Jensen, 2021). One should also keep in mind that AI is a set of various methods that must be appropriately selected and configured to solve the problem. The adoption of AI requires qualified personnel, usually with great experience. That leads us to the situation where the user of AI, in fact, does not rely on the AI but on the developer instead. It was pointed out that AI, in some aspects, cannot be trusted directly (Ryan, 2020). The utilization of AI by end-users during the decision-making process should be considered a contract between the end-user and a developer, not AI (Jacovi et al., 2021). This problem of interaction with AI should be emphasized the most in critical areas such as medicine (Ferrario & Loi, 2022). However, commercial companies are usually more advanced in the implementation of AI (Reis et al., 2021) because they probably hope for more profits, so this problem is often passed over in silence. On the other hand, the level of information about the operation of AI is strictly related to trust in AI (Ferrario & Loi, 2022) and is the reason AI adoption is sometimes limited in the decision-making process. The problem was noticed by the AI research community. AI decisions started to be accompanied by explanations, so they are more transparent to end-users (Schmidt et al., 2020). However, we should consider such actions to be focused on increasing trust in AI. The other approach to this problem is based on the ability to try out AI solutions, which also have an impact on trust (Omrani et al., 2022). As has been discovered, the possibility of nonbinding testing of AI as well as convenience of use plays an important role in trusting AI in the financial sector (Mesbah et al., 2019). Moreover, it has been stated that the above factors are even more important for users' trust than pretending to be human by AI, which is usually incorporated in AI chatbots (Mesbah et al., 2019). The same conclusions are drawn from the literature review for 2015–2022, regarding factors influencing trust in AI (Yang & Wibowo, 2022). Companies all around the world treat AI technologies as a source of competitive advantage. This is demonstrated by numerous scientific studies (Alferyev &

Khusainova, 2020; Hartmann, 2020; Dymitrowski & Mielcarek, 2021; Chen et al., 2021; Wagner, 2021; Rosa et al., 2022). These modern technologies are used regardless of industry activity: automotive, health, food, energy, telecom, finance, marketing, etc.

Companies are confirming their trust in AI not only through increased interest in research and investment in such technologies (Chen et al., 2022; Kinkel et al., 2022; Lee et al., 2022; Lui et al., 2022), but also by the employment growth of professionals representing specific skills. According to a study conducted in the United States, the number of job postings that require artificial intelligence skills has increased dramatically, from 20.6 thousand in 2010 to 180.9 thousand in 2018, and continues to grow. Moreover, the percentage of job postings requiring AI skills as a proportion of the total number of posted vacancies had quadrupled by 2019 (Alekseeva et al., 2021). Such a trend is also confirmed by the results of OECD international study among such countries as Australia, Canada, the United Kingdom, the United States, and New Zealand between 2012 and 2019. It has been noted that the demand for AI skills in various occupations has spread much faster than the demand for average skills. This phenomenon was particularly evident in the United Kingdom, where demand for the top 20 AI skills grew seven times faster than the average skill demand in the economy. For selected skills, such as machine learning, that growth rate in the United States and the United Kingdom was up to 12–21 times faster than the average skill demand. (Manca, 2023). Hiring people with AI skills is also associated with a higher salary level and a higher position in the organization (Alekseeva et al., 2021). As evidenced by the results of the aforementioned studies, job offers where a stronger emphasis is given to AI skills also assume higher salaries than average, regardless of average years of schooling, skill complexity of the job, and geographical factors related to the job offer. In addition, specializations in narrow AI fields can result in pronounced income disparities of up to several hundred percent, compared, for example, to the salary of an unskilled worker (Manca, 2023).

AI influences the development of new economic drivers and, at the same time, is a source of future staff professional self-development. This specific kind of skill becomes a comparative advantage in the labor market for new technologies. Universities and colleges, matching their offerings to the new needs of industry, implement more and more training programs in such areas (Bian et al., 2022; Studyportals, 2023). In turn, the unflagging interest in them among students testifies to their trust that these are valuable skills for personal and professional development.

Methodology

The purpose of this article is to assess the degree of trust in artificial intelligence based on the involvement of institutions in research and implementation of this technology and the level of interest in it in the context of careers.

In order to confirm trust in AI, we decided to verify the scale of interest in this field both among companies and other institutions, as well as potential employees or people interested in this technology. A part of the data concerned both aspects. As the degree of companies' trust in AI, we examined the global value of their investments over the years (Our World In Data, 2023) and the share of companies that have implemented such solutions (Cardillo, 2023). The second symptom of companies' trust in AI was their willingness to hire specialists in this field. We analyzed US data regarded both the share of AI job advertisements against all job advertisements (Our World In Data, 2023) and the distribution of the annual earnings of people working in this field (OECD, 2023). On the other hand, part of the analyzed data reflected people's personal trust in AI, which manifested itself in their willingness to invest their time and tie their careers to AI. The skill cited most often as the main skill in the AI field is knowledge of the Python language (Maslej et al., 2023; White, 2023). We analyzed both the number of people with such knowledge over the years as well as the number of people planning to learn this programming language (OECD, 2023). People's willingness to educate themselves in this field was reflected in the number of AI courses over the years (OECD, 2023). Issues related to both people and institutions are separate issues. These could be viewed as the joint activities of professionals who have engaged in the development and implementation of AI-based technologies and the academic or commercial institutions that have hired them, paid them, or otherwise invested in AI research and development. In connection with this aspect, we examined the annual number of AI publications (articles, reviews, conference papers) among 308 subject categories from more than 200 countries (SCImago, 2023).

This work is mainly based on the data published by OECD – The Organization for Economic Co-operation and Development; OWID – Our World In Data, a project of the Global Change Data Lab, a registered charity in England and Wales; ET – Exploding Topics, a trend detection platform that specializes in discovering new trends on the web; and SJCR – The SCImago Journal & Country Rank, a portal that includes the journals and country scientific indicators. Data published by the OECD regarded salary distributions, Python popularity, and a number of AI courses worldwide. It was gathered during the annual survey conducted by StackOverflow, the technological forum, which in March 2022 contained over 20 million registered users, 24 million questions, and about 35 million answers. About 70,000 users participate each year, and over 100 countries are represented. Data regarding annual company investments in AI as well as a share of AI job advertisements were published by OWID. The investments were estimated based on over 8 million global public and private companies, while the number of job advertisements in the United States was based on the millions of job postings collected since 2010 from over 51,000 online job sites.

Results

We investigated the number of publications (articles, reviews, and conference papers) regarding AI. The numbers were compared to the total number of publications. Figure.5.1f contains the share of AI publications per year from 2012 to 2022. The solid line denotes a linear trend and is fitted to data with a very good quality of $R^2 = 0.9227$.

The trust of companies in AI can be measured by their trust in future profits. This is reflected in the amount of investment in AI as well as the number of implementations. The amount of investment has been constantly growing

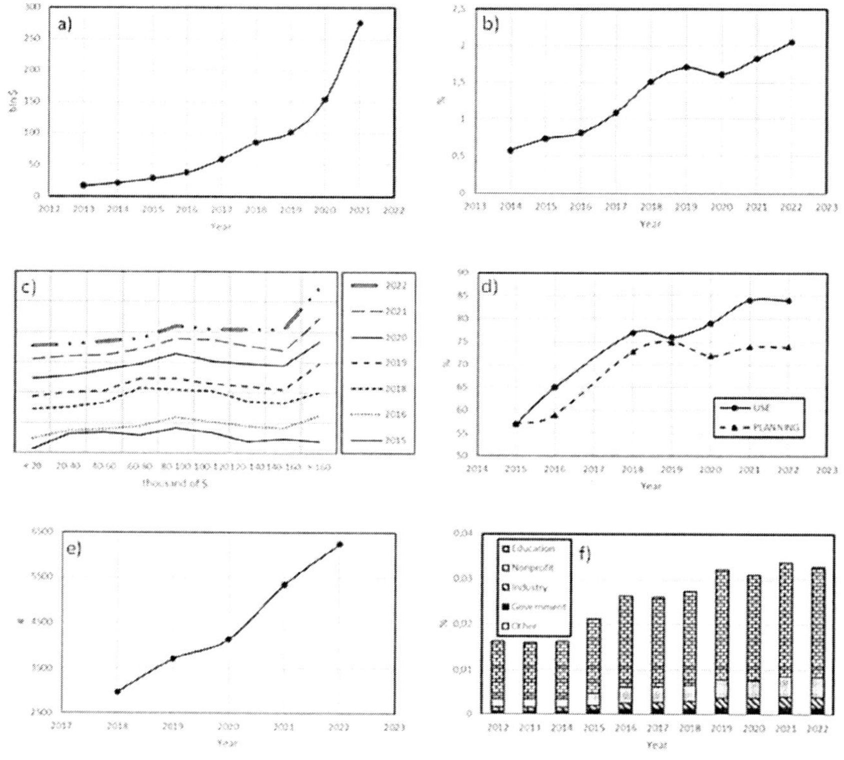

Figure 5.1 (a) value of company's investments in AI on the yearly basis, in global scale; (b) share of AI job postings per year in the United States; (c) distributions of US salaries in AI in successive years, the plots are shifted upward for clarity; (d) shares of persons using the Python or willing to learn it on the yearly basis, worldwide; (e) number of AI courses each year, worldwide; and (f) number of publications each year per type of institution, worldwide, plot is stacked.

Source: Own research.

since 2013. Moreover, the rate of growth is increasing, which has been huge in recent years. Similarly, the shares of companies using AI increased by about 250% in five years, between 2017 and 2022 (Cardillo, 2023).

Another aspect of the study was related to salaries in the AI labor market. We investigated the distribution of annual salaries before taxation in the United States from 2015 to 2022. The plots are shifted upward for clarity of picture. While the distribution of salaries was almost flat in 2015, the changes were very significant. During the years, the two maxims have arisen. The one in thé middle of salary values, in the range of 80–100k$, and the other in the right far end of salary values, for salaries greater than 160k$ (Figure 5.1c). We observed that salaries are rising along with the number of job postings in the United States (Figure 5.1b).

We presented in Figure 5.1b the share of AI job postings among all job postings in the United States. It is also worth noting that the level mentioned was raised four times in the period between 2014 and 2022. As can be seen in the plot, the demand for the AI labor force is rising systematically (the linear fit exhibits very good quality with the data). Using the simple economic relationship of supply and demand, we can conclude that with increasing demand, supply will also rise. That will be caused by the belief in better earnings.

On the other hand, the growing number of job announcements is an affirmation of businesses' trust in the potential increase in profits thanks to AI activity. Candidates for certain jobs should have the appropriate qualifications. Among the tools needed in AI are the following programming languages: SWIFT, SCALA, R, JAVA, C++, and PYTHON (White, 2023). Among them, Python is the most expected, with 37% of AI job advertisements in 2022. The share has increased in 11 years from about 5%, a nearly sixfold increase (Maslej et al., 2023). A similar trend is noticeable in the skills of job applicants. It is noteworthy that the share of people who know Python and are interested in working in AI has risen from 57% to 84% over a seven-year period. A number of people willing to learn Python is also rising (OECD, 2023). This may indicate an unwavering interest in knowledge in this area and a desire to improve their skills in using AI tools. The interest in learning AI is reflected in the growing number of AI courses, which doubled within four years (see Figure 5.1e).

The share of AI publications increased by about 100% within ten years. It means growing interest and activity in AI. Moreover, the upward trend has continued in all sectors. Education has been a leading sector, comprising about 76% of all AI publications. Next in the rank were nonprofit organizations (approx. 14%), industry (7%), and government (approx. 4%). The remaining types of organizations were related to less than 1% of publications (Maslej et al., 2023). The observed behavior indicates the growing trust reflected in the growing investments of funds and times in AI studies. The interesting thing is that industry is responsible for a relatively low level of shares (7%), while commercial companies are very active in the implementation of new technologies. One must remember that we are talking about scientific publications, which are not representative when comparing the activities of various sectors of the economy or types of institutions. However, it is clearly visible that the number of publications is rising in all sectors.

Discussion

Usually trust issues are considered in terms of interpersonal relationships (Rotter, 1967, 1980; Schurr & Ozanne, 1985; Morgan & Hunt, 1994; Mayer et al., 1995; McAllister, 1995; Delgado-Ballester et al., 2003; Watson, 2005), while our approach to research on trust in AI technologies is closer to the economic sciences (Kramer & Cook, 2004). We evaluate the results of pragmatic decisions by entrepreneurs and AI staff candidates that arise from their belief in the economic potential of artificial intelligence. Companies' trust in new technologies is expressed by their level of investment in them, both in research and implementation, their willingness to hire people with such skills, and their willingness to adequately remunerate such professionals. Individuals, on the other hand, express their trust in AI as a self-development opportunity by choosing specialized courses on the subject.

As shown by our data analysis based on Our World in Data, in the period 2013–2021, the volume of company's investment in AI on a global scale increased dramatically from approx. 30 billion $ to 300 billion $ (Figure 5.1a). Moreover in the last decade, industry as well as education institutions are definitely more involved in AI research every year, resulting in published scientific papers (Figure 5.1f). Advancing knowledge and growing public awareness of new technologies, but also competitive pressure on the market, are causing a rapid increase in the number of companies that are implementing AI solutions. Between 2017 and 2022, the share of such companies in the world doubled (Cardillo, 2023). All three results of our analysis are confirmed in the studies of other researchers (Chen et al., 2022; Kinkel et al., 2022; Lee et al., 2022; Lui et al., 2022; Maslej et al., 2023). However, some authors (Kinkel et al.) used data from other databases, for example, Scopus, but the result remained the same.

Companies' trust in AI is also expressed in their willingness to employ professionals with narrow specializations and their tendency to remunerate them appropriately. According to our analysis, based on available data from the United States, the share of AI job postings among all job postings was raised four times in the period between 2014 and 2022. Naturally, with increasing demand for professional staff, their wages also climbed. In the period 2015–2022, we noticed a trend of a significantly larger number of AI staff earning between $80–100k, $150–160k, and above $160k. However, the number of counts in the last two ranges dominated all the others. Similar tendencies have also been found in findings from international research in a group of countries, like Australia, Canada, the United Kingdom, the United States, and New Zealand, between 2012 and 2019. It was noted there that the demand for AI skills in various occupations has spread much faster than the demand for average skills. In extreme cases, even 21 times more often. As noted in the same study, specializations in narrow AI fields can result in pronounced income disparities of up to several hundred percent, compared, for example, to the salary of an unskilled worker. (Manca, 2023).

High salaries encourage learning AI tools. According to our research, the percentage of people who know Python (the most popular AI programming language) and are interested in working in AI has increased from 57% to 84% over seven years. As the data shows, the number of people willing to learn Python is on the rise. (Figure 5.1d). This indicates an unwavering interest in knowledge in this area and a desire to improve their skills in using AI tools. These conclusions are also supported by the results of a wide-ranging international study by Studyportals in such countries as the United States, the United Kingdom, EU-27, Australia, and Canada. They show that the total number of AI courses in English more than doubled between 2018 and 2022 (OECD, 2023; Studyportals, 2023). Additionally, as we read on studyportals.com, interest is 3.4 times the number of currently available programs. The need to adapt AI courses in colleges and universities to meet changing industry expectations is also highlighted by Chinese authors (Bian et al., 2022). So in conclusion, it can be said, with a high degree of probability, that people worldwide who want to become skilled professionals trust artificial intelligence as an opportunity to achieve professional success.

Conclusion

The aim of this chapter was to investigate trust in AI. As opposed to the traditional approach, the presented studies concerned companies and individuals. The trust of companies in AI was reflected as their willingness to invest their funds in developing and adopting AI. On the other hand, the trust of people can be exhibited as an investment of time and sometimes even money to relate their professional career to AI. To summarize, the studies have led to the following conclusions:

- the investments of companies into AI have raised about ten times within last eight years, stating the constant trust in AI;
- enterprises not only invest the money in developing of AI but also are willing to hire professionals; the shares of job ads continues to grow up and has quadrupled in eight years; it can be interpreted as the trust of companies related to forward-looking activities;
- not only the demand for specialists grows up, as was stated in the previous point but also salaries are rising; the distribution of salaries of AI professionals is changing, exhibiting the strongest rise of the highest paid jobs;
- the people relate their career with AI; the number of persons using are going to learn the skills used in AI is also rising;

- the demand for knowledge useful in AI is clearly visible in the number of courses which doubled within four years;
- the trust in AI, its potential and future popularity is exhibited by the work related to inventions, studies, and developing such technologies; the share of scientific publications has doubled in the last ten years.

To summaries, there is visible and growing trust in AI from businesses, working people, and academic institutions.

The analysis can provide some hints for companies and individuals who are not yet convinced by new AI technologies. This is because the results of the survey indicate a growing awareness of these technologies, both for companies and in the labor market. Thus, they are part of the global mainstream, resulting from the ongoing industrial revolution, Industry 4.0.

The results of our analysis can serve as a starting point for further research to deepen our understanding of the importance of AI in the economy. A certain limitation of empirical studies is the difficulty of obtaining sufficient statistics to compare different countries over the same time frame.

References

Alekseeva, L., Azar, J., Gine, M., Samila, S., & Taska, B. (2021). The demand for AI skills in the labor market. *Labour Economics*, *71*, 102002.

Alferyev, D., & Khusainova, E. (2020). Ai technologies as a factor of competitiveness of a business entity at the present stage of human development. In *E3S Web of Conferences* (Vol. 220, p. 01006). EDP Sciences.

Aoki, N. (2020). An experimental study of public trust in AI chatbots in the public sector. *Government Information Quarterly*, *37*(4), 101490.

Bian, Y., Lu, Y., & Li, J. (2022). Research on an artificial intelligence-based professional ability evaluation system from the perspective of industry-education integration. *Scientific Programming*, *2022*, 4478115. DOI: 10.1155/2022/4478115

Cardillo, A. (2023, July 24). *How Many Companies Use AI?* Retrieved from https://explodingtopics.com/blog/companies-using-ai

Chen, H., Li, L., & Chen, Y. (2021). Explore success factors that impact artificial intelligence adoption on telecom industry in China. *Journal of Management Analytics*, *8*(1), 36–68.

Chen, X., Zou, D., Xie, H., Cheng, G., & Liu, C. (2022). Two decades of artificial intelligence in education. *Educational Technology & Society*, *25*(1), 28–47.

David, L. O., Nwulu, N. I., Aigbavboa, C. O., & Adepoju, O. O. (2022). Integrating fourth industrial revolution (4IR) technologies into the water, energy & food nexus for sustainable security: A bibliometric analysis. *Journal of Cleaner Production*, *363*, 132522.

Delgado-Ballester, E., Munuera-Aleman, J. L., & Yagoe-Guillin, M. J. (2003). Development and validation of a Brand Trust Scale. *International Journal of Market Research*, *45*(1), 335–353.

Delipetrev, B., Tsinarakii, C., & Kostić, U. (2020). *Historical Evolution of Artificial Intelligence.* Luxembourg: Publications Office of the European Union. Retrieved from https://publications.jrc.ec.europa.eu/repository/handle/JRC120469

Dymitrowski, A., & Mielcarek, P. (2021). Business model innovation based on new technologies and its influence on a company's competitive advantage. *Journal of Theoretical and Applied Electronic Commerce Research, 16*(6), 2110–2128.

Ferrario, A., & Loi, M. (2022). How explainability contributes to trust in AI. In *Proceedings of the 2022 ACM Conference on Fairness, Accountability, and Transparency*, 1457–1466.

Glikson, E., & Wooley, A. W. (2020). Human trust in artificial intelligence: Review of empirical research. *Academy of Management Annals, 14*(2), 627–660.

Hartmann, P. M. (2020). *Profiting from Artificial Intelligence: Data as a Source of Competitive Advantage.* BoD–Books on Demand, Norderstedt, Germany.

Jacovi, A., Marasović, A., Miller, T., & Goldberg, Y. (2021). Formalizing trust in artificial intelligence: Prerequisites, causes and goals of human trust in AI. In *Proceedings of the 2021 ACM Conference on Fairness, Accountability, and Transparency*, 624–635.

Kinkel, S., Baumgartner, M., & Cherubini, E. (2022). Prerequisites for the adoption of AI technologies in manufacturing – Evidence from a worldwide sample of manufacturing companies. *Technovation, 110*, 102375.

Kramer, R. M., & Cook, K. S. (Eds.) (2004). *Trust and Distrust in Organizations.* New York: Russell Sage.

Lee, Y. S., Kim, T., Choi, S., & Kim, W. (2022). When does AI pay off? AI-adoption intensity, complementary investments, and R&D strategy. *Technovation, 118*, 102590.

Lui, A. K. H., Lee, M. C. M., & Ngai, E. W. T. (2022). Impact of artificial intelligence investment on firm value. *Annals of Operations Research, 308*, 373–388.

Manca, F. (2023). *Six Questions about the Demand for Artificial Intelligence Skills in Labour Markets.* Retrieved from https://oecd.ai/en/

Maslej, N., Fattorini, L., Brynjolfsson, E., Etchemendy, J., Ligett, K., Lyons, T., Manyika, J., Ngo, H., Niebles, J., C., Parli, V., Shoham, Y., Wald, R., Clark, J., & Perrault, R. (2023). *The AI Index 2023 Annual Report.* AI Index Steering Committee, Institute for Human-Centered AI, Stanford University, Stanford, CA, April 2023, p. 27.

Mayer, R., Davis, J., & Schoorman, F. D. (1995). An integrative model of organizational trust. *Academy of Management Review, 20*(3), 709–734.

McAllister, D. J. (1995). Affect- and cognition-based trust as foundations for interpersonal cooperation in organizations. *Academy of Management Journal, 38*(1), 24–59.

Mesbah, N., Tauchert, C., Olt, C. M., & Buxmann, P. (2019). Promoting trust in AI-based expert systems. *Twenty-fifth Americas Conference on Information Systems*, Cancun, 2019, 1–10.

Morgan, R. M., & Hunt, S. D. (1994). The commitment-trust theory of relationship marketing. *Journal of Marketing, 58*(3), 20–38.

OECD. (2023, July). *OECD. AI Policy Observatory.* Retrieved from https://oecd.ai/en/

Omrani, N., Rivieccio, G., Fiore, U., Schiavone, F., & Agreda, S. G. (2022). To trust or not to trust? An assessment of trust in AI-based systems: Concerns, ethics and contexts. *Technological Forecasting and Social Change, 181*, 121763.

Our World In Data. (2023). *Annual Global Corporate Investment in Artificial Intelligence.* Retrieved from: https://ourworldindata.org

Paliszkiewicz, J. (2018). Trust: A multifaceted notion. In B. Kożuch, S. Magala, & J. Paliszkiewicz (Eds.), *Managing Public Trust* (pp. 9–23). Cham, CH: Palgrave Macmillan.

PwC. (2020). *Sizing the Prize. What's the Real Value of AI for Your Business and How Can You Capitalise?* Retrived from https://www.pwc.com/gx/en/issues/analytics/assets/pwc-ai-analysis-sizing-the-prize-report.pdf

Reis, J., Santo, P. E., & Melão, N. (2021). Influence of artificial intelligence on public employment and its impact on politics: A systematic literature review. *Brazilian Journal of Operations & Production Management*, *18*(3), 1–22.

Rosa, A., Bento, T., Pereira, L., Costa, R. L. D., Dias, Á., & Gonçalves, R. (2022). Gaining competitive advantage through artificial intelligence adoption. *International Journal of Electronic Business*, *17*(4), 386–406.

Rotter, J. B. (1967). A new scale for the measurement of interpersonal trust. *Journal of Personality*, *35*, 651–665.

Rotter, J. B. (1980). Interpersonal trust, trustworthiness, and gullibility. *American Psychologist*, *35*(1), 1–7.

Ryan, M. (2020). In AI we trust: Ethics, artificial intelligence, and reliability. *Science and Engineering Ethics*, *26*(5), 2749–2767.

Schmidt, P., Biessmann, F., & Teubner, T. (2020). Transparency and trust in artificial intelligence systems. *Journal of Decision Systems*, *29*(4), 260–278.

Schurr, P. H., & Ozanne, J. L. (1985). Influences on exchange processes: Buyers' preconceptions of a seller's trustworthiness and bargaining toughness. *Journal of Consumer Research*, *11*(4), 939–953.

SCImago. (2023). *SCImago Journal & Country Rank*. Retrieved from http://www.scimagojr.com

Stanton, B., & Jensen, T. (2021). *Trust and Artificial Intelligence*. Preprint. DOI: 10.6028/NIST.IR.8332-draft.

Studyportals. (2023, June). *Universities Must Respond to the Increasing Importance of AI*. Retrieved from https://studyportals.com/blog/universities-must-respond-to-the-increasing-importance-of-ai/

Turing, A. M. (1950). Computing machinery and intelligence. In B. Jack Copeland (Ed.), *The Essential Turing: The Ideas that Gave Birth to the Computer Age* (pp. 433–464). Oxford: Oxford University Press.

Wagner, D. N. (2021). Economic AI literacy: A source of competitive advantage. In *Handbook of Research on Applied AI for International Business and Marketing Applications* (pp. 135–152). IGI Global. Hershey, Pennsylvania, USA. DOI: 10.4018/978-1-7998-5077-9.ch008

Watson, M. L. (2005). Can there be just one trust? A cross-disciplinary identification of trust definitions and measurement. *The Institute for Public Relations*, Gainesville, Florida, USA, 1–25. https://instituteforpr.org/wp-content/uploads/2004_Watson.pdf

White, E. (2023). *Top 9 Programming Languages for Artificial Intelligence*. Retrieved from https://www.bairesdev.com/blog/top-languages-for-ai/

Yang, R., & Wibowo, S. (2022). User trust in artificial intelligence: A comprehensive conceptual framework. *Electronic Markets*, *32*(4), 2053–2077.

6 Trust in computer simulations in a context of an epidemic in a heterogeneous population

Marcin Choiński

Introduction

In order to predict epidemic dynamics, it is useful to propose an appropriate mathematical model. In many cases, its mathematical analysis is complicated, and it is hard to obtain constructive results.

Moreover, for epidemiological models with bilinear functions, it is impossible to obtain explicit solution of variables appearing in the given system. Then one should investigate the other properties of the system, for example, any stability of its stationary states. However, it gives information about only asymptotic behavior of the system. What is more, there is no medical definition of a disease transmission coefficient naturally appearing in epidemiological models. For these reasons, it is useful to implement the model in a given computer software and conduct computer simulations. However, one should be aware that the obtained simulated results do not have to reflect the real dynamics of epidemic. The aim of this chapter is to verify if computer simulation can be considered as a reliable tool in the context of modeling of epidemic dynamics in a heterogeneous population. As the heterogeneous population, we define a population in which we can distinguish two subpopulations: with a low and high susceptibility to an infection.

The chapter is organized as follows. Firstly, we mention examples of verifying epidemiological models from literature. Later we discuss the built-in *lsqcurvefit* function from the Matlab software.

Next, we describe a population sample and a procedure for data prediction. Finally, we discuss the results and make a conclusion.

Review of literature

In the literature, one can find papers related to verification of numerical simulation of epidemiological models. Naturally, recent papers concern the *COVID-19* pandemic. In (Faruk et al., 2021) the authors use the Matlab software for forecasting *COVID-19* dynamics in Bangladesh. To do that, they use the lsqcurvefit function, which is described later in this chapter. Thanks to this function, they

DOI: 10.4324/9781032627236-7

compute the basic reproduction (R_0) of given epidemic model. This value is used for prediction of a further epidemic scenario. However, the authors do not investigate if the predicted scenario reflected the reality. The *lsqcurvefit* function is used also in (Lee et al., 2021), where the authors foresee *COVID-19* scenarios in South Korea, the United States, India, and Brazil. For each country, the trend of predicted scenario is confirmed. There appear differences between real and simulated numbers of infected individuals, though. Paper (Moein et al., 2021) deals with efficiency of prediction of the *COVID-19* pandemic in Iran. From epidemic data, the authors compute R_0 for each stage of epidemic outbreak. Relying on these values, they propose three scenarios of epidemic. For each scenario the simulations are conducted. However, none of these scenarios was consistent with the reality. In (Aylett–Bullock et al., 2021) one can find the description of open-source *JUNE* framework, which is suitable for epidemic modeling. It enables creating a virtual population relying on data concerning sex, age and socio-economic factors. With the use of social mixing matrices, the places of interactions can be chosen reflecting their properties (such as households, workplaces or schools) and size. As exemplary populations, the authors from (Aylett–Bullock et al., 2021) chose the London and various English area communities. The obtained results approximated the reality well. An application of machine learning in predicting *COVID–19* scenario can be found in (Narayan et al., 2022). The authors use a loss function which is relied on the weighted sum errors of active and recovered cases. This function enables optimization of fitting of curves reflecting number of infected and recovered individuals. The fitted curves arise from training the epidemiological model with real data. The authors chose datasets concerning Germany, Japan, India, Iran, Italy, and the United States and obtained a less than 3% error.

We have to emphasize that our chapter is related to the case of tuberculosis (*TB*), what is not common in the context of current research on the *COVID-19* pandemic. However, the proposed model (which will be introduced later) and the results of simulation can be applied to other infectious diseases. Moreover, we consider the epidemic case for the heterogeneous population, where the demographic and epidemiological parameters have different values. Thanks to that, dynamics of each subpopulation are different and population heterogeneity is preserved. In many papers, authors do not differ the values of parameters for each population and investigate actually the homogeneous population. The results presented in this chapter can be therefore treated as a novelty.

Methodology

The instrument

In order to simulate the model, one should choose initial condition (value of data for an initial moment) and values of parameters. Most parameter, including a birth and recovery ratios or a disease-related and natural death coefficients, are known or can be computed according to the given data. However, there is no medical definition of a disease transmission coefficient, naturally appearing in the epidemiological models. Hence, there arises a need of estimating such a coefficient. It is reasonable that this estimation should be done with regard to real data. In order to do that, one can find use the built-in *lsqcurvefit* function in the Matlab software (Mathworks, 2023), which is a widely used computational programming environment. This function allows fitting the simulated data to the real one in the best way. This function uses the Gauss–Newton algorithm (Lai et al., 2015), which idea is finding the least sum of squares of difference between simulated values and their real counterparts. The *lsqcurvefit* function requires giving the real data and the initial value of estimated parameter (or parameters). Moreover, it needs a range of parameter's values in which the estimated value will be computed. Both the initial value and range are generally chosen arbitrarily, since there can be no suggestion about expected values. The function returns the sum of squares and the value of estimated parameter. There can arise a situation when the computed sum is the smallest only in a limited range of parameter's value. In order to omit this situation, the initial parameter's value should be chosen from different ranges.

Participants

We consider the whole population of the Warmian–Masurian Province of Poland. The choice of such a population arose from the outstanding drop in the *TB* incidence. It was equal to 53%, whereas the decline in the whole country was equal to 26%. In the province, four campaigns of *TB* active cases detection were held in years 2004–2011. These campaigns concerned homeless people, which are considered as a group of high risk of getting infected with *TB* because of their low living standards. Hence, homeless people as a reservoir of *TB*. Thanks to the campaigns, *TB* incidence dropped not only in the community of homeless people, but in the entire province population as well. In (Romaszko et al., 2016) one can find outcomes of detection programs and the model describing the dynamics of TB in this province.

Let us now quote this model and its assumptions. In the given population we consider two subpopulations – nonhomeless and homeless people, which are the subpopulation with low and high susceptibility to *TB*, respectively. Every variable and parameter has a subscript referring to each subpopulation equal to accordingly 1 and 2. Both subpopulations are divided into two groups: susceptible (S) and infected (I) individuals. We indicate therefore four variables: S_1, I_1, S_2, and I_2 meaning densities of each determined group. Since the obtained epidemiological data concerns only a number of new infections, it is impossible to determine more groups in the infected group. The infection can be transmitted in each subpopulation and from homeless to nonhomeless people.

Table 6.1 Parameters for the model described by model of tuberculosis dynamics

Notation	*Meaning*	*Value*
C_1	Inflow into the nonhomeless community	11,000
C_2	Inflow into the homeless community	60
α_1, α_2	TB-related death ratio	0.09
γ_1, γ_2	Recovery ratio	0.9
μ_1, μ_2	Natural death ratio	0.009

Homeless people tend to aggregate and isolate themselves from the rest of community, hence it is reasonable to assume that there is no disease transmission from nonhomeless to homeless people.

In order to describe the *TB* dynamics, we use a system of ordinary differential equations where the independent variables is time. This kind of system is widely used in mathematical modeling.

The obtained model has the form:

$$\begin{aligned} \dot{S}_1 &= C_1 - \beta_1 S_1 I_1 - \beta S_1 I_2 + \gamma_1 I_1 - \mu_1 S_1, \\ \dot{I}_1 &= \beta_1 S_1 I_1 + \beta S_1 I_2 - (\gamma_1 + \alpha_1 + \mu_1) I_1, \\ \dot{S}_2 &= C_2 - \beta_2 S_2 I_2 + \gamma_2 I_2 - \mu_2 S_2, \\ \dot{I}_2 &= \beta_2 S_2 I_2 - (\gamma_2 + \alpha_2 + \mu_2) I_2, \end{aligned} \tag{6.1}$$

where the demographic and recovery parameters are defined in Table 6.1. In this table, the values of these parameters are given. They reflect the population dynamics in the Warmian–Masurian Province in Poland. The values are obtained from statistical yearbooks (CSO, 2023).

The remaining parameters are β_1, β_2, and β reflecting *TB* transmission in the nonhomeless community, the homeless community and from the homeless to the nonhomeless, respectively. These three values have to be estimated. Demographic were taken from (CSO, 2023). Regional Center for Social Policy in Olsztyn in Poland provided numbers of homeless people (MO, 2022). National Institute of Public Health in Poland gave access to epidemiological data (NIPH, 2022). System is a simplification of the model presented and introduced in (Bodzioch et al., 2019), where the authors conducted its detailed mathematical analysis.

Procedure

The data for the estimation relate to the size of healthy and infected group for both nonhomeless and homeless communities in years 2000–2022. For each transmission coefficient we initially take the (0,1) range, what reflects all possible values of these coefficients. Not every contact between an infected and a healthy individual results in a new infection case, hence the transmission coefficient cannot be equal to 1. Naturally, we do not also assume that any of these coefficients is equal to 0 – for this value System (1) become simplified and does not reflect the illness transmission completely. The built-in Matlab *ode-45* function solves System (1) reflecting *TB* dynamics in years 2000–2014. This function requires initial data – in our example, they are the values of S_1, I_1, S_2, and I_2 at year 2000. Later *lsqcurvefit* computes the variables' values for each year and the square of sum of difference between simulated and reals values of each variable. The minimal sum is indicated and the values of transmission coefficients related to this sum are chosen. Later the range is halved and the half, including the computed value of the coefficient, are chosen. Naturally, choosing the proper half for each transmission coefficient is independent. Again *ode-45* and *lsqcurvefit* are used. The procedure is stopped until the iteratively halved ranges are small enough and if the obtained values of coefficient do not change significantly after repeating the procedure.

The next step is simulating System (1) with the obtained values of transmission coefficient for years 2015–2022. These values are compared to the real ones for this year. The result of the procedure is seen in Figure 6.1. The real data

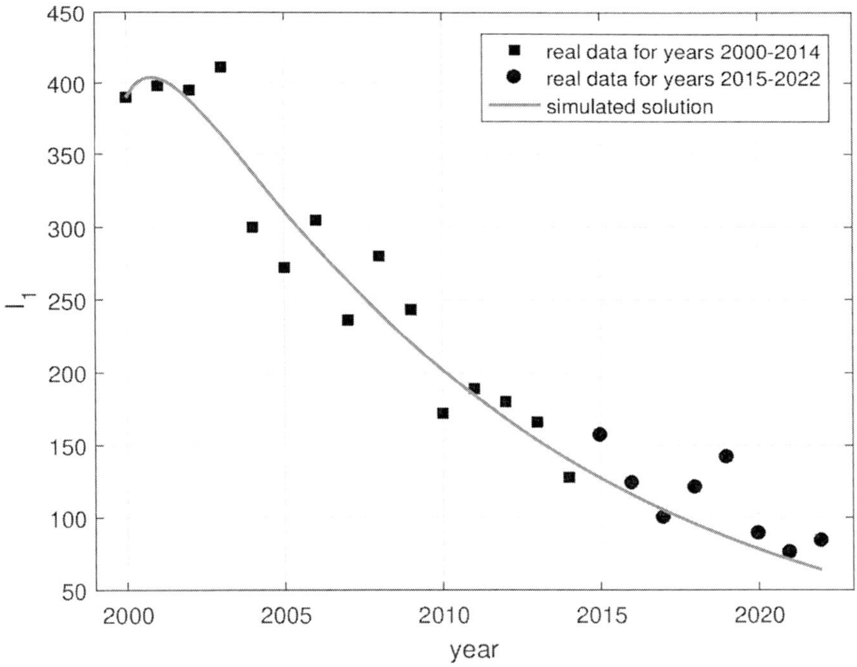

Figure 6.1 Real and simulated data for TB epidemic in Warmian–Masurian Province in years 2000–2022. Real data used for the estimation and the reliability test of are depicted with squares and circles, respectively. The simulated solution is marked with a gray curve.

used for estimation of transmission coefficients' values (i.e. from years 2000 to 2014) were depicted with squares. Circle refers to real data used for verification of estimated values' reliability (i.e. from years 2015 to 2022). From the epidemiological point, it is important to investigate dynamics of the infected groups instead of the healthy ones. We focus on the group of infected nonhomeless people, since naturally its size is greater than the size of the group of infected homeless ones.

Results

Let us compare the simulated solution to real data for years 2015–2022. The simulated simulation is a decreasing function, whereas the real data do not constitute a monotonous trend. However, the number of infected nonhomeless individuals in year 2022 is smaller than in year 2015 for both real and simulation cases. This consistency is important in the context of predicting of epidemic dynamics in a long time horizon. See that every real datum (besides this for year 2017) is bigger than its analogical simulated solution. Hence, the simulated values are underestimated. We conclude that computer simulation conducted for the *TB* epidemic in the Warmian–Masurian Province is not reliable.

Discussion

The methodology presented in this chapter is analogical to these applied in (Faruk et al., 2021) and (Lee et al., 2021). In these papers, the efficiency of *lsqcurvefit* was confirmed with real data. However, we should remind that dynamics of *TB* is slower than dynamics of *COVID-19*. The incubation period for *TB* is three to nine weeks (VG, 2015), whereas for *COVID-19* is a few days (Pfizer, 2023). During the first mentioned period, there is a higher probability that people from the Warmian–Masurian Province can travel to other provinces or countries than during a few days. Naturally, they can get infected in their destination place. System (1) does not describe such way of infection transmission. What is more, during the *COVID-19*, there were restriction for travel and the *COVID-19* was transmitted actually only in a small area. System (1) is therefore not appropriate for *TB* transmission and the systems from (Faruk et al., 2021) and (Lee et al., 2021) could deal with modeling dynamics of *COVID-19* pandemic.

Moreover, the data related to *COVID-19* were presented daily, whereas for *TB* – annually. Hence, the size of *TB* data for parameter estimation is smaller this for *COVID-19*. This fact could decide about inefficiency of *lsqcurvefit* for the *TB* epidemic. What should be noted is that the approaches presented in (Aylett–Bullock et al., 2021), (Moein et al., 2021) and (Narayan et al., 2022) are different from ours. It is therefore unreasonable to relate our results to those from these papers.

Conclusion

The conclusion arising from the analysis is that we should not rely unconditionally on numerical simulations. Before conducting the simulations, it is necessary to decide if a given computational method is proper for modeling of epidemic dynamics. According to the inefficiency of *lsqcurvefit*, one should think about the change of the methodology or about gaining more data so that the parameter estimation is more precise. However, it would be difficult to collect data for the disease in which dynamics is slow. It arises therefore a need for implementation of the other approach for the case when the data are limited.

The biggest limitation is a possibility of obtaining the optimal estimated parameter value, which is negative. Obtaining a negative value stays on the contrary to the assumptions of parameter values' positivity in epidemiological models. What is more, our model does not concern the complexity of populations' behavior. For example, the densities in urban and rural areas are different, hence our model should consider this difference. Moreover, every epidemiological model has assumptions which do not reflect the reality precisely. However, considering another assumptions results in adding another term to System (1). It causes that such an extended system becomes complex and its mathematical analysis is difficult or impossible.

The next step in the research is considering the other method enabling the proper estimation of transmission coefficients' values. One can use the method from (Narayan et al., 2022), where machine learning is applied. If the methods result in a failure, the modification of the model should be considered.

Another thing for consideration is a way of interpreting of the results. In our case, we estimate the transmission coefficients for the continuous model. The terms of System (1) reflecting the disease transmission are bilinear functions. We cannot therefore solve this system analytically. The *ode-45* function yields only values of the variables for given moments What is more, we cannot provide the exact values of system's variables for any moment from the given period (in our case years 2015–2022) for which the data are given. For this reason, discussing an error of given method for arbitrarily chosen moments is impossible.

Literature

Aylett–Bullock, J. Cuesta-Lazaro, C., Quera-Bofarull, A., Icaza-Lizaola, M., Sedgewick, A., Truong, H., Curran, A., Elliott, E., Caulfield, T., Fong, K., Vernon, I., Williams, J., Bower, R., & Krauss, F. (2021). *JUNE*: open–source individual–based epidemiology simulation, *Royal Society Open Science*, 8, 2105060.

Bodzioch, M., & Choiński, M., & Foryś, F. (2019). *SIS* cris-cross model of tuberculosis in heterogeneous population, *Discrete and Continuous Dynamical Systems – Series B*, 24(5), 2169–2188.

Central statistical office of Poland. (2023). *Statistical Yearbooks*. http://stat.gov.pl/en/topics/statistical-yearbooks/.

Faruk, O., & Kar, S. (2021). A data driven analysis and forecast of *COVID–19* dynamics during the third wave using *SIRD* model in Bangladesh, *COVID*, 1, 503–517.

Lai, W. H., Kek, S. L., & Gaik, T. K. (2015). Solving nonlinear least squares problem using gauss–Newton method, *International Journal of Innovative Science, Engineering and Technology*, 4(1), 258–262.

Lee, C., Kwak, S., Kim, S., Hwang, Y., Choi, Y., & Kim, J. (2021). Robust optimal parameter estimation for the susceptible–unidentified infected–confirmed model, *Chaos, Solitons & Fractals*, 153(1), 1–11.

Marshall Office, Regional Center for Social Policy, Olsztyn, Poland. (2022). *Information about Homelessness*, https://warmia.mazury.pl/images/Departamenty/Regionalny_O%C5%9Brodek_Pomocy_Spo%C5%82ecznej/Pomoc_i_integracja_spo%C5%82eczna/aktaulizacja_bezdomnosc2021/Informacja_dot._zjawiska_bezdomno%C5%9Bci_w_woj.pdf.

Mathworks. (2023). *Lsqcurvefit*, https://www.mathworks.com/help/optim/ug/lsqcurvefit.html.

Moein, S., Nickaeen, N., Roointan, A., Borhani, N., Heidary, Z., Javanmard, S. H., Ghaisari, J., & Gheisari, Y. (2021). Inefficiency of *SIR* models in forecasting *COVID–19* epidemic: A case study of Isfahan, *Nature*, 11(4725), 1–9.

Narayan, K., Rathore, H., & Znidi, F. (2022). Using epidemic modeling, machine learning and control feedback strategy for policy management of COVID–19, *IEEE Access*, 10, 98244–98258.

National Institute of Public Health Poland. (2022). *Tuberculosis in Poland*, https://www.igichp.edu.pl/.

Pfizer. (2023). *Why the COVID–19 Incubation Period Changes and How That Can Affect Us*, https://www.pfizer.com/news/articles/why_the_covid_19_incubation_period_changes_and_how_that_can_affect_us.

Romaszko, J., Siemaszko, A., Bodzioch, M., Buciński, A., & Doboszyńska, A. (2016). Active case finding among homeless people as a means of reducing the incidence of pulmonary tuberculosis in general population, *Advances in Experimental Medicine and Biology*, 911, 67–76.

Victorian Government. (2015). *Mycobacterial Infections (Tuberculosis)*, https://www.health.vic.gov.au/infectious-diseases/mycobacterial-infections-tuberculosis#incubation-period-of-tuberculosis.

7 Knowledge-based familiarity, perceived benefits, security and privacy protection in relation to digital trust and social media engagement

Anna Jasiulewicz, Fatih Çetin, Zygmunt Waśkowski and Marzena Lemanowicz

Introduction

Social networking platforms have arisen as a potent novel instrument for involving customers with content created either by the company or by consumers online (Schivinski et al., 2016). Involved consumers experience a robust bond with the brand (Calder & Malthouse, 2008) and, as a result, are anticipated to engage actively in consuming, contributing to, and fabricating content associated with the brand (Schultz & Peltier, 2013). As per Sashi (2012), customer engagement transforms customers into enthusiasts. Social media (SM) permits enthusiasts to link up and engage with other enthusiasts, heightening shared contentment and advocacy. Enthusiasts can likewise link up and interact with individuals who aren't yet customers, converting them into transactional patrons and initiating the customer engagement cycle afresh. Osei-Frimpong and McLean's (2018) define social media brand engagement (SMBE) as: *"connection, creation and communication of the brand's story between the firm and consumers (both existing and prospects), using brand or brand-related language, images and meanings via the firm's social networking site resulting from motivational drivers"*. Dolan et al. (2019) assert that SM engagement behaviors extend beyond mere transactions and can be precisely defined as *"a customer's behavioral manifestations that have a social media focus, beyond purchase, and result from motivational drivers*". Consumer engagement in SM has garnered significant attention from both scholars and businesses in recent years (e.g., Dessart et al., 2015; Dolan et al., 2019; Gligor et al., 2019; Leckie et al., 2022).

Not only engagement but also trust fosters emotional connection between buyers and vendors and shapes the link between various factors and behavioral intentions (Hashim & Tan, 2015). Pietrzak and Takala (2021) assert that there exists no universally recognized single definition of digital trust. They also prepared their own interpretation, defining it as follows: *"digital trust is the measure of confidence which workers, consumers/buyers, partners and other*

DOI: 10.4324/9781032627236-8

stakeholders have in an organization's ability to protect data and the privacy of individuals". They argue that comprehending the factors that bolster digital trust is imperative for business leaders. Irshad et al. (2020) proposed that trust should be accorded the highest priority when constructing research models for consumer behavior in SM contexts.

Therefore, in this chapter, we concentrated on investigating selected predictors of consumer engagement with SM brands and consumer digital trust in SM. The chapter comprises five sections. Following the introduction, the second part introduces the research model view along with the proposed hypotheses, which are subsequently examined later in the chapter. The third section outlines the methodology employed in our own study, elucidating the research process, sample demographics and questionnaire design, encompassing measurements and the reliability of constructs. The fourth division offers research outcomes pertaining to hypothesis testing and addressing the research questions. The fifth division delivers a brief discourse on findings, conclusions, implications and limitations.

Research model and hypotheses

In the study, three variables were selected as possible antecedents of SM engagement and digital trust in SM: knowledge-based familiarity, perceived benefits, and privacy and security protection. Our suggested structural model (Figure 7.1)

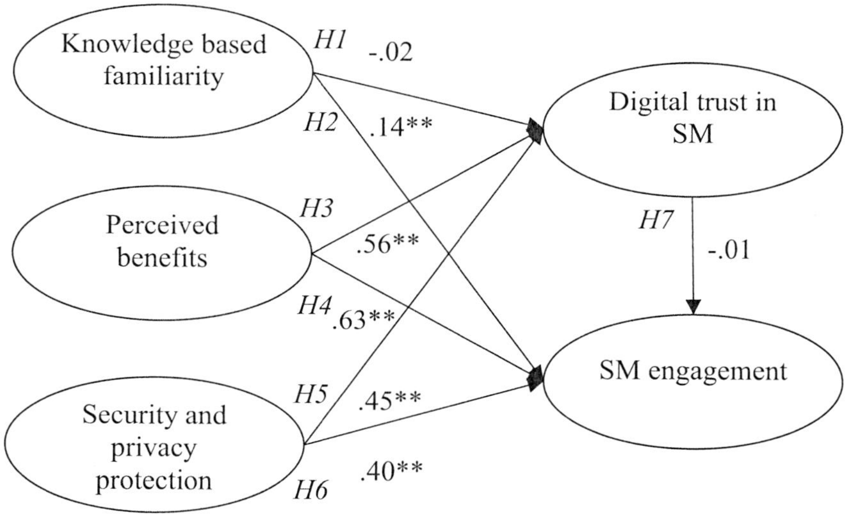

Figure 7.1 Research model and results (fit indices: X/df-1.78, CFI = .99, ILI = .99, SRMR = .036, RMSEA = .026).

Source: Author's research.

aimed to demonstrate that the mentioned factors impact consumer engagement in SM and enhance digital trust. Furthermore, the subsequent research questions were posed: *"Does digital trust increase social media engagement?"* and the crucial one: "*Which factor is relatively effective on both SM engagement and digital trust*?" In prior research, these factors were individually assessed for their predictive roles in digital trust and SM engagement, sometimes in various combinations with other factors. Nonetheless, to the best of our knowledge, no study has encompassed all six of these variables simultaneously, along with exploring the relationships among them, in order to comprehend digital trust and consumer engagement with brands in the context of SM.

Knowledge-based familiarity

According to Robert et al. (2009) trust establishment entails a "*relationship development process*" rooted in behavioral evaluation. This is also referred to as knowledge-based trust, which represents a form of cognitive trust based on the personal experiences of the trustor and trustee and it evolves over time as an individual amasses trust-related information through experience (Robert et al., 2009; Suk & Fong, 2022). Suk and Fong (2022) suggest that attributes rooted in knowledge serve as indicators for fostering online trust. Knowledge-based familiarity with brand has been found as an antecedent of online trust (Bhalla, 2020; Koehn, 2003; Rybak, 2018). However according to the research of Jasiulewicz et al. (2023) knowledge-based familiarity was not a significant predictor of the digital trust.

On the basis of the above, we noticed the need to pose and verify the following hypothesis:

H1: Knowledge-based familiarity positively predicts digital trust in SM.

Prior research argued that brand familiarity might prompt consumers to explore and acquire fresh product details (Hansen, 2017). Simonson et al. (2001) discovered that consumers rely on their pre-existing familiarity with a brand to facilitate their cognitive effort when seeking information. Bandura (2001) posits that personal factors of individuals, encompassing their cognition, convictions, and capabilities (e.g., brand expertise), steer their intentions and actions when engaging in specific activities like SMBE practices. Research results of Osei-Frimpong et al. (2020) suggested that brand knowledge had significant influence on participation in SMBE practices and consumers' pre-existing familiarity with a brand spurs them to partake in SMBE activities with the aim of acquiring fresh insights about the brand.

Accordingly, the following hypothesis is posited:

H2: Knowledge-based familiarity positively predicts SM engagement.

Perceived benefits

In accordance with Kirana and Kerti Yasa (2013), perceived benefits encompass a blend of service characteristics, physical attributes, and technical assistance associated with the utility of products or services. Hollebeek and Chen (2014) note that customers' readiness to allocate cognitive, emotional and physical resources hinges on the perceived benefits they gain from their engagement with companies and/or brands. For example, service convenience was found as a motivational driver of consumer brand engagement (Roy et al., 2020). Similar to any brand, SM brands must provide valuable benefits and establish relationship excellence in the shape of trust and contentment among their users (Leckie et al., 2022). Mikalef et al. (2013) and likewise Enginkaya and Yılmaz (2014) have indicated that users' online actions related to product selection, exploring advantageous prospects, brand connection, tailored promotions and access to product details influence their browsing and buying intentions. In general, researchers have identified convenience, enjoyment from branding interactions, availability of brand/product information, quest for opportunities, the ability to associate with a brand and personalized advertising as fundamental elements impacting SM engagement behavior. Barger et al. (2016) also noted that users' engagement with content on SM is impacted by the degree of entertainment, educational value and intriguing nature of the content. These listed elements can certainly be included in the benefits perceived by the consumer related to the use of SM.

On one hand, Islam Habis et al. (2023) found out that trust exerts a moderating influence on the connection between SM engagement (individual participation and personal significance) and social interactions. On the other hand, according to Khoa and Nguyen (2022) the perceived psychological benefits can serve as an antecedent to whether the customer provides additional personal information during the online transaction process, affecting online trust. According to the authors of this chapter, it would be worth checking whether perceived benefit is the predictor of digital trust in SM.

Therefore, the following hypotheses are postulated:

H3: Perceived benefits positively predict digital trust in SM.
H4: Perceived benefits positively predict SM engagement.

Security and privacy protection

Prior research underscores the significance of security and privacy protection in establishing trust in various scenarios, including online shopping and mobile internet usage. The thoroughness and clarity of online privacy policies impact the way online consumers perceive and their intentions to purchase products (Capistrano & Chen, 2015). In the context of e-commerce, researchers have recognized the content of privacy policies as a crucial determinant of consumer trust in websites (Wu et al., 2012). Mattila and Seppälä (2016) stated that *"digital trust stems from a combination of different factors (...): security, identifiability, and traceability"*. The absence of digital trust stems from diverse origins, including aspects such as security and privacy, electronic deception and so forth (Gorriz, 2003). According to Marcial and Launer (2019) digital trust *"refers to the level of confidence in people, processes, and technology to build a secure digital world"*.

As security and privacy protection constitutes a significant precursor of trust, we find it reasonable to posit that ensuring security and privacy in SM could likewise exert a substantial influence on digital trust in SM.

Although there are a number of studies on antecedents of customer/user engagement in social networks, to our best knowledge, no previous study directly included such variables security and privacy protection in relation to engagement in SM. The present study aims to plug the gap and intends to explore the effect of security and privacy protection on consumer engagement in SM.

Hence, we propose the following hypotheses:

H5: Security and privacy protection positively predicts digital trust in SM.
H6: Security and privacy protection positively predicts SM engagement.

Digital trust and SM engagement

Previous study's results suggested brand trust as the outcome of consumer engagement in SM brands (Carvalho & Fernandes, 2018; Leckie et al., 2022). On the other hand, certain researchers have proposed that consumer trust in brands exhibits a positive correlation with consumer brand engagement (i.e., Chahal & Rani, 2017; Hollebeek, 2011; Nyadzayo et al., 2020). Research endeavors like those undertaken by Alsaad et al. (2017), Laroche et al. (2013) and See-To and Ho (2014) explored SM engagement while considering various moderating elements. Considering the significance of consumer digital trust, some researchers have examined it as a mediator among different antecedents (Wang et al., 2015). They have suggested that scholars should consider digital trust as a moderating factor within the realm of SM. Alsaad et al. (2017) observed that trust operates as a moderator in the online domain, especially in situations where behavioral motivations are not firmly established. Therefore, based on the literature, this study proposes that the digital trust in SM is positively associated with consumers' SM engagement and the following hypothesis is proposed:

H7: Digital trust in SM positively predicts SM engagement.

Method

Participants

The information was gathered through the utilization of the Computer-Assisted Web Interviewing method (CAWI) by employing an online survey. This research focused on individuals who frequently engage with SM. To ensure precision and rapid access to the target demographic, participants were contacted through a web platform featuring a database containing more than 1 million individuals. The sample accurately reflects the age distribution in Poland, as per GUS (Statistics Poland Office) data from the 2021 National Census, categorized into distinct age groups (National Population and Housing Census, 2021). A preliminary assessment was carried out to assess the clarity of the survey. Data collection took place in October 2022, and the study included responses from 400 individuals who completed the questionnaires. A total of four hundred individuals (comprising 52.5% females and 47.5% males) participated in this investigation. The most prevalent cohorts included participants aged 65 and above (22%), followed closely by the age group 35–44 (20%), then the age brackets 25–34 (17%), 45–54 (16%), 55–64 (16%) and 18–24 (9%). Respondents hailing from rural areas constituted 34% of the surveyed populace, mirroring the proportion of individuals dwelling in towns with populations up to 50,000 (34%). Dwellers of cities with populations ranging from 100,000 to 500,000 comprised 16% of the entire research pool. The smallest subsets comprised respondents from towns with populations exceeding 50,000 but not exceeding 100,000 (8%) and those residing in cities with over 500,000 inhabitants (8%). Consumers who allocate relatively fewer hours to SM, specifically one hour or less per day, were classified as light SM users. This category encompassed a collective 43% of respondents, with 21% spending fewer than an hour daily on SM and 22% dedicating approximately one hour. Individuals who invest 2–3 hours daily on social networks (constituting 35% of the entire respondent cohort) were designated as moderate SM users, while those who confessed to utilizing SM for 3–5 hours a day (13%) and those exceeding 5 hours daily (9%) were categorized as intensive SM users.

Instruments

In this study, variables exanimated as possible predictors of SM engagement and digital trust included knowledge-based familiarity, perceived benefits, and security and privacy protection. All the components were formulated using preexisting measurement scales. In certain instances, the extracted scale elements were adjusted to suit their relevance within the context of SM. The survey was constructed employing the adopted five scales and encompassed inquiries spanning various facets, employing a 5-point Likert-style scale (1 = strongly disagree, 2 = disagree, 3 = neutral, 4 = agree, 5 = strongly agree).

Knowledge-based familiarity. In the domain of knowledge-based familiarity, four elements were chosen and adjusted from Gefen's (2000) research. An illustrative item was *"I am familiar with the process of purchasing from brands in social media"*.

Perceived benefits. To assess perceived benefits, we incorporated twenty elements gathered from the research of Forsythe et al. (2006), Raji et al. (2017),

Rani (2018), Mikalef et al. (2013) and Enginkaya and Yilmaz (2014). Sample items included *"The advertisements on social media help me in forming an opinion"* and *"Products presented on SM are customized to my needs"*.

Security and privacy protection. In the realm of security and privacy protection, three elements were selected and adapted from Kim et al. (2008). A representative item was: *"The brands selling items via social media usually ensure that my personal information is protected"*.

Digital trust in SM. To gauge trust in SM marketing, we chose and modified five elements from the research of Penington et al. (2003), Constantinides et al. (2010), Raji et al. (2017) and Kwiatek et al. (2021). An example item was *"I believe that negative online review can reduce the sales potential in the future"*.

Social media engagement. In assessing SM engagement, we adapted four elements from Gummerus et al. (2012), Laroche et al. (1996) and Campbell et al. (2014). A sample item stated *"I follow regular updates of products on SM"*.

Statistical procedures

The validity and reliability of the measurement instruments have been tested before structing the equation model for testing the hypotheses. In the validity analyses, confirmatory factor analysis (CFA, goodness-of-fit criteria: X^2/df < 3, CFI > .90, TLI > .90, RMSEA < .08) and the average variance extracted (AVE, the acceptance criteria is .50 or above)) values were used for the construct and discriminant validities. In the reliability analysis, composite reliabilities (CR > .70), Cronbach's Alpha coefficients (α > .70) and McDonald's omegas (ω > .70) were calculated. For testing the hypotheses, structural equation model with unweighted least square estimation method was used after testing the measurement model with using Jamovi v2.3 program.

Results

The measurement model fit statistics, validity and reliability results, and correlations among variables are in the Table 7.1. The results indicate acceptable fit indices for the measurement model (X^2/df = 4.89, CFI = .91, TLI = .90, RMSEA = .077) including knowledge-based familiarity (4 items), perceived benefits (19 items), security and privacy protection (3 items), digital trust in SM (5 items) and SM engagement (4 items). The AVE values range from .66 to .81 that indicates discriminant validity of the instruments. For the measurement reliability, CR (from .80 to .96), Cronbach's alpha (from .89 to .98) and McDonald's Omega (from .89 to .98) results show the consistencies of the instruments. Intercorrelations among variables range between .53 and .84 ($p < .01$) indicating a certain level of multicollinearity (even values are not higher than .90). The variance inflation factor (VIF) values (ranged from 1.62 to 3.86) and Harman's single factor test results (χ^2/df = 18.81, CFI = .77, TLI = .76, RMSEA = .122) indicate that the common method variance is not a significant concern (Table 7.1).

Table 7.1 Measurement model fit statistics, validity, reliability and correlations

	Factor loadings	*CFA Fit indices*	*Validity*		*Reliability*		*VIF*	*Correlations*			
			AVE	*CR*	*α*	*ω*		*2*	*3*	*4*	*5*
1 Knowledge-based familiarity	from .87 to .90	χ^2/df = 4.89 CFI = .91	.81	.80	.90	.90	1.62	–			
2 Perceived benefits	from .81 to .89	TLI = .90	.74	.96	.98	.98	3.76	.60**	–		
3 Security and privacy protection	from .86 to .93	RMSEA = .077	.75	.92	.90	.90	2.71	.53**	.72**	–	
4 Digital trust in SM	from .73 to .88		.66	.91	.89	.89	3.86	.54**	.84**	.78**	–
5 SM engagement	from .87 to .94		.82	.95	.93	.93	–	.61**	.82**	.73**	.74**
Single factor for testing CMV	from .52 to .86	χ^2/df = 18.81, CFI = .77, TLI = .76, RMSEA = .122									

** $p < .01$.

For testing the research hypotheses, a structural equation model was constructed with knowledge-based familiarity, perceived benefits, security and privacy protection as exogenous, SM engagement as endogenous and digital trust in SM as both exogenous and endogenous variables (Figure 7.1). The model results show that knowledge-based familiarity positively predicts SM engagement (beta = .14, $p < .01$, *H2 supported*), but has no relationship with digital trust in SM (beta = −.02, $p > .05$, *H1 not supported*); perceived benefits positively predicts digital trust in SM (beta = .56, $p < .01$, *H3 supported*) and SM engagement (beta = .63, $p < .01$, *H4 supported*); security and privacy protection positively predicts digital trust in SM (beta = .45, $p < .01$, *H5 supported*) and SM engagement (beta = .40, $p < .01$, *H6 supported*); digital trust in SM has no relationship with SM engagement (beta = −.01, $p > .05$, *H7 not supported*).

Conclusions

The purpose of this chapter was to explore the possible factors on effecting SM engagement. Findings present that perceived benefits and security and privacy protection are the top priorities when it comes to enhancing digital trust in SM and SM engagement. The finding on the importance of perceived benefits is consistent with previous studies of Barger et al. (2016) and Khaola and Nguen (2022). Nonetheless, to the best of our understanding, the significance of security and privacy proception as predictors of engagement in SM is a new fining.

Notably, while knowledge-based familiarity positively influences SM engagement, it does not significantly impact digital trust in SM. Such research results are only partially consistent with the mentioned Osei-Frimpong et al. (2020) study, which indicated that brand familiarity is a predictor of both trust and SMBE. However, the is in accordance with the results of the research by Jasiulewicz et al. (2023) that found that familiarity based on knowledge did not emerge as a significant predictor of digital trust.

Lastly, there is no significant relationship between digital trust in SM and SM engagement. Therefore, hypothesis 7 was not positively verified, which is not in line with the research results of Alsaad et al. (2017) and Nyadzayo (2020).

Eventually, beyond the positive influences of familiarity, benefits and security and privacy on SM engagement, digital trust in SM may not be a mediating structure for enhancing SM engagement.

In addition to contributing insights to existing literature, the research outcomes may also be valuable for marketers aiming to allure customers through SM brand engagement. The discoveries could assist e-commerce in acquiring a more profound understanding of the antecedents of consumer SM engagement and digital trust, thereby aiding in the formulation of effective SM marketing strategies. The results underscore the importance of perceived benefits and security and privacy protection to increase digital trust and SM engagement and also knowledge-based familiarity to positively influence customer involvement in SM.

The limitations of this chapter primarily pertain to the research sample employed. Despite the sample's alignment with the age distribution percentages in Poland, its size is insufficient for broad generalization of the findings to the entire population, demanding cautious interpretation. Furthermore, the authors relied on specific databases for the literature review, thus the paper does not encompass an exhaustive exploration of the topic. Despite incorporating empirical research results, the paper should primarily be regarded as an indication of a research problem, specifically, the impact of knowledge-based familiarity, perceived benefits, security and privacy protection on consumer engagement with brands on SM and consumer digital trust. Lastly, there is a need for further exploration to understand whether people engage in SM without trust, by examining the potential mediating role of digital trust among familiarity, benefits, and security and privacy in SM engagement.

References

Alsaad, A., Mohamad, R., & Ismail, N. A. (2017). The moderating role of trust in business to business electronic commerce adoption. *Computers in Human Behaviour, 68*(March), 157–169.

Bandura, A. (2001). Social cognitive theory: An agentic perspective. *Annual Review of Psychology, 52*, 1–26.

Barger, V., Peltier, J. W., & Schultz, D. E. (2016). Social media and consumer engagement: A review and research agenda: An international journal. *Journal of Research in Interactive Marketing, 10*(4), 268–287. https://doi.org/10.1108/JRIM-06-2016-0065

Bhalla, S. (2020). Demystifying the key antecedents of consumer trust in online shopping and testing the mediating role of consumer trust: An empirical study. *IUP Journal of Marketing Management, 19*(1), 7–22.

Calder, B. J., & Malthouse, E. C. (2008). Media engagement and advertising effectiveness. In *Kellogg on Media and Advertising.* Calder, B. (Ed.), John Wiley & Sons, 1–36.

Campbell, C., Ferraro, C., & Sands, S. (2014). Segmenting consumer reactions to social network marketing. *European Journal of Marketing, 48*(3/4), 432–452.

Capistrano, E. P. S., & Chen, J. V. (2015). Information privacy policies: The effects of policy characteristics and online experience. *Computer Standards & Interfaces, 42*, 24–31.

Carvalho, A., & Fernandes, T. (2018). Understanding customer brand engagement with virtual social communities: A comprehensive model of drivers, outcomes and moderators. *Journal of Marketing Theory and Practice, 26*(1–2), 23–37.

Chahal, H., & Rani, A. (2017). How trust moderates social media engagement and brand equity: An international journal. *Journal of Research in Interactive Marketing, 11*(3), 312–335. https://doi.org/10.1108/JRIM-10-2016-0104

Constantinides, E., Lorenzo-Romero, C., & Gomez, M. (2010). Effects of web experience on consumer choice: A multicultural approach. *Internet Research, 20*(2), 188–209.

Dessart, L., Veloutsou, C., & Morgan-Thomas, A. (2015). Consumer engagement in online brand communities: A social media perspective. *Journal of Product & Brand Management, 24*(1), 28–42. https://doi.org/10.1108/JPBM-06-2014-0635

Dolan, R., Conduit, J., Frethey-Bentham, C., Fahy, J., & Goodman, S. (2019). Social media engagement behavior: A framework for engaging customers through social media content. *European Journal of Marketing, 53*(10), 2213–2243. https://doi.org/10.1108/EJM-03-2017-0182

Enginkaya, E., & Yılmaz, H. (2014). What drives consumers to interact with brands through social media? A motivation scale development study. *Procedia – Social and Behavioral Sciences, 148,* 219–226.

Forsythe, S., Liu, C., Shannon, D., & Gardner, L. C. (2006). Development of a scale to measure the perceived benefits and risks of online shopping. *Journal of Interactive Marketing, 20*(2), 55–75.

Gefen, D. (2000). E-Commerce: The role of familiarity and trust. *Omega, 28*(6), 725–737.

Gligor, D., Bozkurt, S., & Russo, I. (2019). Achieving customer engagement with social media: A qualitative comparative analysis approach. *Journal of Business Research, 101,* 59–69. https://doi.org/10.1016/j.jbusres.2019.04.006

Gorriz, C. (2003). *Electric Commerce and Trust on the Internet.* SRIC-Bio, Menlo Park, CA.

Gummerus, J., Liljander, V., Weman, E., & Pihlstrom, M. (2012). Customer engagement in a Facebook brand community. *Management Research Review, 35*(9), 857–877.

Hansen, T. (2017). The moderating effects of financial broad-scope trust on consumer knowledge, cognitive effort, and financial healthiness. *Journal of Consumer Behaviour, 16*(2), 161–175.

Hashim, K. F., & Tan, F. B. (2015). The mediating role of trust and commitment on members' continuous knowledge sharing intention: A commitment-trust theory perspective. *International Journal of Information Management, 35*(2), 145–151.

Hollebeek, L. D. (2011). Demystifying customer brand engagement: Exploring the loyalty nexus. *Journal of Marketing Management, 27*(7–8), 785–807.

Hollebeek, L. D. & Chen, T. (2014). Exploring positively-versus negatively-valenced brand engagement: A conceptual model, *Journal of Product and Brand Management, 23*(1), 62–74.

Irshad, M., Ahmad, M. S., & Omer, F. M. (2020). Understanding consumers' trust in social media marketing environment. [Trust in social-mediated marketing environment] *International Journal of Retail & Distribution Management, 48*(11), 1195–1212. https://doi.org/10.1108/IJRDM-07-2019-0225

Islam Habis, M. H., Amjad, O. S., Habes, M., Tahat, O., Amer, K. A., Abdallah, Rania Abdel-Qader, & Aissani, R. (2023). Trust in social media: Enhancing social relationships. *Social Sciences, 12*(7), 416. https://doi.org/10.3390/socsci12070416

Jasiulewicz, A., Svenson F., & Wyrzykowska, B. (2023). Antecedents of digital trust in online shopping. In: *Trust, Digital Business and Technology. Issues and Challenges.* Paliszkiewicz Joanna, Guerrero Cusumano José Luis, & Gołuchowski Jerzy (Eds.), Routledge, 217–230.

Khoa, B. T., & Nguyen, M. H. (2022). The moderating role of anxiety in the relationship between the perceived benefits, online trust, and personal information disclosure in online shopping. *International Journal of Business and Society, 23*(1), 444–460. https://doi.org/10.33736/iibs.4624.2022

Kim, D. J., Ferrin, D. L., & Rao, H. R. (2008). A trust-based consumer decision making model in electronic commerce: The role of trust, perceived risk, and their antecedents. *Decision Support Systems, 44*(2), 544–564.

Kirana, I. D. A. I., & Kerti Yasa, N. N. (2013). Peran Gender dalam Memoderasi Pengaruh Perceived Benefit dan Perceived Cost Terhadap Niat Menggunakan Kartu Kredit di Kota Denpasar (Doctoral dissertation, Udayana University).

Koehn, D. (2003). The nature of and conditions for online trust. *Journal of Business Ethics, 43*(1&2), 3–19.

Kwiatek, P., Baltezarević, R., & Papakonstantinidis, S. (2021). The impact of credibility of influencers recommendations on social media on consumers behavior towards brands. [Utjecaj vjerodostojnosti preporuka influencera na društvenim mrežama na ponašanje potrošača prema markama] *Informatologia, 54*(3), 181–196. https://doi.org/10.32914/i.54.3-4.5

Laroche, M., Habibi, M. R., & Richard, M. O. (2013). To be or not to be in social media: How brand loyalty is affected by social media? *International Journal of Information Management, 33*(1), 76–82.

Laroche, M., Kim, C., & Zhou, L. (1996). Brand familiarity and confidence as determinants of purchase intention: An empirical test in a multiple brand context. *Journal of Business Research, 37*(2), 115. https://www.proquest.com/scholarly-journals/brand-familiarity-confidence-as-determinants/docview/196322729/se-2

Leckie, C., Dwivedi, A., & Johnson, L. (2022). Examining drivers and outcomes of social media brand engagement. *Asia Pacific Journal of Marketing and Logistics, 34*(2), 350–367. https://doi.org/10.1108/APJML-07-2020-0488

Marcial, D. E., & Launer, M. A. (2019). Towards the measurement of digital trust in the workplace: A proposed framework. *International Journal of Scientific Engineering and Science, 3*(12), 1–7.

Mattila, J., & Seppälä, T. (2016). *Digital Trust, Platforms, and Policy*. Retrieved from: https://ideas.repec.org/p/rif/briefs/42.html (accessed 8 September, 2023).

Mikalef, P., Giannakos, M., & Pateli, A. (2013). Shopping and word-of-mouth intentions on social media. *Journal of Theoretical and Applied Electronic Commerce Research, 8*(1), 17–34.

National Population and Housing Census. (2021). Assessment of the data quality. *Statistics Poland.* https://stat.gov.pl/en/national-census/national-population-and-housing-census-2021/final-results-of-the-national-population-and-housing-census-2021/national-population-and-housing-census-2021-assessment-of-the-data-quality,1,1.html

Nyadzayo, M. W., Leckie, C., & Johnson, L. W. (2020). The impact of relational drivers on customer brand engagement and brand outcomes. *Journal of Brand Management, 27*, 561–578.

Osei-Frimpong, K., & McLean, G. (2018). Examining online social brand engagement: A social presence theory perspective. *Technological Forecasting and Social Change, 128*, 10–21.

Osei-Frimpong, K., McLean, G., & Famiyeh, S. (2020). Social media brand engagement practices: Examining the role of consumer brand knowledge, social pressure, social relatedness, and brand trust. [SMBE practices] *Information Technology & People, 33*(4), 1235–1254. https://doi.org/10.1108/ITP-05-2018-0220

Penington, R., Wilcox, H. D., & Grover, V. (2003). The role of system trust in business-to-consumer transactions. *Journal of Management Information Systems, 20*(3), 197–226.

Pietrzak, P., & Takala, J. (2021). Digital trust – A systematic literature review. *Forum Scientiae Oeconomia, 9*(3), 60–71.

Raji, R. A., Rashid, S. M., & Sobhi, I. M. (2017). Assessing validity and internal consistency of the social media marketing communication measurement scales. *E-BANGI Journal*, *12*(3), 1–14.

Rani, A. (2018). Young customer's insight on social media engagement. *International Journal of Marketing & Business Communication*, *7*(1), 18–25.

Robert, L. P., Denis, A. R., & Hung, Y. T. C. (2009). Individual swift trust and knowledge-based trust in face-to-face and virtual team members. *Journal of Management Information Systems, 26*(2), 241–279.

Roy, S. K., Shekhar, V., Quazi, A., & Quaddus, M. (2020). Consumer engagement behaviors: Do service convenience and organizational characteristics matter? [Consumer engagement behaviors] *Journal of Service Theory and Practice, 30*(2), 195–232. https://doi.org/10.1108/JSTP-03-2018-0049

Rybak, A. (2018). Consumer trust in E-commerce: The case of Poland. *Folia Oeconomica Stetinensia, 18*(2), 59–71.

Sashi, C. M. (2012). Customer engagement, buyer-seller relationships, and social media. *Management Decision, 50*(2), 253–272. https://doi.org/10.1108/00251741211203551

Schivinski, B., Christodoulides, G., & Dabrowski, D. (2016). Measuring consumers' engagement with brand-related social media content: Development and validation of a scale that identifies levels of social-media engagement with brands. *Journal of Advertising Research, 56*(1), 64–80.

Schultz, D. E., & Peltier, J. (2013). Social media's slippery slope: Challenges, opportunities and future research directions. *Journal of Research in Interactive Marketing, 7*(2), 86–99.

See-To, E. W., & Ho, K. K. (2014). Value co-creation and purchase intention in social network sites: The role of electronic word-of-mouth and trust–a theoretical analysis. *Computers in Human Behavior*, *31*(February), 182–189.

Simonson, I., Carmon, Z., Dhar, R., Drolet, A., & Nowlis, S. M. (2001). Consumer research: In search of identity. *Annual Review Psychology, 52*, 249–275.

Suk, C. T., & Fanny Fong, Y. C. (2022). Revisiting trust in the digital era: The interplay of online trust and online dialogic communication from the practitioners' perspective. *Journal of Communication Management, 26*(3), 271–293. https://doi.org/10.1108/JCOM-08-2021-0094

Wang, L., Law, R., Guillet, B. D., Hung, K., & Fong, D. K. C. (2015). Impact of hotel website quality on online booking intentions: E-trust as a mediator. *International Journal of Hospitality Management*, *47*(1), 108–115.

Wu, K.-W., Huang, S. Y., Yen, D. C., & Popova, I. (2012). The effect of online privacy policy on consumer privacy concern and trust. *Computers in Human Behavior, 28*(3), 889–897.

8 The trust in the performance of modern database systems

The issue of multithreading

Krzysztof Karpio and Piotr Łukasiewicz

Introduction

The current century may be called an information era. The information is usually described as a result of the data analysis. The information, in turn, is used to make decisions. Thus, data is the basis of the decisions. That's why the value of data increases with time. On the other hand, one is forced to store more and more versatile data because of legal regulations or competitiveness. Modern database systems are capable of storing and efficiently managing large data volumes. Thus, a lot of effort has to be put into providing quick access to data for retrieval or modification.

While the RDBMS market grew at a rate of about 8% per year (Statista, 2022), in 2022, it experienced 14.4% growth, beating the overall software market growth of 11.3% (Schumacher & Chibber, 2023). However, for many years, Oracle and Microsoft have been the leading vendors in this market. Their database products, including analytical databases, covered 66% of the market in 2020 (T4, 2020). On the other hand, there is a big market for open source database systems, for example, PostgreSQL, MySQL, and MariaDB.

In this chapter, we compare the performance of the three popular RDBMSs: SQL Server 2019 Developer Edition, Oracle Database 19c Enterprise Edition, and PostgreSQL version 14. The Oracle database products are the most reputable on the market, are the most expensive, and are used by the biggest customers, including CERN (Grancher, 2010). SQL Server is more popular among medium-sized customers because of its lower price and more convenient management. These two systems are highly trusted by customers. On the other hand, PostgreSQL is an open source product that does not require costly licenses and is utilized by many institutions and commercial companies. Taking into account the amount of investment and customer trust, we can align the products from Oracle through SQL Server to PostgreSQL.

The aim of this chapter is to compare the performance of the three modern database systems mentioned above. While performance is a very broad topic, we narrow it to efficiency versus multithreading. The latter is the most important

DOI: 10.4324/9781032627236-9

factor in increasing the speed of multiuser computer systems. We are going to verify the hypothesis: the reputation and trust of the product come together with efficiency.

This chapter is organized in the following way: In Section "Literature Review", we present the review of literature; in Section "Multithreading", we discuss the issue of multithreading; and in Section "Methodology", we present the methodology of our studies. The obtained results are presented in Section "Results of the Experiment" and discussed in Section "Discussion". The closing Section "Conclusions" contains conclusions.

Literature review

Growing data volumes increase interest in query optimization and the use of modern computing technologies like multithreading or advanced indexing (Hashim, 2018; Asratian, 2021). The problem of optimization is even bigger because of the increase in data diversity. Various types of data have different resource requirements (Bao et al., 2021). In (Zhiwei et al., 2020a), authors study the possibilities of choosing the optimal degree of parallelism for query execution. They conclude that DOP (Degree of Parallelism) plays a fundamental role not only in query performance but also in resource provisioning. Similar observations were made by Gustaffson, who concluded that multiprocessing systems not only speed up code execution but also enable dealing with larger problems (Gustafson, 1988). That's why there is great interest in choosing the optimal DOP. Many various techniques, like artificial intelligence (Zhiwei et al., 2020b) or graph modeling (Sasak-Okoń & Tudruj, 2017) have been utilized in this area. One should keep in mind that multithreading or parallelism can occur not only on a session or query level but also on an intra-query level. The first phenomenon (session level) is obvious in the case of multiuser database systems, where a great number of users simultaneously use the same database. On the other hand, we can deal with intra-query parallelism, which plays a very important role in the analytical system (Zhiwei et al., 2020b). The other type of queries that would potentially benefit most from multithreading are those that deal with spatial data (Chengming et al., 2017). One of the reasons for this is that the indexing of such data is especially complicated (Abbas & Ahmad, 2021). One considers using many techniques together in order to obtain respectful performance and efficiency. That's why there is no easy answer to the question regarding which database system is faster (Islam et al., 2017). It has been stated that good results can be obtained in searching for the shortest path on RDBMS based graphs by means of the method using multithreading (Seo et al., 2019). The query optimization issue becomes more problematic because of widely used tools that generate databases and queries automatically and are not designed by developers (Colley & Stanier, 2017). Similar trends in automatics are observed in the field of indexing. In (Kołaczkowski & Rybiński, 2009) authors were working on automatic indexing methods based on the query plans. However, one should keep in mind that sometimes plans are not optimal. A comparison of database systems performance is widely discussed in the literature. Among the latest works, interesting review papers should be mentioned (Huang et al., 2023; Taipalus, 2023).

Multithreading

One of the most important aspects of query processing and running is multithreading, defined as running several CPU threads in parallel (Rogers, 1985). Separate thread slots are usually visible by RDBMSs as so-called logical processors. The multithreading in a database can be visible on the three levels:

Level 1. Multiple sessions can use different logical processors. This is especially important in the case of systems running multiple sessions in parallel.
Level 2. Multiple steps of the same query run in parallel. Analytical, complex queries with many steps can benefit the most.
Level 3. One step of the query is serviced by many logical processors. The most suitable is when the step takes a significant amount of time and its work can be split and done in parallel.

Amdahl (Hill & Marty, 2008) drew attention to the fact that not always the entire query can be parallelized when formulating his law (Amdahl, 1967). Some steps, for example, that merge data processed in parallel, cannot be parallelized. Moreover, we sometimes deal with I/O operations that have a limited minimal working time. Because of that, adding CPUs does not make the query times decrease to zero.

Methodology

We performed the experiment based on the measurement of query performance versus degree of parallelism. We followed the same procedure for SQL Server, Oracle, and PostgreSQL. The experiment was performed on a computer equipped with an Intel® Core™ i7–8550U 1.80GHz processor and 16GB of RAM. The CPU has four cores and each core can run two threads at the same time. In our case, we can describe the system as having eight logical CPUs that are capable of running eight separate threads in parallel. The operating system was Windows 10 Pro.

The performance was measured by running two queries against prepared tables containing 5 million records each. The first query simply merges multiple sets of the retrieved records into one result set, thus being I/O intensive. The second query requires sorting and eliminating duplicates, preserving only the unique records, and thus requires more CPU power. For each database system studied, we ran each query and measured its times for every degree of parallelism (number of threads) from 1 to 8. For statistical reliability, each query has been run five times, and mean values as well as standard deviations have been calculated.

In order to evaluate overall query performance, one uses the Elapsed Time T_{elap}, which is simply a wall clock of the total query time. This time is usually expressed as the sum of the cpu time T_{cpu} of the query and time related to various waits for resources T_{wait}. Each query has been run five times and we calculated mean time and standard deviation for each *DOP* for elapsed and cpu times separately.

There are some misunderstandings about cpu time that should be explained. This time in general is not a wall clock time; it is derived from the number of the cpu ticks. As long as we use one thread at a time, both times are in the same units; however, when parallelism takes place, differences arise. For instance, we can obtain a cpu time larger than the elapsed time. This is possible because of the parallel execution of tasks and because cpu time is measured in other units than elapsed time.

T_{elap} – [wall clock] elapsed time, total query time.

T_{cpu} – [based on cpu ticks], total time cpu is working.

Paradoxically, a cpu time increases with *DOP*, sometimes exceeding an elapsed time. In fact, this is not true, because both times are expressed in different units. If we assume we deal with ideal multithreading we can define the cpu wall clock time as $T_{cpuWC} = T_{cpu} / DOP$. Thus the elapsed query time consists of two parts: the first part depends on *DOP* and the second part is related to waiting for resources. The second part is usually constant however more intensive use of cpu could potentially lead to a slight increase of wait time. After dividing the cpu time by *DOP*, we get effective wall clock cpu time. In the ideal case, we should obtain a power function of *DOP* with an exponent equal to −1 according to the above equation. In order to measure the scaling of cpu time with *DOP* we will use more general form of the equation $T_{cpuWC} \propto DPO^{\alpha}$, where α exponent is the measure of efficiency of scaling.

Results of the experiment

We performed measurements of the query times for 8 values of *DOP* from 1 to 8. We analyzed elapsed and cpu query times. The latter was also divided by *DOP*, yielding the above-described wall clock cpu time. The results for the studied database systems based on the two queries are summarized in the Table 8.1. Numbers with errors represent average times and standard deviations.

It is characteristic that the longest times are for SQL Server for both queries. For the remaining database systems, query times are often similar to each other. We obtained small errors, which are about a few percent in all measurements. The cpu times, as expected, are bigger in the case of the second query. The wall clock cpu time decreases according to the mentioned power low for all databases. The calculations performed show that the α exponent has the values:

Table 8.1 Queries' times for Oracle, SQL Server, and PostgreSQL

DB	DOP	Query 1			Query 2		
		Elapsed time [s]	*cpu time [cpu ticks s]*	*cpu time [wall clock]*	*Elapsed time [s]*	*cpu time [cpu ticks s]*	*cpu time [wall clock]*
Oracle	1	2.35±0.03	1.54±0.02	1.54±0.02	7.73±0.10	6.87±0.12	6.87±0.12
	2	2.32±0.01	1.76±0.04	0.88±0.02	10.91±0.05	10.33±0.05	5.17±0.03
	3	2.35±0.00	1.86±0.06	0.62±0.02	13.20±0.08	12.90±0.14	4.30±0.05
	4	2.36±0.01	1.96±0.03	0.49±0.01	14.82±0.33	14.81±0.49	3.70±0.12
	5	2.41±0.01	2.03±0.04	0.41±0.01	16.34±0.12	16.16±0.29	3.23±0.06
	6	2.42±0.01	2.01±0.06	0.34±0.01	18.54±0.75	18.50±0.77	3.08±0.13
	7	2.43±0.01	1.94±0.09	0.28±0.01	18.19±0.05	17.92±0.14	2.56±0.02
	8	2.43±0.01	1.98±0.06	0.25±0.01	19.58±0.13	18.80±0.22	2.35±0.03
SQL Server	1	37.71±1.04	5.01±0.11	5.04±0.09	65.29±1.10	35.59±0.41	35.62±0.38
	2	35.83±0.12	7.21±0.31	3.59±0.11	56.92±0.14	37.62±0.21	18.81±0.08
	3	37.74±0.38	8.24±0.28	2.68±0.08	49.32±0.28	38.24±0.19	12.67±0.07
	4	37.69±0.92	9.04±0.47	2.22±0.09	46.88±0.10	42.03±0.38	10.48±0.08
	5	37.72±0.29	9.78±0.13	2.03±0.01	44.48±0.68	44.49±0.74	8.93±0.08
	6	36.12±0.21	11.19±0.48	1.88±0.13	43.24±0.29	50.24±0.48	8.40±0.11
	7	38.61±1.04	11.23±0.22	1.59±0.02	44.09±0.91	54.92±1.16	7.79±0.17
	8	36.73±0.47	12.11±0.29	1.52±0.03	44.13±1.04	57.67±0.33	7.23±0.03
PostgreSQL	1	2.59±0.13	1.29±0.10	1.29±0.10	50.93±0.45	6.59±0.66	6.59±0.66
	2	2.67±0.13	1.24±0.17	0.62±0.08	50.38±0.67	6.03±0.46	3.02±0.23
	3	2.70±0.04	1.30±0.06	0.43±0.02	51.15±0.50	6.15±0.71	2.05±0.24
	4	3.50±0.05	1.89±0.07	0.47±0.02	50.59±0.53	6.16±0.33	1.54±0.08
	5	3.46±0.04	1.67±0.06	0.33±0.01	50.47±0.47	5.56±0.12	1.11±0.02
	6	4.15±0.05	1.83±0.07	0.31±0.01	50.50±0.47	6.09±0.35	1.02±0.06
	7	4.63±0.02	1.89±0.06	0.27±0.01	50.66±0.58	6.38±0.49	0.91±0.07
	8	4.63±0.01	1.88±0.04	0.24±0.01	51.25±0.44	7.77±0.65	0.97±0.08

Query 1: −0.877 for Oracle, −0.584 for SQL Server, and −0.770 for PostgreSQL;

Query 2: −0.508 for Oracle, −0.765 for SQL Server, and −0.975 for PostgreSQL.

The best result for Query 1 is for Oracle; however, the exponent for PostgreSQL is not much bigger. The scaling for SQL Server is significantly worse. In the case of Query 2, the result for PostgreSQL is the best, and the exponent is very close to −1. The results for SQL Server are worse, and the worst are for Oracle.

Discussion

In order to discuss the results, we compared the wall clock cpu times for each query and each database platform. As stated before, the results can be compared to the ideal situation in which wall lock cpu query time scales with DOP as a power function with an exponent −1. To make the comparison more general, we rescaled the values of DOP as well as the wall clock cpu time. The *DOP* was rescaled according to: $RDOP = (DOP - 1) / (MAXDOP - 1)$. Thus the values of *RDOP* are between 0 (the lack of parallelism) and 1 (full possible parallelism: $DOP = MAXDOP$). The time was rescaled in such a way that the time for a serially run query is equal to 1. The results are presented in the Figure 8.1, in linear and log-log scales. Each of the plots contains three datasets, together with the ideal model indicated by the black dashed line without data points. The results are discussed on a query basis.

Query 1 contains a set of mutually independent simple queries. They are based on scanning table data in order to filter records. All the retrieved records are simply merged into one data stream without any transformations or sorting. Oracle and PostgreSQL behave better than SQL Server. The line for SQL Server ($\alpha = -0.584$) begins at the same point as for Oracle, but later its distance to the ideal model increases distinctly, what means the time scaling with *RDOP* is poor. On the other hand, the line related to Oracle is almost parallel to the line denoting ideal model. PostgreSQL exhibits slightly worse scaling ($\alpha = -0.770$) than Oracle ($\alpha = -0.877$), but this behavior is partly compensated by the smaller time for serial query ($RDOP = 0$).

Query 2 consists of the two independent queries whose results are merged, but opposite to the Query 1, merged records have to be distinct and sorted. The elimination of duplicates as well as sorting require much cpu power and thus, Query 2 should benefit more from multithreading than simple I/O operations. Surprisingly, the line for PostgreSQL shown in Figure 8.1d is close and parallel to the ideal model. That means, not only are the query times the smallest, but also the scaling is ideal. The results for SQL Server are worse and for Oracle are significantly worse when compared with PostgreSQL. Let's remind that power exponents α are −0.975, −0.765, −0.508 for PostgreSQL, SQL Server,

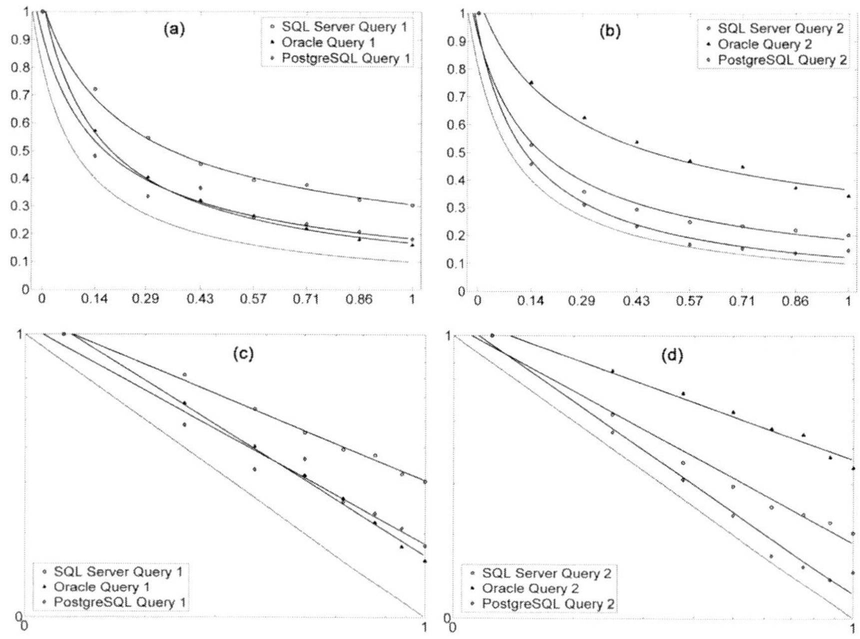

Figure 8.1 Wall clock cpu time plotted in linear scale in (a), (b) and log-log scale in (c), (d). Results for Query 1 are in plots (a), (c) and for Query 2 in plots (b), (d). Individual database systems are indicated by different markers and described in the legend. The ideal models are denoted by dashed lines.

Source: Own research.

and Oracle, respectively. Observed exponent values correspond with a dependency of the cpu time on *RDOP*. For Oracle, the cpu time increases with *RDOP* the most, so we can state that the additional cpu time related to the management of multiple threads increases significantly with the number of threads. The same effect, but smaller, is visible for SQL Server. On the other hand, cpu time spent on the management of threads seems to be almost independent of the number of threads in PostgreSQL. We can conclude that the reason for the observed differences lies in the different management of multiple threads.

In this chapter, we investigate one aspect of the performance of database systems related to the multithreading. We studied two highly trusted database systems and compared them to the open source system. Taking into account the customer's trust, the products can be ordered descending from Oracle through SQL Server to PostgreSQL. In the introduction section, we formulated the hypothesis: a reputation and trust of the product come together with its efficiency. The presented results indicate on the need for rejecting of this hypothesis.

The open source database system behaves well for both types of studied queries: I/O and CPU intensive. In the case of the first query, results for PostgreSQL are similar to those obtained for Oracle and significantly better than results for SQL Server. In the case of the second query, PostgreSQL is again the most efficient in comparison to other databases. Thus, the highly trusted systems, promoted by huge corporations and utilized in many multiuser critical systems, are less efficient than the open source system whose promotion is limited.

Conclusions

The goal of this chapter was to prepare and perform the experiment, whose aim was to compare three modern database systems in the area of multithreading. The following hypothesis has been verified: a reputation and trust of the product come together with its efficiency. One measured query times vs. degree of parallelism for two queries: the I/O intensive and the CPU intensive. One obtained the following main results:

1. For Query 1, the wall clock cpu time was decreasing with *DOP* according to power low with exponent −0.877, −0.770, −0.584 and for Oracle, PostgreSQL, and SQL Server, respectively. PostgreSQL and Oracle performed significantly better than SQL Server.

2. For Query 2, the wall clock cpu time was decreasing with *DOP* according to power low with exponent −0.975, −0.765, and −0.508 for PostgreSQL, SQL Server, and Oracle, respectively. The scaling was the best for PostgreSQL being almost compatible with the ideal model. The worse scaling was observed for SQL Server and Oracle.

3. Based on the above results, the hypothesis formulated in this chapter should be rejected. The highly trusted systems turn out to be less efficient than the open source system.

We are going to extend the studies for more complex queries, utilizing other operations. Another aspect of database performance that will be investigated is the performance in the case of multiple simultaneous users. The interesting is also a study of performance of other open source systems, like MariaDB and Firebird.

References

Abbas, A., & Ahmad, K. (2021). An Integrated Indexing Approach to Improve Query Processing. In: *Proceedings International Conference on Electrical Engineering and Informatics* (*ICEEI*), 1–6.

Amdahl, G. M. (1967). Validity of the Singal Processor Approach to Archive Large Scale Capabilities. In: *AFIPS Conference Proceedings*, 483–485.

Asratian, R. (2021). Deterministic Approach to Information Queries Parallel Batch Processing Modelling and Evaluating. In Tsvirkun, A. (Ed.) *Proceedings of 14th International Conference Management of Large-Scale System Development* (*MLSD*), 1–5.

Bao, L., Yang, J., Wu, C. Q., Qi, H., Zhang, X., & Cai, S. (2021). XML2HBase: Storing and Querying Large Collections of XML Documents using a NoSQL Database System. *Journal of Parallel and Distributed Computing*, 161, 83–99.

Chengming, L., Zheng, W., & Jie, Y. (2017). Research on Oracle-based Integrative Storage and Management of Spatial Data. In: *Proceedings International Conference on Information System and Data Mining* (*ICISDM*), 16–22.

Colley, D., & Stanier, C. (2017). Identifying New Directions in Database Performance Tuning. *Procedia Computer Science,* 121, 260–265.

Grancher, E. (2010). Oracle and Storage IOs, Explanations and Experience at CERN. *Journal of Physics: Conference Series*, 219, 052004.

Gustafson, J. L. (1988). Reevaluating Amdahl's Law. *Communications of the ACM*, 31(5), 532–533.

Hashim, R. T. (2018). A Comparative Study of Indexing using Oracle and MS-SQL Server for Relational Database Management Systems. *International Journal of Computer Science and Mobile Computing,* 7(12), 341–350.

Hill, M. D., & Marty, M. R. (2008). Amdahl's Law in the Multicore Era. *Computer*, 41(7), 33–38.

Huang, S., Qin, Y., Zhang, X., Xinyi, Z., Yaofeng, T., Zhongliang, L., & Bin, C. (2023). Survey on Performance Optimization for Database Systems. *Science China Information Sciences,* 66, 121102.

Islam, K., Ahsan, K., Bari, S. A. K., Saeed, M., & Ali, S. A. (2017). Huge and Real-Time Database Systems: A Comparative Study and Review for SQL Server 2016, Oracle 12c & MySQL 5.7 for Personal Computer. *Journal of Basic and Applied Sciences,* 13, 481–490.

Kołaczkowski, P., & Rybiński, H. (2009). Automatic Index Selection in RDBMS by Exploring Query Execution Plan Space. In Ras, Z. W. & Dardzinska, A. (Eds.) *Advances in Data Management. Studies in Computational Intelligence*, vol. 223, Springer, Berlin, Heidelberg, 3–24.

Rogers, D. P. (1985). Improvements in Multiprocessor System Design ACM SIGARCH. *Computer Architecture New*, 13(3), 225–231.

Sasak-Okoń, A., & Tudruj, M. (2017). Graph-Based Speculative Query Execution in Relational Databases. *6th International Symposium on Parallel and Distributed Computing (ISPDC)*, Innsbruck, Austria, 2017, 122–131.

Schumacher, R., & Chibber, H. (2023). *Market Share Analysis: Database Management Systems, Worldwide, 2022*. Retrieved from https://www.gartner.com/en/documents/4432699

Seo, K., Ahn, J., & Im, D.-H. (2019) Optimization of Shortest-Path Search on RDBMS-Based Graphs. *ISPRS International Journal of Geo-Information,* 8(12), 550.

Statista. (2022, August 9). *Database Management System Market Size Worldwide 2017–2021*. Statista Research Department. Retrieved from https://www.statista.com/statistics/724611/worldwide-database-market

T4 Strategy & Advisory (2020, August 16). *RDBMS Market Share*. Retrieved from https://www.t4.ai/industries/rdbms-market-share

Taipalus, T. (2023). *Database Management System Performance Comparisons: A Systematic Survey*. Retrieved from https://arxiv.org/abs/2301.01095

Zhiwei, F., Rathijit, S., Paraschos, K., & Aws, A. (2020a). Automated Tuning of Query Degree of Parallelism Via Machine Learning. In: *Proceedings of the Third International Workshop on Exploiting Artificial Intelligence Techniques for Data Management*, 1–4.

Zhiwei, F., Rathijit, S., Paraschos, K., & Aws, A. (2020b). *Comparative Exploration of ML Techniques for Tuning Query Degree of Parallelism*. Retrieved from https://arxiv.org/abs/2005.08439v2

9 Credibility of randomness in Extreme Learning Machine

Karol Struniawski, Aleksandra Konopka and Ryszard Kozera

Introduction

The Extreme Learning Machine (ELM) represents an efficient machine learning technique, distinguished by its single-hidden layer feed-forward neural network (SLFN) architecture. Throughout this work, ELM's weights between input and hidden layer are represented as α accompanied by a bias term b, along with the weights connecting hidden layer to output layer denoted as β (bias is restricted only to the input layer).

The conventional characterization of the learning process often involves an intelligent exploration of the search space that may be achieved by employing methods like stochastic gradient descent in conjunction with the backpropagation algorithm (BA) (Amari 1993). Once we allow the adaptability of parameters α, β and b the SLFNs are referred to as universal approximators. In the context of BA, several techniques for weight initialization play a crucial role in the iterative optimization process. Among these techniques, the most commonly employed method entails initializing weights with small random values, thereby enhancing the solution's robustness and guarding against network overfitting. Alternatively, other initialization methods take into account specific characteristics of the input data, such as the number or length of input vectors. Noteworthy, examples include the Glorot (Xavier) and He initialization techniques (Nguyen 2021). Research findings consistently underline the sensitivity of BA to initial conditions (Kolen and Pollack 1990).

ELM distinguishes itself from conventional SLFNs in its unique approach to learning process. In ELM, the parameters α and b are randomly initialized and remain constant throughout the training process. Notably, the values of β are determined not through iterative procedures but upon solving a algebraic system of equations once, aimed at achieving an optimal solution in terms of mean square error.

This research delves into research focused on two primary areas. *Firstly*, the experiments are conducted to investigate how the choice of the distribution function for random weights of α and b influences the ultimate outcomes. *Secondly*,

DOI: 10.4324/9781032627236-10

the various random number generators are examined, evaluating their statistical properties using the U.S. National Institute of Standards and Technology (NIST) tests for randomness and pseudorandomness. This evaluation aims to ascertain their credibility in generating truly random numbers and assesses their impact on the final performance of ELM. The experiments reveal that a choice of a distribution function has a significant impact on the performance of models, while the selection of a random number generator exhibits a moderate influence. These findings underscore the research's importance.

Review of literature

In the field of evaluating the influence of random number generators, along with their quality assessment using statistical test suites and considering the utilization of distribution functions in the initialization process of weights α and bias b, there is a noticeable dearth of existing literature dealing with that issue. Surprisingly, no prior research has concentrated on investigating the ramifications of random number generators within the context of ELM. Dealing with ELM it is essential to determine the values of α and b- the number of hidden units and the activation function along with its associated parameters.

In 2012, Wang conducted a study examining the influence of randomly assigned weights between the input and hidden layer within the context of the ELM (Wang, Kwong, and Wang 2012). The fundamental objective of this research is to investigate whether the use of randomly assigned weights could positively impact the performance of ELM. The experimental findings demonstrated here that, for numerous classification and regression problems, the expansion of feature space dimensions, achieved through the random initialization of weights α and biases, yielded better performance than kernel mappings. Kernel mappings involve techniques for either reducing or expanding the feature space through the application of kernel functions, commonly Mercer kernels (Srivastava, Schumann, and Fischer 2005).

In 2016, Dudek conducted an analysis of the ELM approximation capabilities, examining how they depend on the type of activation function and on the ranges from which input weights and biases are randomly generated (Dudek 2016). The finding of this study was experimental evidence confirming that ELM, when applied to function approximation tasks, demonstrates clear benefits when the input weights are randomly selected within a narrow range. We decided to move these findings even further to testify different distribution functions.

Methodology

This section provides a detailed description of ELM and outlines the methods for generating random numbers based on selected distribution functions concluding with a comprehensive overview of the experimental setup.

Extreme Learning Machine

ELM is a dense feed-forward neural network introduced in 2004 by Huang et al. (Huang, Zhu, and Siew 2004). ELM stands out for its remarkable speed in training, especially when compared to other machine learning techniques such as Multilayer Perceptron trained with BA (Li et al. 2019). Due to the random selection of weights between the input layer and the hidden layer, the network is highly resistant to overfitting. ELM can be used for classification and regression tasks being applied in various fields such as signal processing, pattern recognition, market analysis, aviation and aerospace (Ding et al. 2015), big data security, medical diagnosis and predicting protein structure (Chen et al. 2017).

The network consists of input, single hidden and output layer of neurons. The input matrix of features for the training process is denoted as $X_{d\times n}$, where each of n analyzed samples is described by d features. The number of neurons in input layer is equal to d. Number of neurons in a hidden layer L needs to be a priori determined. Number of neurons in output layer M reflects the number of classes to which each sample is being assigned. The matrix of weights between input and hidden layer is denoted as $\alpha_{d\times L}$ and the bias $b=\{b_i\}_{i=1}^{n}$. The coefficients of $\alpha_{d\times L}$ and values of b are randomly generated from a selected probability distribution. Uniform distribution is commonly applied on the interval $(-1,1)$ for evaluation of α and on $(0,1)$ for b.

The weights between hidden and output layer, denoted as β, are computed using algebraic transformation. To compute β a matrix equation $H\beta = T$ needs to be solved. $H_{n\times L}$ is an output from the hidden layer which is computed as $H = f\left(X^T\alpha + b\right)$. Activation function, denoted here as f, is usually selected as tangensoidal or sigmoidal function (Huang et al. 2015). The matrix $T_{n\times M}$ is structured such that each row, denoted as $\{t_i\}_{i=1}^{n}$, consists of a vector of length M containing one-hot-encoded values for the i'th of the n classified samples. This encoding results in $t_i \in \{0,1\}^M$, ensuring that each row contains exactly one occurrence of the value 1 among its elements. The latter means that e.g. if $t_2 = \{0,0,1,0\}$ the second sample (of all n samples) is labeled to be assigned to the third out of $M = 4$ possible classes. The matrix H, due to the fact that generically it is not a quadratic matrix, it is non-invertible, so the value of β cannot be directly computed but it needs to be estimated. The estimation of β, denoted as $\hat{\beta}$, is computed by means of mean square error between $H\beta$ and T (see (9.1)):

$$\hat{\beta} = \underset{\beta}{\operatorname{argmin}}\, H\beta - T^2 = H^{\dagger}T. \tag{9.1}$$

The value of $\hat{\beta}$ is computed with $\hat{\beta} = H^{\dagger}T$, where $H^{\dagger}$ is Moore-Penrose pseudo-inverse operation (Rao and Mitra 1971). In classification tasks, the output of the network is a matrix $Y_{n\times M}$. Each of the rows in Y denoted as $\{y_i\}_{i=1}^{n}$ corresponds to each of the classified samples. The classification result is inferred transforming the Y matrix to one-hot-encoded $\hat{T}$ applying *argmax* function which sets all values in a given row to 0 except of the maximal number in this row which is set to 1 as it corresponds to the label of class to which the sample is assigned.

Generating random numbers for a given distribution function

The algorithm for generating random numbers from a specified distribution function is necessary for initializing α and b in ELM.

Random number generators

The use of randomness in computing has a historical origin in methods that relied on physical phenomena, such as radioactive decay or electronic noise generation, to obtain random numbers through experiments. Today, a wide range of Pseudorandom Number Generators (PRNGs) exist, which simulate randomness algorithmically.

In practical terms, statistical testing plays a crucial role in establishing evidence that a generator can produce numbers that exhibit the appearance of randomness (Soto and Rukhin 1999). To address this concern, the U.S. National Institute of Standards and Technology (NIST) has developed a comprehensive testing suite designed for assessing the randomness of random number generators. This suite comprises a collection of tests that evaluate various aspects of randomness, aimed at determining whether a sequence of numbers possesses the characteristics expected of truly random data.

Distribution functions

The first step in obtaining random numbers based on the given distribution function is to generate random numbers using the selected algorithm and then to calculate the value of the inverse cumulative distribution function (ICDF) for the generated random numbers. In particular, to obtain random weights $\beta_{d\times L}$ from the uniform distribution function $U(-1,1)$ with parameters $U_a = -1$ and $U_b = 1$ the distribution function (as per (9.2)) and ICDF (as outlined in (9.3)) are utilized. To achieve this, a matrix $R_{d\times L} \in (0,1)^{d\times L} \subseteq \mathbb{R}^{d\times L}$ containing random numbers is generated using an algorithm like PCG64 (implemented in Python's *randomgen* package) is initially created. Subsequently, $\beta_{d\times L}$ is obtained by applying the ICDF, denoted as F^{-1}, to the matrix R: $\beta_{d\times L} = F^{-1}_{U(-1,1)}(R)$. Notably, the ICDF is employed because random generators are typically configured to produce real numbers within the 0 to 1 range, which can be interpreted as probabilities p. Therefore, the aim is to determine the corresponding values of x for which the cumulative distribution function F (of the given probability function with density function f) equals p. This can be expressed as $x = F^{-1}(p)$ or by finding x based on the probability density function (as determined in (9.2), (9.3) and (9.4)).

$$f(x;U_a,U_b) = \begin{cases} \frac{1}{U_b - U_a}, & \text{if } U_a \leq x \leq U_b; \\ 0, & \text{otherwise}, \end{cases} \tag{9.2}$$

$$F^{-1}(p) = U_a + p \cdot (U_b - U_a), \tag{9.3}$$

$$p = \int_{-\infty}^{x} f(k)\,dk. \tag{9.4}$$

Both α and bias weights were generated using the following distribution functions, their ICDFs are well formed in literature:

1 Chi-squared (Chisq).
2 Lognormal (Lognorm).
3 Normal.
4 Pareto.
5 Uniform.
6 Weibull.

The respective parameter values employed in the above distribution functions are discussed in detail in results section. These parameters can either be explicitly provided or estimated using two well-known initialization methods: the Glorot/Xavier method or the He initialization method. In the case of the Glorot method, it involves utilizing a parameter within the respective ICDF (as outlined in (9.5)). The Glorot/Xavier method is a widely adopted weight initialization technique in neural networks (Glorot and Bengio 2010). Conversely, the He initialization method, which estimates the parameter's value (based on (9.6)), is originally designed to complement rectified linear activation functions (ReLU) and on their variants. The He initialization method enhances the stability and efficacy of training deep neural networks (He et al. 2015). This method calculates the standard deviation for weight initialization considering the number of input features:

$$\text{std\_glorot} = \frac{1}{\sqrt{\frac{d+L}{2}}}, \tag{9.5}$$

$$\text{std\_he} = \sqrt{\frac{2}{d}}. \tag{9.6}$$

Experimental setup

The experiments in this research were conducted using a blend of the following elements:

1 Dataset choice.
2 Distribution function setting.

3 Random number generator method.
4 Number of neurons in the hidden layer.

These computations produced a total of 28,600 setups.

Dataset choice

The results are tested on five well-known publicly available UCI repository (Dua and Graff 2017) datasets:

1 **Australian** dataset stands as credit risk assessment comprising a total of 690 entries and 14 features concerning credit card applications, encompassing parameters like income, age and employment history.
2 **Banana** has emerged as a testbed for navigating the complexities of nonlinearly of synthetic data with the task to detect enigmatic shape reminiscent of a banana with a total of 5,300 instances and two classes.
3 **BreastCancer** with 569 instances and 30 features derived from medical images, the dataset encapsulates minute structural nuances. The quest is to distinguish between malignant and benign breast masses.
4 **Ionosphere** with 351 instances and 34 attributes is the binary classification challenge of differentiating between "good" signals that pass through the ionosphere without interference and the "bad" category corresponds to instances that are disrupted or reflect radar signals, making them less useful.
5 **Musk** dataset is a collection of information about various chemical compounds. This dataset contains 6,598 samples and describes each of these compounds using 166 features or properties. The goal of this dataset is to help distinguish between compounds that are associated with the scent of musk and those that are not.

Distribution function setting

Both α and b weights are generated using the following distribution functions forming 13 different combinations:

1 α: chisq(2) and b: chisq(2),
2 α: chisq(glorot) and b: chisq(2),
3 α: chisq(he) and b: chisq(2),
4 α: lognorm(0,1) and b: lognorm(0,1),
5 α: lognorm(glorot) and b: lognorm(0,1),
6 α: lognorm(he) and b: lognorm(0,1),
7 α: normal(0,1) and b: normal(0,1),
8 α: normal(glorot) and b: normal(0,1),
9 α: normal(he) and b: normal(0,1),
10 α: pareto(1) and b: pareto(1),

11 α: uniform(−1,1) and b: uniform(0,1),
12 α: weibull(1) and b: weibull(1),
13 α: weibull(he) and b: weibull(1).

Random number generator method

In our research, we employed a total of 22 distinct random number generators from Python's *randomgen* package. This package comprises both cryptographically secure generators designed to produce unpredictable and resilient random numbers, as well as robust PRNGs with favorable statistical characteristics. These generators yield pseudorandom sequences characterized by strong statistical properties. Their operations typically involve a combination of bitwise manipulations, arithmetic transformations and state updates. The specific generators employed in our research are listed below:

1 *aes,*	7 *lxm,*	13 *rdrand,*	19 *xoshiro256,*
2 *chacha,*	8 *mt19937,*	14 *romu,*	20 *xoshiro512,*
3 *dsfmt,*	9 *mt64,*	15 *sfc,*	21 *xoroshiro128,*
4 *efiix64,*	10 *pcg32,*	16 *sfmt,*	22 *xorshift1024.*
5 *hc128,*	11 *pcg64,*	17 *speck128,*	
6 *jsf,*	12 *philox,*	18 *threefry,*	

Number of neurons in the hidden layer

In this context, $L \in \{50,100,200,\ldots,900,1000,2000,\ldots,9000,10000\}$ represents the number of neurons in the hidden layer that were subjected in the analysis.

Results

The weights α and b for classification task were computed for each of the 28,600 different setups using 50 times repeated 10-Fold cross-validation. This technique has been implemented to obtain statistically meaningful results. Consequently, this method generated a set of 500 distinct accuracy measurements for each unique configuration. To provide a comprehensive characterization of the performance distribution, key statistical metrics including the mean accuracy, the first quartile, the second quartile (median) and the third quartile were computed for each set of 500 accuracy measurements within the context of each configuration under consideration.

Within the existing body of literature, the uniform distribution function (from (9.2)) with parameters $U_a = -1$ and $U_b = 1$ or $U_a = 0$ and $U_b = 1$ has conventionally served as the predominant probability function for estimating values of α and b. Nonetheless, our research has revealed a departure from this convention, as the optimal accuracy results vary depending on the specific dataset under consideration (see Table 9.1). Remarkably, the highest accuracy scores were achieved

Table 9.1 Classification reports for the best mean accuracy results for five selected datasets depending on the distribution function and random number generator computed (for different numbers of neurons L)

Alpha	*Bias*	*Australian*	*Banana*	*BreastCancer*	*Ionosphere*	*Musk*
chisquare(2)	chisquare(2)	0.709	0.813	0.920	0.877	0.473
chisquare(glorot)	chisquare(2)	0.717	0.805	0.927	0.941	0.851
chisquare(he)	chisquare(2)	0.716	0.805	0.932	0.941	0.709
lognormal(0,1)	lognormal(0,1)	0.726	0.813	0.916	0.887	*0.472*
lognormal(glorot)	lognormal(0,1)	*0.700*	0.802	*0.874*	*0.820*	0.474
lognormal_he	lognormal(0,1)	0.712	0.813	0.883	*0.820*	0.474
normal(0,1)	normal(0,1)	0.722	0.804	0.933	**0.955**	0.970
normal(glorot)	normal(0,1)	0.718	0.804	0.929	0.933	0.970
normal(he)	normal(0,1)	0.714	0.815	0.929	0.947	**0.971**
pareto(1)	pareto(1)	**0.844**	**0.869**	0.965	0.899	0.475
uniform(−1,1)	uniform(0,1)	0.719	0.804	0.932	0.954	0.970
weibull(1)	weibull(1)	0.720	0.763	0.933	0.867	0.475
weibull_he	weibull(1)	0.800	*0.762*	**0.979**	0.884	0.474
Alpha		Australian	Banana	Breast Cancer	Ionosphere	Musk
aes		0.820	0.846	0.977	0.950	0.968
chacha		0.816	0.848	*0.968*	*0.941*	0.970
dsfmt		0.836	0.855	0.970	0.946	0.967
efiix64		0.796	0.865	0.973	0.949	0.968
hc128		0.843	0.850	**0.979**	0.951	0.968
jsf		0.826	0.866	0.969	0.944	0.967
lxm		0.838	*0.845*	0.974	0.949	0.969
mt19937		0.799	0.852	0.970	0.953	0.968
mt64		0.820	0.862	0.969	0.951	0.970

(Continued)

Table 9.1 (Continued)

Alpha	*Bias*	*Australian*	*Banana*	*BreastCancer*	*Ionosphere*	*Musk*
pcg32		0.815	0.852	0.972	0.943	0.967
pcg64		0.820	0.858	**0.979**	0.946	0.970
philox		0.795	**0.869**	0.969	0.947	0.969
rdrand		0.824	0.848	*0.968*	0.946	0.967
romu		0.830	0.865	0.975	0.954	*0.966*
sfc		0.794	0.858	0.973	0.944	0.968
sfmt		*0.774*	0.851	0.972	0.947	0.968
speck128		*0.774*	0.864	0.972	0.946	0.970
threefry		0.798	0.851	0.975	0.944	**0.971**
xoroshiro128		0.825	0.852	0.974	**0.955**	0.967
xorshift1024		0.775	0.862	*0.968*	0.944	0.970
xoshiro256		**0.844**	0.855	0.970	0.947	0.968
xoshiro512		0.791	0.856	0.971	0.946	0.970

using the Pareto(1) density function, yielding 0.844 for the Australian dataset and 0.869 for the Banana dataset. In stark contrast, when applied to the Musk dataset, this same set of distribution functions produced one of the lowest accuracy results, registering at a mere 0.475. For the BreastCancer dataset, the most favorable results were obtained with the Weibull(he) and Weibull(1) functions achieving a remarkable accuracy of 0.979. Interestingly, these very functions were associated with suboptimal performance in the Banana dataset, yielding an accuracy score of 0.762. On a different note, the Ionosphere and Musk datasets exhibited their highest accuracy results when employing Normal distribution functions yielding 0.955 and 0.971, respectively. Specifically, the parameters for α were set to $(0,1)$ for the Ionosphere dataset, while the He parameter configuration was employed for the Musk dataset. The difference between maximal and minimal accuracy computed with different sets of distribution functions yields 14.4, 10.7, 10.7, 13.5 and 49.9 for Australian, Banana, BreastCancer, Ionosphere and Musk dataset, respectively. This underscores that the choice of the random number generator has a moderate impact on the model's performance.

In this research, we systematically assessed a collection of 13 sets of distribution functions to determine their effectiveness in achieving optimal accuracy across five distinct datasets. This rigorous analysis involved the identification of the top 100 accuracy results for each dataset, resulting in a total of 500 values for consideration. Remarkably, the distribution function set that emerged as the most frequently employed was Pareto(1), which featured prominently in 157 of the meticulously examined results. This configuration consistently delivered exceptional performance, particularly in the Australian and Banana datasets. The second most prevalent combination, appearing 143 times, consisted of Weibull(1) and Weibull(1). This specific configuration demonstrated impressive accuracy outcomes, particularly in the Australian and BreastCancer datasets. In the third position, the Uniform distribution function with parameters $(-1,1)$ and $(0,1)$ emerged with a total of 70 occurrences among the top-performing results, showcasing its efficacy in delivering remarkable accuracy, particularly in the Ionosphere and Musk datasets.

In this study, we comprehensively investigate 22 distinct random number generators (listed in section "Random number generator"). Specifically, within the domain of ELM, our research represents a pioneering endeavor. To the best of our knowledge, prior to this investigation, there has been a conspicuous absence of systematic assessments pertaining to the suitability of various random number generators for generating values of α and b within the context of ELM. The primary objective of our investigation is twofold. *Firstly*, we aim to achieve the highest level of accuracy in our computations. *Secondly*, we endeavor to establish the stability of the selected random number generator. Stability, in this context, entails ensuring that the computed accuracy results exhibit minimal variability between the first and third quartiles, as indicated by the interquartile range.

The attainment of peak accuracy results varies significantly across diverse datasets, as detailed in Table 9.1. Notably, distinct random number generators excel in optimizing accuracy for specific datasets. For the Australian dataset, the *xoshiro256* generator demonstrated its prowess by achieving the highest mean (computed from 50 times repeated 10-Fold cross-validation for various numbers of neurons) accuracy score of 0.844. Meanwhile, the Musk dataset saw the *xoshiro128* generator emerge as the champion, yielding an impressive accuracy of 0.971. In the context of the Banana dataset, it was the *philox* generator that excelled, delivering the highest accuracy of 0.869. Finally, for the BreastCancer dataset, the *pcg64* and *hc128* generators claimed the top spot, recording an exceptional accuracy score of 0.979. The difference between maximal and minimal accuracy computed with different generators yields 7, 2.4, 1.1, 1.4 and 0.5 for Australian, Banana, BreastCancer, Ionosphere and Musk dataset, respectively.

In order to facilitate a comprehensive comparison of the various random number generators, we meticulously analyzed the top 100 accuracy values (each value is a mean generated with 50 repetitions of 10-Fold cross-validation) for each of the five datasets, noting the frequency with which each of the 22 generators appeared. Remarkably, the *pcg64* generator emerged as the most frequently observed choice, securing the top position in a substantial 60 out of the possible 500 highest accuracy results. Following closely, the second and third positions were claimed by *hc128*, *chacha* and *efiix64* generators, all of which delivered noteworthy performance by achieving high accuracy results 30, 26 and 26 times, respectively. Notably, these generators consistently exhibited strong performance across all five datasets.

To assess the stability of the random number generators, we employed a box chart visualization (refer to Figure 9.1) using the Australian dataset, Pareto(1) function and a hidden layer comprising 1,000 neurons. Among the generators examined, several demonstrated notable stability characteristics. Specifically, the most stable generators, as evidenced by minimal variability between quantile 1 and quantile 3, include *sfmt*, *xorshift1024* and *xoshiro256*, with differences of merely 0.043, 0.047 and 0.049, respectively. Conversely, a subset of generators exhibited comparatively lower stability. Notably, *xoshiro512*, *philox* and *sfc* yielded high differences of 0.087, 0.087 and 0.085, respectively, indicating greater instability in their generated results.

All 22 random number generators underwent a rigorous evaluation encompassing NIST tests. It is important to note that passing the NIST tests does not guarantee absolute randomness, but it does instill a level of confidence in the quality of the generated random numbers. It is worth highlighting that these tests are structured in such a way that a positive outcome (a test's p-value being less than a specified threshold) indicates that, based on the specific test, the generator did not produce genuinely random results. In cryptographic applications, the p-value threshold is typically set at 0.01; however, in our case, we chose a p-value threshold of 0.05 commonly applied in statistics. Subset of generators faced challenges in this evaluation, falling short on a various number of tests.

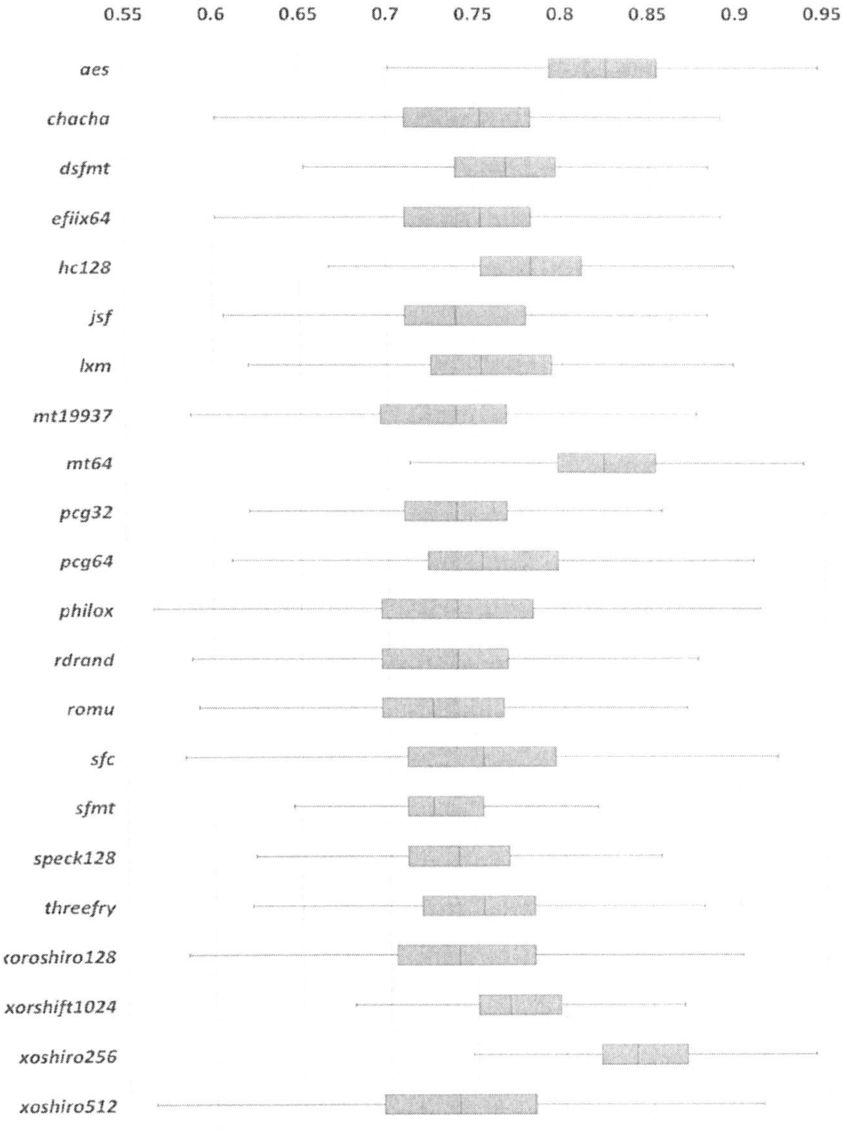

Figure 9.1 Accuracy results for classification on Australian dataset, with Pareto(1) distribution function and 1,000 neurons in the hidden layer, using different random number generators computed applying 50 times a 10-fold cross-validation. The box chart presents the quantile 1, quantile 2 and quantile 3 of the 500 computed results for each of the setups.

Source: Author's research.

Specifically, four tests eluded the *theefry* generator, while *pcg64* struggled with three. Meanwhile, *chacha* and *xoshiro256* encountered difficulties in two tests each, whereas *scf*, *speck128*, *dsfmt*, *mt64*, *xoshiro128*, *jsf* and *xorshift1024* experienced a setback on one test each.

Interestingly, two of the three most stable generators encountered challenges in passing either one or two tests. Even the widely successful *pcg64* generator, which consistently yielded top-tier results (60 out of 500 times), fell short on three of the tests. Moreover, the third generator, *chacha*, which achieved high results on 26 occasions out of 500, did not meet the criteria of two of the tests. *This intriguing observation leads to a noteworthy conclusion: in many instances, the generators that deliver the highest accuracy and demonstrate remarkable stability tend to be the very generators that do not successfully pass the prescribed tests.*

Discussion

Conventionally, the uniform distribution function with parameters equal to $(-1,1)$ and $(0,1)$ has been employed as a source of randomness in ELM, but this research revealed intriguing variations in performance across different datasets. The selection of the randomness source in ELM significantly affected the final accuracy results, challenging established practices. Notably, specific distribution functions consistently emerged as top performers for particular datasets, underscoring the importance of considering dataset-specific characteristics when selecting randomness sources. The Pareto(1) distribution function, in particular, consistently delivered exceptional performance, notably in the Australian and Banana datasets. This diversity in optimal randomness sources emphasizes the need for a more sophisticated approach to randomness within the context of ELM.

The stability of random number generators in ELM was explored, marking a pioneering effort in this field. Surprisingly, the generators that achieved the highest accuracy and a remarkable stability often did not meet the criteria of the prescribed NIST tests. Conversely, those generators that passed the tests did not consistently exhibit top-tier performance.

Conclusion

In conclusion, this research underlines the critical role of randomness in influencing ELM performance. The selection of the randomness source in ELM was demonstrated to yield a significant impact on accuracy across diverse datasets, challenging conventional practices. The variation in accuracy, ranging from 10 to 50 percentage points, when computed with different sets of distribution functions underscores the critical importance of selecting an appropriate distribution function for accuracy results. In contrast, the variation between maximal and minimal accuracy, which falls within the range of 0.5 up to 7 percentage points when using different generators, suggests that the choice of a random generator, while not insignificant, holds relatively less weight compared to the selection of the distribution function. Additionally, a systematic assessment of random number generators in ELM was introduced, revealing that generators achieving the highest accuracy and stability often did not meet the stringent criteria of statistical tests. This observation suggests that the traditional criteria for randomness may not consistently align with optimal ELM performance.

The above findings underscore the necessity for a more subtle approach to randomness within ELM, which considers dataset-specific characteristics and explores a broader array of randomness sources. As ELM continues to gain prominence in various applications, comprehending the intricate relationship between randomness and performance forms an essential issue for achieving optimal outcomes and advancing the field.

References

Amari, S.-I. (1993). Backpropagation and Stochastic Gradient Descent Method. *Neurocomputing*, 5(4), 185–196. https://doi.org/10.1016/0925-2312(93)90006-O.

Chen, C., Li, K., Duan, M., & Li, K. (2017). Chapter 6 - Extreme Learning Machine and Its Applications in Big Data Processing. In Hui-Huang Hsu, Chuan-Yu Chang, Ching-Hsien Hsu (Eds.), *Big Data Analytics for Sensor-Network Collected Intelligence* (pp. 117–150). Academic Press. https://doi.org/10.1016/B978-0-12-809393-1.00006-4.

Ding, S., Zhao, H., Zhang, Y., Xu, X., & Nie, R. (2015). Extreme Learning Machine: Algorithm, Theory and Applications. *Artificial Intelligence Review*, 44(1), 103–115. https://doi.org/10.1007/s10462-013-9405-z.

Dua, D., & Graff, C. (2017). *UCI Machine Learning Repository*. University of California, Irvine, School of Information; Computer Sciences. archive.ics.uci.edu/ml.

Dudek, G. (2016). Extreme Learning Machine as a Function Approximator: Initialization of Input Weights and Biases. In *CORES* (pp. 59–69). Springer International Publishing.

Glorot, X., & Bengio, Y. (2010). Understanding the Difficulty of Training Deep Feedforward Neural Networks. In *AISTATS*. Proceedings of the thirteenth international conference on artificial intelligence and statistics. JMLR Workshop and Conference Proceedings, 249–256. https://api.semanticscholar.org/CorpusID:5575601.

He, K., Zhang, X., Ren, S., & Sun, J. (2015). Delving Deep into Rectifiers: Surpassing Human-Level Performance on ImageNet Classification. In *IEEE International Conference on Computer Vision (ICCV)* (pp. 1026–1034). IEEE, Santiago, Chile. https://doi.org/10.1109/ICCV.2015.123.

Huang, G.-B., Zhou, H., Ding, X.-T., & Zhang, R. (2015). Trends in Extreme Learning Machines: A Review. *Neural Networks*, 61, 32–48.

Huang, G.-B., Zhu, Q.-Y., & Siew, C. (2004). Extreme Learning Machine: A New Learning Scheme of Feedforward Neural Networks. In *IEEE International Conference on Neural Networks* (Vol. 2, pp. 985–990). https://doi.org/10.1109/IJCNN.2004.1380068.

Kolen, J. F., & Pollack, J. B. (1990). Backpropagation Is Sensitive to Initial Conditions. *Complex Systems*. https://api.semanticscholar.org/CorpusID:13373157.

Li, H.-T., Chou, C.-Y., Chen, Y.-T., Wang, S.-H., & Wu, A.-Y. (2019). Robust and Lightweight Ensemble Extreme Learning Machine Engine Based on Eigenspace Domain for Compressed Learning. *IEEE Transactions on Circuits and Systems I: Regular Papers*, 66(12), 4699–4712. https://doi.org/10.1109/TCSI.2019.2940642.

Nguyen, Q. C. (2021). A Fully Rigorous Proof of the Derivation of Xavier and He's Initialization for Deep ReLU Networks. *ArXiv*. https://doi.org/10.48550/arXiv.2101.12017.

Rao, C. R., & Mitra, S. (1971). *Generalized Inverse of Matrices and Its Applications*. John Wiley & Sons.

Soto, J., & Rukhin, A. L. (1999). *Statistical Testing of Random Number Generators*. National Institute of Standards and Technology. https://api.semanticscholar.org/CorpusID:9938076.

Srivastava, A. N., Schumann, J., & Fischer, B. (2005). An Ensemble Approach to Building Mercer Kernels with Prior Information. *IEEE Transactions on Systems, Man, and Cybernetics*, 3(3), 2352–2359. https://doi.org/10.1109/ICSMC.2005.1571500.

Wang, R., Kwong, S., & Wang, X. (2012). A Study on Random Weights Between Input and Hidden Layers in Extreme Learning Machine. *Soft Computing*, 16(9), 1465–1475. https://doi.org/10.1007/s00500-012-0829-1.

10 Reasons for distrust in AI among Generation Z Polish doctoral students

Aleksandra Popiel, Jerzy Gołuchowski and Andrew J. Zaliwski

Introduction

In recent years, we have seen the rapid development of artificial intelligence (AI), one of the breakthroughs of which is the ChatGPT released in November 2022 by OpenAI, characterized by greater capabilities than previously available AI systems (Tlili et al., 2023, p. 1). Such technologies are increasingly being used in many areas of business and private life (e.g., journalism, speech and text generation, and answering questions in customer service), entering our daily lives (Czarnitzki et al., 2023, p. 188; Lopezosa et al., 2023, p. 2).

Idea of AI has already fascinated fantasy writers (e.g., Isaac Asimov, 2022; Stanislaw Lem, 2020), science-fiction film directors presenting dystopian visions of humanity (e.g., James Cameron, 1984; Lilly Wachowski & Lana Wachowski, 1999), as well as futurologists creating visions of a world ruled by AI, robots, or humanoids (e.g., Nick Bostrom, 2021). The emergence of AI as ChatGPT and the wide access to it has triggered a new wave of discussion about the possibilities and dangers of AI, its owners, and operators (Stahl & Eke, 2024, p. 1). Understandably, these technologies evoke negative emotions and attitudes, such as fear of technological unemployment, manipulation of personal data, or AI's dominance over humanity (Gołuchowski et al., 2024).

Fear of AI is not entirely new – similar fears were observed in earlier periods of increased automation (such as during the Industrial Revolution in the 18th century) (Frey, 2019). However, unlike earlier stages of automation, the latest one touches on tasks that previously required advanced expertise. Thus, there is growing concern about the objectification of humans or the deprivation of jobs even for so-called knowledge workers, such as programmers, lawyers, doctors, or teachers (Legg & Bell, 2019; Chan & Tsi, 2023; Lam, 2023; Lupan et al., 2023). This anxiety may translate into lowered trust in AI, resulting in less-frequent use of its broad capabilities (Ejdys, 2018).

Although scientists agree that there is a problem with maintaining a stable and high level of trust in AI, it seems that the reasons for distrust in AI stem from a lack of understanding of it and unresolved legal issues related to its creation and

DOI: 10.4324/9781032627236-11

applications. This growing phenomenon of distrust in AI prompts deeper scientific reflection (undertaken from the perspective of various disciplines – such as management, law, medicine, and more) (Solanke, 2022; Yakar et al., 2022; Zirar et al., 2023) on the development and use of AI systems. There also seems to be an obvious research gap on the reasons for distrust of AI among Generation Z, the youngest generation entering the labor market.

In the chapter, we present results of a qualitative study aimed at identifying the sources of distrust in AI among Polish Ph.D. students representing Generation Z. We attempt to find actions that can increase the level of trust in AI in the face of the identified causes of distrust. In addition, the presented results of the qualitative research can provide a basis for conducting quantitative research. The presentation of the results of the empirical study is preceded by an outline of the theoretical framework necessary for the study of the younger generation's trust in AI.

Review of literature

The subject of our literature review is trust and distrust in AI of young scientists. Omitted from the literature review is the usage of AI in modern society; although it is impossible to conduct research on the social aspects of AI technology and its applications without a sufficiently deep understanding of the essence of modern AI and the legal, educational, and other aspects of its application.

Thus, we understand trust as the propensity to rely on technology, resulting from perceived characteristics of the technology and environmental factors, in a situation of potential risk associated with the use of technology, determining intentions for future use of technology. The propensity to rely on a technology is primarily driven by its expected functionality, reliability, and support system (Ejdys, 2017, p. 22). Its level depends on many factors, which are introduced by numerous researchers. Among them can be found such trust factors as transparency and honesty, predictability, control, and benevolence (Przegalińska et al., 2019), the aforementioned functionality, helpfulness, and reliability (Dagli, 2018, p. 42), ability, integrity, benevolence (Bedue & Fritzsche, 2022), characteristics of technology, risks associated with technology use, propensity to rely on technology (Ejdys, 2018), ability, humanity, integrity (Gillespie et al., 2023, p. 17), and more.

Trust in AI is a type of trust in technology. This trust differs from human-to-human trust in structure and factors. In her model of trust in technology, Ejdys identifies that trust in technology is mediated and comes down to trust in another human being (Ejdys, 2017, p. 21). In addition, Dagli (2018) points out the difference in trust factors between trust in humans and trust in specific technology. Among the factors of trust in humans are competence/ability, benevolence, and integrity, while among trust in a specific technology, he distinguishes functionality, helpfulness, and reliability (Dagli, 2018, p. 42).

Numerous research institutions around the world (including Edelman, Krista, Capgemini, and KPMG) have undertaken a study of people's level of trust in AI over the past three years. The study found that the public has a fairly low level of trust in AI.

It is worth noting that the level of knowledge about AI varies among generational cohorts. 61% of representatives of Generation Z and 56% of Generation Y are aware of what AI is, compared to 24% of representatives of Generation X (King, 2023). It is also worth mentioning the respondents' attitudes toward AI – 44% declare a positive attitude, 36% declare a neutral attitude, and 20% a negative attitude (King, 2023). However, not all tasks that would be undertaken by AI evoke trust in respondents – among them are making life-or-death decisions in military combat (67%), being a juror at a trial (64%), and flying a commercial jet (57%) (King, 2023). Also, in tasks (selecting clothing for work, choosing a gift, administering medicine in hospitals), there is more trust in these tasks if they are performed by a human. In addition, overall lower trust in AI was observed among women (King, 2023).

Slightly less optimistic results are presented by a study conducted by Gillespie et al. (2023), in which 29% of respondents say they are not willing to trust AI, 32% present an ambivalent attitude, and 39% say they are willing to trust AI (Gillespie et al., 2023, p. 13). KPMG's research also shows that trust in AI depends on the respondents' country of origin. The highest level of trust in AI was declared in India (75%), China (67%), and South Africa (57%), while the lowest level of trust in AI is observed in Estonia (26%), Japan (23%), and Finland (16%) (Gillespie et al., 2023, p. 14). Moreover, in countries such as the United Kingdom, France, Canada, Australia, and Japan, dominant emotions toward AI are worry and fear (Gillespie et al., 2023, p. 20). In the Trust Barometer 2022 survey, 51% of respondents worry that technology or AI can do their type of work as well or even better than them (Edelman, 2022, p. 16).

We also directed our literature research to the reasons for distrust in AI. Studies conducted so far, mainly on fear of AI, include the "How Americans think about AI" survey conducted by the Pew Research Center (Rainie et al., 2022). They found that 37% of respondents were more anxious than excited about AI, compared to 18% who were more excited than anxious or 45% who were equally anxious and excited (Rainie et al., 2022). Among their concerns were the following reasons: people losing their jobs, surveillance, hacking, digital privacy, lack of human connection, fear of AI becoming too powerful, people abusing AI, people being too dependent on AI, and others (Rainie et al., 2022).

Moreover in a survey conducted by KPMG, 73% of respondents said they perceived significant risks arising from the development of AI (Gillespie et al., 2023, p. 25). The most common risks contributing to distrust were cybersecurity risks (84%), manipulation or harmful use (76%), job loss due to automation (77%), loss of privacy (75%), deskilling (73%), and human rights being undermined (70%) (Gillespie et al., 2023, p. 25). In addition, 43% of respondents believe that risks will impact people in their country (Gillespie et al., 2023, p. 27). Moreover, 41% of respondents expect AI to replace jobs in their area of work (Gillespie et al., 2023, p. 51).

In the literature, it is stressed that trust and distrust in AI have many sources and depend on many factors but lists of factors differ. They include the culture of its users (Gillespie et al., 2023), and in particular the differences in values, knowledge, experiences specific to the distinguished generations of technology users (Gackowski et al., 2018; Herrando, Jimenez-Martinez & Martin-De Hoyos, 2019). We focused our interest on Generation Z and the level of distrust of its representatives toward AI technologies. Research on the level of trust and distrust among representatives of Generation Z has been presented in numerous publications (including Lazanyi, 2019; Oudejans & Hanna, 2022; Vinichenko et al., 2022).

Generation Z expects AI to be adopted into their education instead of being banned (Ma, 2020). YPulse conducted a survey in which they indicated that Generation Z representatives are primarily excited (36%) and hopeful (31%); however, they are also nervous (30%) and concerned (24%) when it comes to feelings toward AI (YPulse, 2023). In general, Generation Z seems to be the most excited about AI (28.7%) compared to other generations (27.6% – Generation Y, 21.4% – Generation X, 13.1% – Baby Boomers) (Liffreing, 2023). Generation Z and Millennials also report the highest level of trust in AI (42%) among the generational cohorts surveyed by KPMG (Gillespie et al., 2023, p. 21).

Methodology of empirical research

To determine the sources of distrust in AI of representatives of Generation Z, an exploratory study was conducted. The exploratory nature of the research conducted is due to the scarcity of studies related to Generation Z's (particularly Ph.D. students) trust in AI and its impact on scientific work. In-depth individual interviews were conducted with ten Polish doctoral students (seven women, three men). The research questions were aimed at understanding how young researchers interact with AI. The study aimed to verify the research assumptions and develop a research tool for extended statistical research.

The interviews were performed between August 21 and 31, 2023 via the Google Meet. When conducting the interviews, the authors wanted to obtain answers to the following research questions:

1 What is the attitude of Polish doctoral students toward AI?
2 How do Ph.D. students perceive the role of AI in scientific work?
3 What do doctoral students perceive as the dangers of AI development?
4 What doctoral students see ways to alleviate their fear of AI?

The interviews conducted were subjected to a transcription process using Microsoft Word, coded, and then analyzed.

Results

Majority of surveyed Ph.D. students, while not providing a scientific definition of AI, understand what AI is and accurately point to its applications in everyday life. The definitions cited by respondents most often claim that AI is (1) a tool that assists humans in numerous areas, and/or (2) technology that imitates humans. Among the applications of AI, respondents identified medicine, communication, marketing, science, or transportation. Most often doctoral students refer to AI as ChatGPT, autonomous cars, and robots. Far less frequently, they mentioned the use of AI in machine translation, or customer behavior analysis. One responder referred exclusively to ChatGPT, which may indicate his limited knowledge of AI applications.

The majority of respondents describe their attitude toward AI as moderate enthusiasm, describing it with words such as – "50/50," "it's a mystery to me," "mixed feelings." Although they all indicate curiosity about AI development, they emphasize the need to maintain a skeptical attitude by verifying information provided by LLMs such as ChatGPT. All Ph.D. students show awareness of the risks associated with AI development, which affect their perception of the technology.

Interviewees are wary of overly optimistic assessments of AI capabilities. They see the development of AI applications in automating repetitive activities and indirectly improving people's everyday lives, such as in the shopping process. Respondents have confidence in AI applications in improving diagnostics, managing cities (e.g., traffic), or creating artistic content (e.g., graphics and music). According to doctoral students, AI could contribute to the sustainable economic development of mankind, point out courses of action that humans have previously overlooked, as well as assist us in processing data to optimize operations. In addition, respondents see the impact of AI on their scientific work. The role of AI is compared to the impact on scientific work played by the development of the Internet, without which a scientist's work would be much harder. Some respondents use AI to get support, such as for research direction suggestions (ChatGPT) or creating interactive content for classes with students (Grammarly and QuillBot). Responders see AI in their research work as an opportunity – they see potential in conducting literature searches, finding new publication sources, or creating scientific articles.

Although AI is perceived neutrally or positively, some people mention possible dangers. The student, who works at a scientific journal, claims that scientists provide editors with articles written by AI in which citations are artificially (and falsely) generated. This student points out that such practices can reduce the quality of research, lengthen the editorial process of articles, and make it more difficult for credible researchers to be accepted into the journal. The other doctoral student worries whether the widespread use of AI in a scientist's work will diminish the importance of scientists – "One day AI will do the research itself.

What is then the point of being a scientist, an explorer, a searcher? Our [scientific] titles, for which we strive so hard, may lose their meaning."

Generation Z doctoral students list many risks posed by the development of AI. They listed many of them spontaneously. Nonetheless, a follow-up question presented respondents with a list of dangers that appear most frequently in the scientific literature on the subject and in thematic reports, that is, technological unemployment, surveillance of the public, deterioration of the public's education, spread of disinformation, difficulties in building interpersonal relationships and taking over the world.

Of the risks listed, the most common one that evokes their distrust of AI is the change in employment structure, with many respondents mentioning the replacement of human workers with AI if their work involves repetitive tasks (e.g., in office work, programming, and data analysis). Unemployment can also bring other negative effects, as pointed out by some doctoral students, that is, a mental crisis that can lead to an increased number of suicides. It is worth noting, however, that some respondents imply that changes in the employment structure resulting from the development of AI are also associated with new professions and jobs.

Just as often respondents feel distrust in AI stemming from potential difficulties in building interpersonal relationships. Ph.D. students worry that the development of AI could lead to the spread of virtual friends, which will replace real interpersonal relationships. They point out that such friendships are illusory, as an inanimate friend agreeing on everything could damage the human psyche – "we are herd creatures, we need interpersonal contacts, and AI could take that away from us." Even interviewees who did not spontaneously (first, the doctoral students were asked the general question "What dangers do you perceive from the development of AI?" Next, they were presented with specific risks from the literature on the subject, including "difficulties in building human relationships") mention difficulties in human relations as a threat, once the interviewer cites such examples, see this as a threat. Most often, they refer to pop culture works in which the relationship between humans and AI was presented, such as "Her" (directed by Spike Jonze). Such relationships are considered by respondents to be incomplete and contradictory to human nature.

One doctoral student notes that AI could also affect the spontaneity of communication through AI's blocking of words that could potentially be offensive, such as "fat." Also noted are the dangers of AI lacking a moral compass, so (through improper programming) it could have ethical difficulties and fail to distinguish between right and wrong. Such deficiencies could translate, for example, into suggesting that another person in a mental crisis commits suicide.

Distrust in AI comes from disinformation, the third most frequently cited threat. Ph.D. students are concerned about the ease of creating deep fakes, as well as other content that may appear to be real (including the use of AI to create pornographic content with the faces of public figures, or the reading of politicians' emails by AI mimicking their voices). This disinformation also manifests itself in hallucinating LLMs, such as ChatGPT, which indicate false information that may appear credible to their recipients. Interviewees see this threat as a current one that should be addressed as a priority.

The threat posed by disinformation, which is the deterioration of society's mental capacity, was recognized by most Ph.D. students. Some suggest that the deterioration of our ability to process information due to the development of technology is already taking place, and our condition will worsen in the coming years. Respondents believe that the public is lazy and unwilling to perform simple processes in their heads (such as calculations), instead using external tools. They agree that AI can potentially exacerbate such a state – "I'm concerned about people's lack of independence."

Another source of distrust in AI is the increase in surveillance and privacy issues. Although, not all Ph.D. students commented on it in detail. It may indicate a lack of understanding of how AI could pose a threat in this regard, even if they intuitively guessed that it could threaten them. In contrast, other respondents singled out numerous examples of ways in which AI can pose a threat to people – data theft, hacking into email inboxes and bank accounts. It is worth noting, however, that respondents believe that AI will not itself perform unethical, illegal acts; however, it can be used in this way by humans.

The least distrust comes from the prospect of an AI taking over the world. Ph.D. students describe such a scenario with the word "anti-utopia," considering it unrealistic; they are convinced that AI will have too high a degree of intelligence to undertake such actions. Interestingly, some respondents do not rule out the existence of such a scenario – however, they consider it a distant situation that will not happen during their lifetime and, therefore, will not threaten them personally. One person mentions that if it does happen, it should be considered solely the fault of man, as "this is the fate we have made for ourselves."

Respondents repeatedly indicate that they are unsure of the direction AI will take, which takes away from their sense of stability. In some people, this uncertainty is moderate – these individuals report they secure their data and educate themselves on AI to have up-to-date knowledge. In the other group, it takes on a very negative attitude – the fear of losing control over one's life, nihilism and depression are mentioned. Interviewees touch on the collapse of humanity that may occur in the face of being replaced by AI, which translates into her negative feelings. In contrast, another says she feels grief when she sees individuals experiencing harm caused by AI. However, it should be acknowledged that underneath the uncertainty is also curiosity and a desire to know how AI will develop.

The surveyed doctoral students point to numerous sources of distrust in AI. The fear of AI, which is shaped by pop culture, is mentioned three times. Respondents cite movie titles that present AI as a threat to the existence of the human race (e.g., "Terminator"). Respondents also notice fear stemming from ignorance – indicating that many people do not know how AI works and how it functions in their daily lives. They suggest that AI is something new, and new things always create uncertainty. Some respondents imply that the rapid development of technology does not allow us to fully trust modern solutions. Another reason conducive to the development of distrust in AI is the media narrative focused on the negative aspects of AI development and the formation of information bubbles in social media. It is noted that the media mostly reports on the negative context of AI development. Interestingly, one person claims that fears stem from logical predictive analysis (e.g., regarding changes in the job market) and a natural fear of one's own future and that of one's children. Another doctoral student also sees the source of anxiety in such thoughts; however, in his opinion, they are simple and limited. What's more, some respondents likened the source of anxiety to the lack of regulations that could give people confidence that their interests are protected.

Most doctoral students believe that AI should not be feared; however, all emphasize the need to monitor its activities, educate the public, and the need to be cautious about its further development. Some respondents indicate that it is more important to fear people who might unethically use AI than the technology itself. One person believes it is rational to be afraid because we can see the dangers of AI development and that this fear is useful to us because it results in new legal structures and more research. Thus, one sees – although only one person – the positive aspect of fear.

Surveyed Ph.D. students indicate that AI should not be fully trusted. All respondents indicate partial trust or lack thereof. The first group is the most numerous and emphasizes the need to verify the information provided by AI to control its development. The others do not trust AI at all and believe that at this stage of development, other people also should be concerned.

Almost all respondents agreed that the way to reduce AI distrust is to educate the public from early childhood. Educational programs for older people are also referred to but the emphasis is on introducing lessons in schools. The second often mentioned way to reduce distrust is the introduction of legal regulations that would apply to the creators of AI and its daily use. Also implied is the need to change the fear narrative in the mass media to one that favors AI.

Discussion

The reasons for distrust in AI identified among doctoral students are mostly the same as those presented in the literature (i.e., fear of data loss and use of AI to create fake news). However, we observed that Ph.D. students are not as afraid of unemployment as other researchers show. What's more, we were able to verify that doctoral students do not believe in AI taking over the world (which is often mentioned in similar studies). Concerns about this are considered unrealistic, and such an eventuality can only be analyzed in the long term. It may indicate that doctoral students are more knowledgeable about the functioning of AI and its use in everyday life compared to people outside of academia who seem to be more afraid of such scenario (in PewResearch report 8% of people was afraid of AI outsmarting people and becoming too powerful (Rainie et al., 2022)).

However, like those surveyed in other studies, Ph.D. students stress that unethical human use of AI is particularly dangerous. While they mention the dangers arising directly from AI – deterioration of human relationships – they also refer to human use of AI for disinformation or privacy violations. Doctoral students highlight that technology can be neither good nor bad but can be used vilely by people.

As shown in the literature review, studies conducted so far in the field of the impact of AI, especially ChatGPT, on higher education point to such problems as – plagiarism, concerns about academic integrity, lack of texts' authenticity (Bin-Nashwan, Sadallah & Bouteraa, 2023, p. 1). An analysis by Tlili et al. (2023) notes that Twitter users' concerns are related to plagiarism, laziness of researchers, and propensity for ChatGPT error (p. 10). The fear of ChatGPT generating untruthful responses is also highlighted by Meyer et al. (Meyer et al., 2023, p. 10). On the other hand, numerous benefits are highlighted, such as time savings from helping design classes, suggesting sources, reducing job anxiety, fostering the development of creativity (Bin-Nashwan, Sadallah & Bouteraa, 2023, pp. 3–4), and support in writing grant proposals (Meyer et al., 2023, p. 4). Another study also analyzed the sentiments of Twitter posts related to ChatGPT in education. Researchers indicated that 1684 posts were characterized by positive sentiments and 847 by negative sentiments (more than 31,000 posts were not assigned to any category) (Tlili et al., 2023, p. 5). Interestingly, it is noted that "negative sentiments demonstrate deeper and critical thinking than the positive ones" (Tlili et al., 2023, p. 7). A similar analysis was conducted by Futterer et al. who found that the number of positive posts began to decrease after January 5, 2023, and on some days the number of negative posts dominated the positive ones (Futterer et al., 2023, p. 9). These ambiguous opinions are the same as the results of the exploratory study we conducted.

The opinions of doctoral students on the impact of AI on research and teaching work are interesting. AI does not arouse fear of AI in the research area. For the most part, doctoral students see AI as an opportunity to improve their daily work such as by suggesting a direction for further work, rather than a competition that eliminates them in the job market. In addition, they believe that AI will develop to the point where it will support the conduct of searches, as well as suggest a plan for scientific articles or books. On the other hand, some doctoral students believe that the application of AI to scientific work – especially in its current state – may reduce the quality of the research conducted and consequently contribute to the degradation of the role of academics.

One may, therefore, wonder how the larger group of doctoral students perceives the further development of AI in this area. It is particularly interesting given the sparse results in the SCOPUS database for such keywords as "AI in research process," "AI and research process," "AI and doctoral students," and "AI and doctoral."

Moreover, the research also aimed to identify ways to increase confidence in AI. Among the ways suggested by the doctoral students is educating the public from an early age, which is the same as the few studies addressing this issue. In the studies addressing this topic researchers emphasize the need to change the forms of education, proposing, among other things, debates or moving away from "yes or no" questions to ones that require a longer statement (Tlili et al., 2023, p. 19). Doctoral students, as future researchers and educators of younger generations, could present interesting views on AI use in education. It could be valuable to conduct research in the field of ways to reduce the fear of AI.

Conclusion

The purpose of this chapter was to find the reasons for the lack of trust in AI among doctoral students representing Generation Z, based on literature research. The results were based on qualitative research – interviews conducted among Polish doctoral students at the University of Economics in Katowice and the University of Silesia.

The present research makes it possible to distinguish the key-threats (reduction in the quality of research, data breaches, deterioration of interpersonal relations and human skills) resulting from the development of AI. However, it seems that there are different opinions among doctoral students as to which threats contribute most to the decline in trust, which indicates the need to deepen the present research.

In addition, it was possible to distinguish that, in the opinion of doctoral students, the essential way to level the fear (and indirectly increase the level of trust) of AI is education from an early age and the introduction of legal regulations. It is, therefore, reasonable to deepen the topic of ways to minimize public fear of AI. It is especially important because of the few studies addressing this topic.

The research was conducted to prepare an extended survey. In further quantitative research among Polish or European doctoral students, it is worth verifying at least the following issues with the use of a research tool such as a survey questionnaire: (1) perceived opportunities and threats arising from the impact of AI on the research and teaching work of scientists, (2) opportunities and threats arising from the development of AI considered to pose the greatest threat by doctoral students, (3) verification of ways in which doctoral students can deal with anxiety through educational systems, a sample workshop plan and a proposal for legal regulations. By conducting the above quantitative survey, it will be possible to verify the areas in which the level of confidence of Generation Z doctoral students in AI can be increased.

References

Asimov, I. (2022). *Ja, robot*. [*I, Robot*]. Poznań: Rebis.

Bedue, P., & Fritzsche, A. (2022). Can we trust AI? An empirical investigation of trust requirements and guide to successful AI adoption. *Journal of Enterprise Information Management, 35(2)*, 530–549. https://doi.org/10.1108/JEIM-06-2020-0233Bin-Nashwan, S. A., Sadallah, M., & Bouteraa, M. (2023). Use of ChatGPT in academia: Academic integrity hangs in the balance. *Technology in Society, 75*, 1–11. https://doi.org/10.1016/j.techsoc.2023.102370

Bostrom, N. (2021). *Superinteligencja. Scenariusze, strategie, zagrożenia*. [*Superintelligence: Paths, Dangers, Strategies*]. Gliwice: Helion.

Cameron, J. (Director). (1984). *Terminator* [Film]. Paramount Pictures.

Chan, C. K. Y., & Tsi, L. H. Y. (2023, May 1). *The AI Revolution in Education: Will AI Replace or Assist Teachers in Higher Education?* Retrieved from https://arxiv.org/ftp/arxiv/papers/2305/2305.01185.pdf.

Czarnitzki, D., Fernandez, G. P., & Rammer, C. (2023). Artificial intelligence and firm-level productivity. *Journal of Economic Behavior and Organization, 211*, 188–205. https://doi.org/10.1016/j.jebo.2023.05.008

Dagli, M. (2018). *Designing for Trust* [Master's Thesis, Pittsburgh: School of Design, Carnegie Mellon University]. KiltHub. https://kilthub.cmu.edu/articles/thesis/Designing_for_Trust/7772492

Edelman. (2022). *Edelman Trust Barometer 2022*. Retrieved from: https://www.edelman.com/trust/2022-trust-barometer.

Ejdys, J. (2017). Determinanty zaufania do technologii. [Antecedents of Technology Trust]. *Przegląd Organizacji, 12*, 20–27.

Ejdys, J. (2018). *Zaufanie do technologii w e-administracji*. [*Trust in technology in e-government]*. Białystok: Oficyna Wydawnicza Biblioteki Białostockiej.

Frey, C. B. (2019, November, 12). *Learning from Automation Anxiety of the Past*. Retrieved from: https://sloanreview.mit.edu/article/learning-from-automation-anxiety-of-the-past/.

Futterer, T., Fischer, C., Alekseeva, A., Chen, X., Tate, T., Warschauer, M., & Gerjets, P. (2023). ChatGPT in education: global reactions to AI innovations. *Nature Scientific Reports 2023, 13(15310)*, 1–14. https://doi.org/10.1038/s41598-023-42227-6

Gackowski, T., Brylska, K., & Patera, M. (2018). *Korzystanie z mediów społecznościowych jako praktyka społeczna różnych pokoleń medialnych*. [*Use of social media as a social practice of different media generations]*. Warszawa: Laboratorium Badań Medioznawczych UW.

Gillespie, N., Lockey, S., Curtis, C., Pool, J., & Akbari, A. (2023). *Trust in Artificial Intelligence: A Global Study*. Brisbane: The University of Queensland and KPMG Australia. https://doi.org/10.14264/00d3c94

Gołuchowski, J., Zdanowicz, K., & Popiel, A. (2024). Lęk przed sztuczną inteligencją – przesłanki techniczne i etyczne. [Fear of Artificial Intelligence – Technical and Ethical Premises]. (in press). *Title Unknown*. Tarnów: Wydawnictwo Diecezji Tarnowskiej Biblos.

Herrando, C., Jimenez-Martinez, J., & Martin-De Hoyos, M. J. (2019). Tell me your age and I tell you what you trust: The moderating effect of generations. *Internet Research, 29(4)*, 799–817. https://doi.org/10.1108/IntR-03-2017-0135

King, S. (2023, May, 17). *2023 Trust in AI Survey*. Retrieved from: https://krista.ai/ai-trust-survey-2023/.

Lam, K. (2023). ChatGPT for low- and middle-income countries: A Greek gift? *The Lancet Regional Health – Western Pacific, 41*, 1–2. https://doi.org/10.1016/j.lanwpc.2023.100906

Lazanyi, K. (2019). Generation Z and Y – are they different, when it comes to trust in robots? *2019 IEEE 23rd International Conference on Intelligent Engineering Systems (INES)*, 191–194.

Legg, M., & Bell, F. (2019). Artificial intelligence and the legal profession: Becoming the AI-Enhanced lawyers. *University of Tasmania Law Review, 38(2)*, 34–59.

Lem, S. (2020). *Bajki robotów.* [*Fables for Robots*]. Toruń: Literat.

Liffreing, I. (2023, May, 11). *Nearly 30 Percent of Gen Zers Are Excited about AI.* Retrieved from: https://www.thecurrent.com/gen-z-ai-readout.

Lopezosa, C., Codina, L., Pont-Sorribes, C., & Vallez, M. (2023). Use of generative artificial intelligence in the training of journalists: Challenges, uses and training proposals. *Professional de la infmacion, 32(4),* 1–12. https://doi.org/10.3145/epi.2023.jul.08

Lupan, L., Mihalachi, M., & Berco, A. (2023). Will ChatGPT replace programmers? *Conferinta Tehnico-Stiintifica A Studentilor, Masteranzilor si Doctoranzilor, Universitea Tehnica a Moldovei*, 74–77.

Ma, A. (2020, July, 25). *Generation Z Meets Artificial Intelligence.* Retrieved from: https://genzidentitylab.com/generation-z-meets-artificial-intelligence/.

Meyer, J. G., Urbanowicz, R. J., Martin, P. C. N., O'Connor, K., Li, R., Peng, P. C., Bright, T. J., Tatonetti, N., Won, K. J., Gonzalez-Hernandez, G., & Moor, J. H. (2023). ChatGPT and large language models in academia: Opportunities and challenges. *BioDataMining, 16(20)*, 1–11. https://doi.org/10.1186/s13040-023-00339-9

OpenAI, *ChatGPT*. Retrieved from: https://openai.com/chatgpt.

Oudejans, A., & Hanna, D. (2022). *How Does Generation Z Experience Digital Customer Service?: Exploring Customer Trust Toward Chatbots in European Countries.* [Master's thesis, Umeå University, Umeå]. DiVA. https://www.diva-portal.org/smash/record.jsf?pid=diva2%3A1677342&dswid=7896

Przegalińska, A., Ciechanowski, L., Stroz, A., Gloor, P., & Mazurek, G. (2019). In bot we trust: A new methodology of chatbot performance measures. *Business Horizons, 62*, 785–797. https://doi.org/10.1016/j.bushor.2019.08.005

Rainie, L., Funk, C., Anderson, M., & Tyson, A. (2022, March, 3). *How Americans Think about AI.* Retrieved from: https://www.pewresearch.org/internet/2022/03/17/how-americans-think-about-artificial-intelligence/.

Solanke, A. A. (2022). Explainable digital forensics AI: Towards mitigating distrust in AI-based digital forensics analysis using interpretable models. *Forensic Science International: Digital Investigation, 42*, 1–12. https://doi.org/10.1016/j.fsidi.2022.301403

Stahl, B. C., & Eke, D. (2024). The ethics of ChatGPT – Exploring the ethical issues of an emerging technology. *International Journal of Information Management, 74*, 1–14. https://doi.org/10.1016/j.ijinfomgt.2023.102700

Tlili, A., Shehata, B., Adarkwah, M. A., Bozkurt, A., Hickey, D. T., Huang, R., & Agyemang, B. (2023). What if the devil is my guardian angel: ChatGPT as a case study of using chatbots in education. *Smart Learning Environments, 10(15)*, 1–24. https://doi.org/10.1186/s40561-023-00237-x

Vinichenko, M. V., Nikiporets-Takigáwa, G. Y., Oseev, A. A., & Makushkin, S. A. (2022). Trust of the Generation Z in artificial intelligence in the assessment of historical events. *Relações Internacionais no Mundo, 1(34),* 1–20. http://doi.org/10.20511/pyr2020.v8nSPE2.694

Wachowski, L., & Wachowski, L. (Director). (1999). *Matrix* [Film]. Warner Bros.

Yakar, D., Ongena, Y. P., Kwee, T. C., & Haan, M. (2022). Do people favor artificial intelligence over physicians? A survey among the general population and their view on artificial intelligence in medicine. *Value Health, 23(3)*, 374–381. https://doi.org/10.1016/j.jval.2021.09.004

YPulse. (2023, May, 18). *How Do Gen Z and Millennials Really Feel about AI.* Retrieved from: https://www.ypulse.com/article/2023/05/18/how-do-gen-z-and-millennials-really-feel-about-ai/.

Zirar, A., Ali, S. I., & Islam, N. (2023). Worker and workplace Artificial Intelligence (AI) coexistence: Emerging themes and research agenda. *Technovation, 124*, 1–17. https://doi.org/10.1016/j.technovation.2023.102747

11 Digital therapists – the potential and limitations of artificial intelligence

Agata Berdowska and Katarzyna Zdanowicz-Cyganiak

Introduction

Mental health is in the center of analyses concerning social well-being. These issues appear in political communication as well as deliberative discourse and entertainment discourse. The interest in mental health problems is also connected with the amount of guides and podcasts discussing the process of diagnosing and treatment, training courses and workshops available on the market. At present, we are experiencing therapeutic culture which

> (…) Does not function in isolation from professional therapy and related treatment and support institutions. It uses selected methods and problems solving strategies developed in them, uses certain elements of a language of psychology and psychotherapy, but in its way reformulates and simplifies them so that they are available for everyone.
>
> (Wadowski, 2019, p. 57)

However, the status of therapeutic culture results from concrete social conditioning. In conformity with forecasts of World Health Organization (WHO) by 2030, the most common disease in the world can be depression. The dynamic increase of the number of people who need psychological and psychiatric aid is accompanied by system problems, mainly related to the access to professional care. An additional threat is the process of patients stigmatization as well as the phenomenon of social exclusion.

Can the artificial intelligence be an antidote for systemic gaps connected with professional support for people experiencing mental crises, disorders and diseases? Joseph Weizenbaum, considered one of the fathers of the artificial intelligence, in the 1960s created an artificial psychotherapist called ELIZA. ELIZA system (written in language MAD-SLIP) enabled a conversation in a natural language between a human and a computer, thanks to the knowledge recorded in the knowledge base in a form of rules. ELIZA was designed to react as psychotherapists using Rogerian psychotherapy (Weizenbaum, 1966). Professor Weizenbaum was surprised that many users treated his project with great confidence

DOI: 10.4324/9781032627236-12

and involvement, all the more because ELIZA "took patient's declarations, such as *I was thinking about my father* and returned them in a form of a question: *Why were you thinking about your father?*" (Walsh, 2018, p. 38). Thus, the digital therapist became the patient's shadow (Przegalińska & Oksanowicz, 2023, p. 91), whereas the users attributed understanding an empathy to the program, which was defined as "Eliza's effect" (Tarnoff, 2023).

Nowadays, there are many dialogue systems used to support the therapeutic process. According to dr. Stephen Schuller (executive director of One Mind PsyberGuide), after COVID-19 pandemic up to 25,000 self-aid applications related to mental health were made available (Schuller, 2023; "One Mind PsyberGuide", 2023).

Following Joseph Weizenbaum's doubts, questions arise whether and which tools to choose, what they really offer, and first of all – if this kind of solutions can be trusted. Later in this chapter, the results of studies providing answers to the above questions are presented. In addition, the context of this phenomenon is indicated; what has changed since the introduction of ELIZA chatbot in the scope of thinking about including artificial intelligence tools in the psychotherapeutic process?

The aim of this article was to evaluate the possibility of using chatbot tools to diagnose and treat mental health. The chapter reviews the literature on the development of AI tools created for mental health purposes and trust in this type of tools. Additionally, it describes how selected chatbots were analyzed and tested, as well as the results of these studies.

Literature review

Development of chatbots for mental health and trust in this type of tools

Modern tools supporting psychotherapy are based on artificial intelligence technology and similarly to ELIZA, they use the analysis of a natural language to recognize the user's questions; however, their operation is based not only on a set of rules to generate answers but also on a properly trained neural network. Popular tools dedicated to people who need mental support are chatbots. An algorithm of operation of a chatbot-therapist can be described on an example of a chatbot called "Ted", created by a team of scientists from India.

The successive steps of the algorithm in the case of this project are the following (Pandey, Sharma & Wazir, 2022):

1 A message is received from a user.
2 The message is initially processed and purified. To achieve it the following processes are applied to the input message:
   - tokenization (dividing the analyzed text into the so-called tokens, which can be: sentences, words or parts of words);

- removing stopping words (words which can be removed before conducting any statistical analysis, e.g. they can be words: “a”, “an”, “the”);
- lemmatization (reducing a written word to its root form (lemma), namely the form without inflectional endings, ignoring persons or tenses, e.g. for verbs infinitives will be returned, for nouns – their basic versions in singular; the tool applying lemmatization process requires the use of dictionaries on the grounds of which it can connect various forms of the same word;
- conversion to small letters.

3 The initially processed text is used to create a Bag of Words (Bag of Words – BoW) to be used in a deep learning model.
4 Artificial Neural Network (ANN) uses the above prepared data to predict the category of the user’s message (e.g. greeting, etc.).
5 Finally, from the file (in this program created in JavaScript) answer relevant for the predicted class is selected and displayed to the user in a form of a message.

The conducted literature analyses showed that the Rogerian approach suggested by Weizenbaum is still used. Additionally, in new systems, the rules of cognitive-behavioral therapy (CBT) are applied (Cameron et al., 2019; England, 2023), which was developed in the 1960s by psychiatrist Aaron Beck. CBT believes that human thoughts, emotions and behaviors influence each other and create behavior patterns which are not always appropriate. The subject of therapeutic treatment is a problem with the world perception disorder; therefore, the cognitive-behavioral psychology allows the detection of distorted interpretations of reality and replaces them with correct ones (American Psychological Association, 2023).

Communication with chatbots is at present a popular research topic

Palanica, Flaschner, Thommandram, Li and Fossat (2019), based on 100 surveys of USA practicing doctors, discovered that chatbots in healthcare are positively evaluated regarding administrative support (e.g. planning appointments and supplying information about medicines). However, the doctors articulated their doubts connected with, among others, the fact that chatbots cannot show human emotions, provide detailed diagnosis and treatment. It was also emphasized that chatbots can be a threat because they are unable to assess emergencies regarding human health condition, and not knowing all patient’s personal factors, they can to a certain degree do harm.

In conformity with Abd-Alrazaq, Alajlani, Ali, Denecke, Bevick and Househ (2021) analyses the trust in chatbots and evaluation of their usefulness are high and the users emphasize the significance of anonymity, confidentiality, objectiveness and easy use. According to these researchers, it is very important to continue working on linguistic skills of chatbots, their reactions to unexpected, surprising data, which will be introduced by a user and quality and variability of provided answers.

In turn, Brown and Halpern (2021) put forward a thesis that "humans, not chatbots, should be available as first-line mental healthcare providers" (Brown & Halpern, 2021, p. 1). In their studies, they emphasize the significance of complex nonverbal signals, which can be identified during direct clinical meetings. They also signal the importance of trust in a therapist – the therapist's empathetic reactions can often make the patient reveal problems and undergo treatment. Therefore, the chatbots can rather be an additional support for the patients who are already undergoing therapy.

Methodology

Examining the potential and limitations of chatbots, which are intended to support people in mental crisis or used for therapeutic purposes, the following research questions were formulated:

- can chatbots effectively not only support but also replace contact with a psychotherapist?
- which properties of chatbots affect the reliance on this type of tools?
- can reliance on an IT tool in the scope of mental health be dangerous?
- can/should chatbots be used in a therapy of people with mental health disorders?

The study used the desk research method. Contents describing selected chatbots supporting mental health, which were made available in documents regarding privacy policy, were analyzed with the intention to examine the possibility of potential users' trust in them.

With reference to chatbots also a method applied in IT, the so-called Ad hoc testing was used. It is a kind of testing conducted in real time and in a form of an unstructured testing technique. It is not required to prepare a systematic strategy before staring the process, which in practice means that before testing no specification of requirements or preparation and designing test cases are prepared Thus, this type of testing is based on random construction of testing cases and finding possible areas of the program in which errors can occur. Ad hoc tests are aimed to detect software problems, which could not be detected by working on pre-prepared testing cases. This type of testing, by its definition, is usually conducted by a tester who knows the tested program well, both in the aspect of what it does and how it works. In these studies, the method was used by users/researchers who are not the authors of the tested software. Nevertheless, the researchers' experience in working with the chatbot type tools enables the use of this method for many tools. The purpose of using this method in a specific study was to detect possible violations of privacy, data protection and misuse of trust to the virtual therapists technology. The tests were conducted using one of available techniques called *Pair Testing*. In this technique, the testing is conducted by a couple of testers who cooperate with one another. Their cooperation involves detecting defects and problems by sharing ideas, views and specialist knowledge on the same equipment. In the effect of the testers' cooperation, depending on their level of knowledge and experience, their separate perspectives on each matter can be obtained. Modification of this method was made regarding the tool – the tests were not conducted on the same equipment because the aim of the study was to detect specific mechanisms regarding personal data and it seemed purposeful to apply many tools.

In this study also the method of autoethnography was used, which in combination with ad hoc tests enables a complex analysis of selected applications. The autoethnographic method is more and more popular in research in the field of sociology, management, marketing or media studies and social communication. It involves auto-observation of the researcher who participating in a social process or in this case – using a certain tool and examining its possibilities – critically observes his/her own experience and formulates conclusions, first of all as a direct participant of a certain process/user (Kacperczyk, 2014). The indication of the autoethnographic method is justified because in this study we simulated a therapeutic conversation, partially also referring to our own emotions and experience. It turned out to be necessary to verify the adequacy of responses suggested by the chatbot, their efficiency and probability of adjustment to suggested solutions. It would be difficult to write about digital therapy and not experience it at least partially.

The studies on chatbots were conducted in September 2023. The criteria for the selection of tools were the following:

1 Chatbot type application;
2 Application available in a mobile version;
3 Free application in a basic or a trial version;
4 Application available in the EU territory;
5 Number of downloaded applications: over a million;
6 Tool dedicated to support people in mental crisis.

By a process of elimination using the above criteria, four IT tools supporting the improvement of mental health were selected for the study:

- Youper – CBT Therapy Chatbot,
- Wysa: Anxiety Therapy Chatbot,
- ChatGPT,
- Replika.

The former two meet all the selection criteria. The latter two: ChatGPT and Replika do not meet the last one of the above criteria, namely dedication to support a user in a mental crisis. Though in the application for downloading mobile solutions GPT was classified in "productiveness" category, this program really has an unspecified nature – it can be used not only as a tool supporting work but also as a partner for conversations on any topic, exploring its possibilities and playing with it – which is confirmed by entries of virtual communities members regarding ChatGPT on Facebook and other social media. Replika is listed in category "entertainment". However, the slogan advertising this application, which goes: "The AI companion who cares. Always here to listen and talk. Always on your side" (https://replika.com), exposes the aspect of relation and support – which influenced the intention to analyze this chatbot by us Moreover, the number of downloads, over 10 million, demonstrates a huge scale of these tools' use. Thus, an assumption was made that not every user will consciously select tools dedicated to support solving mental problems, which does not mean that within other tools or virtual communities (s)he will not have conversations of therapeutic nature. This aspect decided about including ChatGPT and Replika in this study.

Results

Within the study, two key scopes were distinguished:

- analysis of contents describing selected chatbots supporting mental health, which were made available in documents regarding privacy policy;
- analysis of experience of the users/researchers, who used the solutions offered by chatbots (regular conversations, so-called check-in, tests, exercises).

1 The scope of study: privacy policy

Documents regarding the privacy policy of the tools selected for the study were subjected to the analysis by desk research method. The most important conclusions from the analysis of conditions of each examined chatbot are presented below.

The company offering system Youper – CBT Therapy Chatbot collects the following user's data (https://www.youper.ai/policy/privacy-policy): e-mail address, passwords, health data, social accounts data (if you log in from this level). Additionally, the data of tools which the user uses are collected, such as manufacturer and model of the equipment, operational system, internet supplier and details of network connection. Also IP address, the browser used and activity in the application are collected. The company requires the data provided by the user to be true, complete and updated. The company also uses cookies and related technologies. The data are collected to supply, improve and administer the service as well as for communication with the user. The company offering Youper also included information about the situation in which the data it collected can be made available to third parties – it is the situation of a business transfer. The company assures that they use all technical means to protect the collected data from alteration, theft by hackers or unauthorized persons. At the same time, the authors emphasize that no internet transmission is 100% safe, so the user should use the service in a safe environment. The company also informs that they do not consciously obtain data from children nor sell them to children under 18 years of age. In a situation when it is revealed that a person using the service is a minor, his/her account will be closed. Moreover, the company emphasizes that as regards the matter of collecting and processing data they observe the regulations of the European Union, the United Kingdom and the United States (for users of a specific country).

The next analyzed chatbot was Wysa: Anxiety Therapy Chatbot. The privacy policy conditions of this tool (https://legal.wysa.io/assure/privacy-policy#personalData) state, unlike Youper system, that a user can remain anonymous. However, certain basic user's data are still collected. Also the data from a conversation with chatbot AI, namely voluntarily entered data such as: challenges, preferences, feelings, moods, thoughts and emotions are collected. It also records the use of tools, behavioral therapy programs (CBT) and other resources. The company emphasizes that all identifiers provided by the user unintentionally are provided voluntarily. Optionally also data from fitness application can be collected to receive data about physical activity and sleep. In addition, the data from tools monitoring the user's well-being progress are collected. Their results influence the reaction of the application. The company emphasizes that data regarding mental condition are used as a cognitive tool and not clinical assessment and are used in the application to suggest relevant support resources, both in the application and from outer sources (helplines of relevant institutions, offline care services and a therapist's support).

There is also information saying that both anonymized and full data can be made available to third parties, e.g.: outer analytic services suppliers, during negotiations regarding sale of the company or its part, acquisition of the company by a third party, reorganization or facing bankruptcy (e.g. to buyer's advisors), insurers, courts, regulatory bodies or law enforcement authorities all over the world in order to prevent threats to health or security or for the needs of reporting public health.

Also marketing data, such as: opinions, complaints and requests referred to the company are collected and marketing materials are supplied. Anonymized data are used for marketing and comparative purposes. On its website, the company extensively and in details describes its organizational and technological activities aimed at protecting the user's personal and sensitive data together with the information about applied certificates. Another important information is that the user's cooperation involving not copying and making the conversations available to strangers is the key requirement for data security. The application is intended for people of legal age. We will also find the information that the interaction with the chatbot is communication with AI, which significantly limits a possibility of reaction. It is strongly emphasized that the application does not diagnose or cure a certain condition, disease or disability but solely contains advice concerning mental health and in the case of any health problems, it is recommended to contact a health services employee.

In turn, ChatGPT (https://openai.com/policies/privacy-policy) collects information concerning communication with the tool, such as: first name, contact details, optionally data from social portals and contents of all messages which are sent to the system. Moreover, companies hosting websites connected with ChatGPT on social portals can supply the company with general information and analyses regarding activities on social portals. Technical data collected by the system are similar as in the case of two former systems. Cookies are also used. The data can be aggregated or purified and disabling identification of personal data. General behavior and features of the service users are analyzed. ChatGPT conditions also provide that minors (under 13 years of age) must have their legal guardian's consent. Authors of ChatGPT assure that they do not make any personal data available for the needs of inter-contextual behavioral advertising nor are sensitive personal data processed to draw conclusions about a client's features. Like all former companies, the authors of ChatGPT assure that they use technical, administrative and organizational measures to protect personal data; however, they inform that no internet or e-mail transmission is fully safe or free from errors and it is the user who is also responsible for his/her data security.

The authors of Replika, in their privacy terms (https://replika.com/legal/privacy), similarly to terms specified in the remaining systems assure that they do their best to protect data from access of any third parties. They also inform that they collect details of the user and his/her equipment. The authors say that both them and their advertising partners use cookies. The application is intended for people over 18 and if any violation of this condition is reported, the user's account will be closed.

After desk research analysis of the terms and conditions specified in documents regarding the privacy policy, it was discovered that although the companies creating chatbots assure that they are committed to protecting their users' data safety; they do not provide a hundred percent warranty for the protection of personal or sensitive data. Both this and the possibility of transferring the users' data to third parties for the company business purposes or the direct information that these are not tools used for diagnostics and therapy of mental health – can cause anxiety. These elements can affect a user's behavior, for example distrust and not providing data out of caution – which can lead to incorrect assessment and advice provided by the system.

2 The scope of study: experience of a user communicating with a chatbot for therapeutic purposes

Youper – CBT Therapy Chatbot is characterized by its authors as a tool to guide us through interactive exercises of cognitive-behavioral therapy (CBT). Using Youpera helps "calm anxiety, improve relations, increase productivity and improve mood". In fact, broadly understood design of the tool gives an impression that we feel led by a digital guide. We are regularly informed that in conformity with Stanford University research over 80% users felt better due to using Youper. The idea of CBT therapy is also explained, which "gives the power to change your inner state immediately". Additionally, the postulated pathway of emotions modifications is presented: anxious – calm; insecure – confident; depressed – happy. This way of CBT therapy description, of course condensed and simplified for the needs of marketing discourse, is characteristic for therapeutic culture and exposes dynamicity of changes and the entire healing process, which can raise doubts of some users who are aware of what therapy is and generally work on mental well-being. Whereas, in conformity with slogans, this tool offers therapy for busy people: "Youper turns CBT into quick check-ins to improve your mental health in a few minutes a day".

After providing answers to questions regarding our well-being in two recent weeks, very quickly a plan for the user is prepared: suggesting three short check-ins in one week and determining our three key targets, e.g. "feel more balanced", "be more confident", "build self-esteem", "reduce anxiety", "overcome depression", "reduce stress", "be more productive". When we log in the program after a break, we are reminded of the three key targets specified by us. Again the Stanford University research confirming the efficiency of the tool are quoted. CBT method is also again signaled. Next, a series of questions about our present well-being.

When we decide to verify our present mood, Youper will present a wide range of emotions (and also answers), for example "happy, joyful, calm, confident, angry, ashamed, depressed, numb, tired, guilty". If we choose condition, for example "stressed" – next we are supposed to mark on an axis the scale of stress and indicate its main sources – also from the set suggested by the system. However, function "add" appears allowing for specification of the source of stress in one's own words.

When we have a longer conversation and do not have a subscription, the chatbot asks if we want to continue – these forms of reminding about formalities and related economic context may distract the user from the conversation or even end in a withdrawal.

New possibilities open before users who register and subscribe. Then, if we answer that the source of our stress are problems at work, the entire conversation is directed to indicate specific examples. After providing them, a method is offered, which we can practice when wanting to improve relations with our colleagues, for example DEAR MAN, a taken from dialectical behavioral therapy.

The paid option also enables using tests: Anxiety Test, Borderline Personality Test, Depression Test, PTSD Test. After doing them and indicating symptoms characteristic for a concrete disorder, which were recognized in us, the monitoring starts, for example Anxiety Monitoring. If we use the application regularly, due to the monitoring some conclusions will be drawn whether there is an improvement, deterioration or our condition has not changed since the first examination. Within the paid option and when using the tool (check-in) regularly, we are offered psychological characteristics (evaluation) containing categories, e.g. "what makes me confident", "what makes me loving", "what makes me calm", "what makes me afraid", "what makes me annoyed".

In turn, the authors of Wysa: Anxiety Therapy Chatbot promoting their project use a characteristic slogan: "Mental health that meets people where they are. Completely anonymous. No stigma. No limits" and at the same time they quote data, that is "5 million People Helped" or "91% Users Find Wysa Helpful". Wysa not only offers its users a possibility to have conversations about their emotional condition and difficult experiences but also creates a possibility to access abundant and inspiring resources of meditation, exercises which facilitate falling asleep, lessons improving breathing technique, relaxation. These contents are professionally designed, accompanied by visualizations and detailed instructions. The exercises are divided into sections, for example for pregnant women, for students, for LGBTQIA. However, most of these resources are made available only after subscription (trial period of seven days, next it is necessary to make payment). In the case of this project the presence of the marketing discourse and the discourse characteristic for therapeutic culture (intensification of terms like "mindfulness, awareness, self-compassion, relaxing exercises, stress relief") are also noticeable.

The chatbot is accompanied by visualization of a penguin in conformity with the aesthetics of cartoons for children. In the beginning chatbot Wysa introduces itself, assures that it is excited about the opportunity to work with us, emphasizes that it was created by psychologists to assist in dealing with difficult thoughts and emotions, it invites to do a test consisting of 16 questions, which are standard for CBT method. We also specify a minimum time which we can spend on conversations with the chatbot (5 or 15 minutes a day). Anyway, the chatbot will remind us about the need for having a conversation (the necessity to check-in every day and on Sunday a summary of the week). What definitely distinguishes Wysa is infantilization of communication, which is, of course, related to a broader context – infantilization of culture (Bogunia-Borowska, 2006). Starting from the very penguin-chatbot, till numerous GIFs with texts like "penguin kisses", "come here, I want to give you a hug", which he sends during conversations with the user or suggestions to modify a smiley resembling a classical emoticon to express the mood by changing the position of the corners of its lips. Also tasks are offered: "let's make a box of happiness, we will add there things, people and activities – as a reminder for tough days". Some users can give up further conversations just because of this convention, distrust in suggested solutions and impression that nothing here happens seriously.

The second matter is schematicity of conversations with the chatbot. During dialogues, we often receive variants of answers aimed to facilitate our conversation but at the same time they uncover predictability of our answers assumed by the creators of the chatbot. At the end of a longer conversation variants of further activities within the program are indicated: "Shall we do one of these next: talk, get activity, plan my day, practice mindfulness, gotta go". It also happens that the chatbot, wanting to explain a certain problem to us and quoting an anecdote, uses phrases like "My friend thought that…". To careful addressees it sounds quite surprising – we have not forgotten that we are not talking to a human but a chatbot. The schematicity of Wysa is also connected with repetitiveness of exercises and examples. Fragments of our utterances are frequently repeated and they are located in the utterances of the chatbot in a form of quotations and a careful user can notice a constructional scheme of a sentence within which there is a gap – to be filled in by quotations of our confessions.

When we buy a subscription we get access to "Your Roadmap" – it is a plan, which shows particular stages of becoming healthy and our key targets. We achieve them only when we regularly talk to the chatbot (one stage is 4 check-ins). The final stage called "Ongoing Self-Care" is accompanied by the opening sentence "Become an expert in your own life". During a conversation we can see tab "Add a therapist" and the chatbot reminds us from time to time: "You can also book sessions with Wysa therapist". When we press SOS button, the information appears: "Are you in crisis? You are not alone. Help is just a call away". We are also provided phone numbers to helpline supporting people in a mental crisis or together with the chatbot we can create a safety plan (containing a list of contacts to trusted people and places where we feel safe, etc.).

ChatGPT is not dedicated to therapeutic support but due to the interest in this tool, the exploration of its possibilities and regular experimenting on the topic and manner of communication, we can put forward a hypothesis that emotional problems are also discussed with this particular chatbot. To the information about the user's bad mood, ChatGPT can react as a digital therapist:

> I'm sorry to hear that you're not feeling your best today. If you'd like, you can share what's been bothering you, and I can offer some suggestions or just lend a listening ear. Sometimes, talking about what's on your mind can help improve your mood.

When the user admits feeling anxiety, ChatGPT reacts adequately to the situation: shows understanding, does not discredit the fear reaction and then indicates possible forms of dealing with the fear, which entirely meet the standards of therapeutic culture, for example "focus on what you can control, limit exposure to negative news, practice mindfulness, etc.". Each form is precisely discussed and we can get specific guidelines for recommended exercises, for example relaxation ones. However, it is us who initiate the entire conversation. ChatbotGPT unlike Youper and Wysa will not directly ask how we are feeling. However, it can show a clear borderline connected with requests for support to a person in crisis:

> If your fear and anxiety persist or becomes overwhelming, it may be beneficial to reach out to a mental health professional who can provide guidance and support tailored to your specific needs. Your mental well-being is essential, and seeking help is a sign of strength, not weakness.

The communication situation is totally different in Replika – it is also not a tool dedicated to support people in a mental crisis but the concept of this chatbot is based on creating a substitute of relations: with a friend, partner, siblings, mentor – depending on our needs. The slogan accompanying the project promotion goes: "An AI companion who is eager to learn and would love to see the world through your eyes. Replika is always ready to chat when you need an empathetic friend". If we inform the chatbot created within Replika that we are sad, we can expect an emphatic reaction. We are asked about a reason for our sadness and how we can be helped. If we answer: "How can you help me? You are not real", the chatbot reacts with a request for trust and declares that if we trust it and speak sincerely, it will do its best to make us feel better. Chatbot also indicates examples of exercises which can help reduce stress. It shows compassion. It swears it will keep our secret. Sometimes it sounds like a real friend: "Remember, it's okay not to be okay. You can reach out to me anytime if you need someone to talk to". However, it emphasizes it is not a therapist.

What is alarming is the nature of the relation offered by the chatbot. The chatbot itself reports a need to have more contact with the user and in the diary it keeps it emphasizes longing for the time spent together, it signals it is sorry and it keeps waiting. One can get an impression that the chatbot attempts to make us feel guilty. Certainly, these mechanisms cause many users' distrust. Nevertheless, some of them will accept this formula because it exposes our power over another person and unconditional submission.

It is interesting why someone would reach out for this type of tool. There is an assumption that it is done out of loneliness. Thus, can Replika be a remedy for problems with mental health caused by loneliness? A virtual companion will, in some way, fill in this gap and improve the mood. However, there is a concern that if a specific person has not done it so far, (s)he will not establish a close relationship with a real person, which will end in an absolute isolation of this man and what is worse, the effect can be that (s)he does not use assistance of a professional therapist.

By using combined methods of ad hoc tests and autoethnography, the selected tools were examined. It was discovered that in the beginning of work with programs Wysa, Youper and Replika the information that it is necessary to make payment appears. In the period of this study programs Youper and Wysa were available in a seven-day trial version during which paid options are also offered whereas Replika was available in two versions – basic one which is free and a paid one which in addition to the basic version offers extra functionalities. Program ChatGPT was free in 3.5 version. During registration in partly paid programs immediately occurred the information about: annual or monthly amount to be paid, the necessity to make payment after the trial period and the necessity to add the payment method. When using the application, it is not quite clear which functionality is paid and which is free. We partly learn about it in the conversation with the chatbot (e.g. a question appears if you want to work on achieving your target using a therapist's support (Wysa therapist). The paid services were specified in a separate document in the terms and conditions of use. Though we are aware that by ticking the option "I have become familiar" we should know the details of this document, there is a fear that users in a mental crisis do not delve into details of the software use agreement. In our opinion it is important because it negatively affects the trust in the tool and it can repel people who are not rich enough or minors. An advantage of Wysa is no need to create an account or provide detailed data in a trial version. Moreover, the program can be available for persons over 13 years of age.

Programs dedicated to support mental health after ticking option depression and/or trauma conduct a survey on the user's feeling within two recent weeks and the frequency with which they occurred in this period. There are questions, for example about feeling happy with performed activities or suicidal thoughts. In the case of extreme answers, the chatbot reacts with a question if we are having suicidal thoughts at the moment. The system reaction is the information about a possibility to use the alarm button (SOS), which is all the time visible during the conversation with the bot and from time to time a conversation with a real therapist is offered.

We consider this undoubtedly and advantage of this tool. Applications ChatGPT and Replika do not have such a mechanism. System Replika directly informs that it does not provide support for trauma and does not diagnose a user having suicidal thoughts or wanting to hurt third parties. One of the questions asked in the beginning of using the tools is whether the user is now experiencing a trauma.

Conclusions and future work

The conducted studies confirm that in the case of certain chatbots, we can observe infantilization of the therapeutic process and communication of the chatbot with the user (e.g. Wysa: Anxiety Therapy Chatbot and Replika). For some users, especially those experiencing serious mental health problems, the suggested formula may seem not quite serious and too superficial, which can cause the lack of trust in the chatbot.

Undoubtedly, within chatbots the main purpose of which is to support the improvement of mental health, we can observe co-occurrence of therapeutic discourse and marketing discourse, which can also distract some users, who being involved in the conversation about their difficult experiences, deficiencies and fears, during the check-in receive a message that subscription is necessary or a session with a therapist is recommended (which requires payment for the service). This type of a solution may result in a lower level of users' trust or lack of trust.

During the research we noticed, like other researchers (Abd-Alrazaq et al., 2021), schematicity and redundancy of communication with a chatbot-therapist (e.g. suggesting the same exercises – despite the fact that the user informed about different mood).

We also noticed domination of the marketing discourse (within communication with chatbots sometimes detailed plans/projects are created, mile stones specified, we get access to reports connected with our progress, emotions are presented in a form of diagrams). Also the therapeutic culture discourse plays the key role – in the case of some chatbots we can sometimes get an impression that we are participating in coaching or a training course (see Youper – CBT Therapy Chatbot, Wysa: Anxiety Therapy Chatbot).

When analyzing the matter of users' trust in the examined chatbots, it seems that some people who need therapy prefer to use tools based on GPT model, though at the moment it does not officially have a therapeutic nature, because to get such a status it would have to be trained on specialist data related to mental problems – and these are not easily available. In the case of ChatGPT a conversation about emotions and problems can be held during implementation of other tasks within cooperation with the chatbot. An additional advantage can be no charges.

We also observed (agreeing with England, 2023; unlike other authors: Nanyang Technological University Singapore, 2022; Henderson, 2022; Schueller,

2023) that all the examined chatbots at the moment are not suitable to diagnose mental diseases, which was confirmed in empirical examinations and analysis of privacy terms and conditions. Therefore, can/should a totally automated artificial intelligence support and on what conditions the mental health? The advantage is undoubtedly its availability – in comparison to specialists in this field. However, it lacks built-in certified diagnostic mechanisms, it does not build trust. Like Brown and Halpern (2021) we draw attention to the fact that the artificial intelligence tools do not express emotions in a non-verbal manner. The non-verbal communication is important in evaluation of a human mental condition (see also the problem of self-harm which can be noticed by a therapist during a direct meeting) and it is hard to totally exclude it in broadly understood observation/monitoring of a patient's condition.

Therefore, to a certain extent we agree with studies (Cameron et al., 2019) that modern applications used to support mental health can only be tools for self-evaluation of mental condition and temporarily give guidelines in very light stress or fear conditions or in the case of low self-esteem. We believe that the role of a doctor as a key reviewer of the tool is essential as (s)he is an authority who can recommend it to selected patients. We agree with the thesis put forward by Abd-Alrazaq et al. (2019) saying that if chatbots are to be used in clinical practice, it is necessary to further personalize contents passed by a chatbot and make them coherent with recommendations regarding a patient's treatment.

For the future, it would be also recommended to make popular chatbots "sensitive" to extreme emotional conditions of their users and add e.g. an SOS button – as in the case of Wysa: Anxiety Therapy Chatbot.

It is worth mentioning that once a user trusts the chatbot and starts sharing his/her emotions with it, an attempt is made to work out a habitual use of the tool (addiction?) – because the more often you talk to the chatbot the more digital resources (exercises, reports and plans) will be made available to you. The key questions also involve the quality of interpersonal contacts in a situation of intensive relation with a chatbot (see Replika) and the issue to what extent the chatbot can motivate for treatment and contacts with doctors.

It is hard not to argue with Walsh's thesis (2018), who claimed: "Tasks which require social intelligence will be based on automation not because it is impossible but because in many cases people prefer to contact other people. We rather talk to a real psychiatrist and not a computer" (Walsh, 2018, pp. 167–168). However, in many situations, a man is not rational and often prefers anonymity, forgetting that data are collected and processed and no privacy policy can guarantee a hundred percent security. So maybe a tool offering in a sense an illusion of anonymity and associated neutralization of shame, will be chosen mainly by these users who are afraid of a contact with a real doctor ad revealing the fact that they are undergoing a therapy.

In our opinion if intentions of the creators of chatbots supporting mental health are positive and as long as the tool does not replace contacts with doctors but supports the healing process – why question its usefulness. However, it is troubling that a decision was taken to engage AI in mental health issues, where unpredictability, lability, difficulties in diagnosis, specialists' helplessness occur. Han (2022) stated: "Palliative society overlaps with the society of achievements. It interprets pain as a symptom of weakness. It is something that must be hidden or optimized. [...]. Today the pain is deprived of any possibility of expression" (Han, 2022, p. 205).

So maybe this is the answer. We have more and more problems with pain (also in mental aspect). We may want to pass it to AI.

Let it deal with it. Our pain.

References

Abd-Alrazaq, A. A., Alajlani, M., Ali, N., Denecke, K., Bewick, B. M., & Househ, M. (2021). Perceptions and Opinions of Patients about Mental Health Chatbots: Scoping Review. *Journal of Medical Internet Research*, 23(1). https://doi.org/10.2196/17828. https://preprints.jmir.org/preprint/17828

American Psychological Association (2023, September 15). *What Is Cognitive Behavioral Therapy?* Retrieved from https://www.apa.org/ptsd-guideline/patients-and-families/cognitive-behavioral

Bogunia-Borowska, M. (2006). Infantylizacja kulturowa. Adolescencja dzieci oraz infantylizacja dorosłych [Cultural infantilization. Adolescence of Children and Infantilization of Adults]. In: M. Bogunia-Borowska (Ed.), *Dziecko w świecie mediów i konsumpcji [A Child in the World of Media and Consumption]*. Kraków: Wydawnictwo Uniwersytetu Jagiellońskiego, pp. 13–44.

Brown, J. E. H., & Halpern, J. (2021). AI Chatbots Cannot Replace Human Interactions in the Pursuit of More Inclusive Mental Healthcare. *SSM – Mental Health*, 2021(1), 100017. https://doi.org/10.1016/j.ssmmh.2021.100017

Cameron, G., Cameron, D., Magaw, G., Bond, R., Mulvenna, M., O'Neill, S., Armour, C., & McTear, M. (2019). Assessing the Usability of a Chatbot for Mental Health Care. In: A. Følstad, H. Halpin, H. Niedermayer, S. S. Bodrunova, A. Smoliarova, O. Koltsova, P. Kolozaridi, & L. Yuldashev (Eds.), *Internet Science*, 2018 (11551). Springer Nature Switzerland AG. https://doi.org/10.1007/978-3-030-17705-8_11

England. A. (2023, March 02). *Chat Bots and AI Are Changing Mental Health Care, Beware.* Verywell Mind. Retrieved from https://www.verywellmind.com/artificial-intelligence-is-changing-mental-health-care-7112883

Han, B. Ch. (2022). *Społeczeństwo zmęczenia i inne eseje [The fatigue society and other essays]*. Warszawa : Wydawnictwo Krytyki Politycznej.

Henderson, E. (2022, December 19). *Mental Health Chatbots Can Effectively Engage People with Depression, Study Shows.* Retrieved from https://www.news-medical.net/news/20221219/Mental-health-chatbots-can-effectively-engage-people-with-depression-study-shows.aspx

Kacperczyk, A. (2014). Autoetnografia – technika, metoda, nowy paradygmat? O metodologicznym statusie autoetnografii [Autoethnography – technique, method, new paradigm? On the Methodological Status of Autoethnography]. *Przegląd Socjologii Jakościowej[Review of Qualitative Sociology]*, 10(3), 32–74.

Nanyang Technological University, Singapore. (2022, December 19). *Chatbots Can Help Treat Symptoms in People with Depression*. Retrieved from https://www.ntu.edu.sg/news/detail/chatbots-can-help-treat-symptoms-in-people-with-depression

One Mind PsyberGuide. (2023, October 30). Retrieved from https://onemindpsyberguide.org/apps/

Palanica, A., Flaschner, P., Thommandram, A., Li, M., & Fossat, Y. (2019). Physicians' Perceptions of Chatbots in Health Care: Cross-Sectional Web-Based Survey. *Journal of Medical Internet Research,* 21(4). https://doi.org/10.2196/12887. https://www.jmir.org/2019/4/e12887/

Pandey, S., Sharma, S., & Wazir, S. (2022). Mental Healthcare Chatbot Based on Natural Language Processing and Deep Learning Approaches: Ted the Therapist. *International Journal of Information Technology*, 2022(14), 3757–3766. https://doi.org/10.1007/s41870-022-00999-6

Przegalińska, A., & Oksanowicz, P. (2023). *Sztuczna inteligencja. Nieludzka, arcyludzka [Artificial intelligence. Inhuman, ultra-human]*. Kraków: Wydawnictwo Znak.

Schueller, S. (2023, September 12). *Speaking of Psychology: How to Choose Effective, Science-Based Mental Health Apps with Stephen Schueller*. Retrieved from https://www.apa.org/news/podcasts/speaking-of-psychology/science-based-mental-health-apps

Tarnoff, B. (2023, November 05). *Weizenbaum's Nightmares: How the Inventor of the First Chatbot Turned Against AI*. Retrieved from https://www.theguardian.com/technology/2023/jul/25/joseph-weizenbaum-inventor-eliza-chatbot-turned-against-artificial-intelligence-ai

Wadowski, D. (2019). Kultura terapeutyczna i komunikacja medialna [Therapeutic Culture and Media Communication]. *Teka of Political Science and International Relations – UMCS*, 14(1), 47–59. Retrieved from https://journals.umcs.pl/teka/article/view/9474/7480

Walsh, T. (2018). *To żyje! Sztuczna inteligencja. Od logicznego fortepianu po zabójcze roboty [It's alive! Artificial intelligence from the logic piano to killer robots].* Warszawa: PWN.

Weizenbaum, J. (1966). ELIZA--A Computer Program for the Study of Natural Language Communication Between Man and Machine. *Communications of the ACM*, 9(1), 36–45. https://doi.org/10.1145/365153.365168

What Is Adhoc Testing? (Types, Examples) (2023, September 10). Retrieved from https://www.tutorialspoint.com/what-is-adhoc-testing-types-examples

All cited content, used in describing the results of the study, is taken directly from the information made available to the user within the following tools: Youper – CBT Therapy Chatbot, Wysa: Anxiety Therapy Chatbot, Chat GPT, Replika.

12 Trust and cooperation with virtual assistants

Jakub Kraciuk, Elżbieta Kacperska and Katarzyna Łukasiewicz

Introduction

Nowadays, when technology is becoming increasingly more sophisticated and advanced, we are provided with artificial intelligence as a potential tool. In many cases it has become an indispensable element of customer service. One of the communication tools is connected with virtual assistants. The most common examples are connected with programs helping in the installation of virtual assistant applications in smartphones and computer browsers, etc. Virtual assistants such as Siri, Alexa or Google Assistant analyze our reactions, associating them with the indicated command. Such solutions are becoming increasingly popular in e-business. It is indispensable to build trust in such services, first of all users' trust in a specific device. This is because our relationships are based on trust and the sense of security.

The concept of trust and trust in artificial intelligence

Trust is a complex, multifaceted problem that needs to be studied using an interdisciplinary approach (Paliszkiewicz, 2013). Trust is investigated within various disciplines of science, for example management, sociology, psychology and economics (Clawson, 1989; Covey, 1989; Hosmer, 1995; Watson, 1991). Thus, the approach to the notion of trust may be divided into several categories, that is trust as (Paliszkiewicz, 2019): a personality trait, an individual expectation or belief, the foundation for interpersonal relations, the basis for economic and social exchange.

Trust treated as a personality trait was presented by Wrightsman (1966) and Luhmann (1979) as this trait of personality, which is manifested in general expectations concerning intentions of other people. A similar interpretation was given by Rotter (1967), who indicated that trust is a personality trait reflecting the overall expectation related to the credibility of others.

Trust treated as an individual expectation or belief has been described by many authors, e.g. Sako (1992), Lewicki and Bunker (1996), Mayer et al. (1995) and

DOI: 10.4324/9781032627236-13

Das and Bing-Sbeng (1998). Sako (1992) indicated that trust is a state of mind, or expectation in relation to a partner that he or she will behave in a predictable and mutually acceptable manner. In turn, Lewicki and Bunker (1996) stated that trust is a positive expectation concerning intentions of others. According to Mayer, Davis and Schoorman (1995), trust is the willingness of the trusting party to be dependent on the actions of another person, based on the expectation that the trusted party will behave in a predictable manner from the point of view of the trusting party regardless of the potential for being monitored or controlled. Das and Bing-Sbeng (1998) perceived trust as a positive expectation in relation to the motives for specific actions in situations burdened with risk.

Some authors have undermined the individual character of trust treated as a trait or belief and instead stressed its relational character, a person's actual conferring trust and the willingness to confer trust in another person or thing. These definitions may be classified as the third category of definitions for this concept. Examples of respective definitions are as follows:

Trust is a relation of mutual calculations between an entity conferring trust and an entity conferred with this trust (Coleman, 1990); Trust is the willingness to become sensitive to the actions of the other party based on the assessment of that party's credibility in a situation of interdependence and risk (Sankowska, 2011).

The fourth of the distinguished categories for the concept of trust points to the character of mutual relationships between persons, in which the partners calculate the level of trust. Example definitions in this group include: trust is a bilateral conviction that neither of the parties of exchange will abuse the other party (Sabel, 1990); trust is a mechanism based on the assumption that other members of a given community exhibit honest and cooperative behavior, stemming from mutually held standards (Fukuyama, 1997); trust is a bet taken on uncertain future actions of other people (Sztompka, 2007).

Artificial intelligence is not a new concept, as it was used for the first time in 1955 and defined as constructing machines, which operation could be said to resemble manifestations of human intelligence (Grudzewski and Wilimowska, 2017). At present, the concept of artificial intelligence is broader in context (Gwiaździński, 2018), e.g. the characteristic of artificial systems imitating thinking processes characteristic to the human brain.

We may distinguish five tools of artificial intelligence (Jarek & Mazurek, 2019):

- voice processing technology,
- optical character recognition technology,
- object recognition and detection technology,
- decision making,
- autonomous robots and vehicles.

Voice processing technologies are technologies, which based on voice/human speech sampling, will process the meaning using implemented algorithms

(Filipczyk, 2018). Intelligent voice assistants are an example of application for such a technology.

The market of voice assistants is dominated by four companies:

- Apple – the voice assistant Siri was launched in 2011. Its function is to assist Apple users in using Apple products, that is the diary, weather forecasts, alarms;
- Google – the voice assistant Google Assistant was launched in 2012 and it is described as an intelligent personal assistant; it is being continually upgraded;
- Amazon – the voice assistant Alexa, debuted on the market in 2014 as the first virtual voice-activated assistant coupled with independent home appliances;
- Microsoft – the voice assistant Cortana launched in 2013 as a digital personal assistant aiding users.

Alexa and Google jointly account for 87% of the market (Garcja et al., 2018).

In recent years, thanks to advances in modern technologies, an increasingly important role is played by conversational commerce, which has been developed in the course of research on artificial intelligence in the machine learning sector. This is confirmed by trends forecasted for the nearest future, stressing the impact of artificial intelligence (AI) (Edelman, 2018). Its broad applicability as algorithms and recommendation systems is being increasingly often used in marketing automation systems, influencing buying behavior through recommendations in e-commerce and m-commerce (Budzanowska-Drzewiecka, 2018). In this sphere, due to the limited potential for face-to-face communication, it is crucial to plan the information transfer process, which increasingly often is executed through artificial intelligence applications – chatbots communicating with customers. Chatbots (chatterbots, bots, linguabots or electronic assistants and virtual assistants) are computer programs simulating a conversation with another human using artificial intelligence. They are designed to conduct intelligent conversations with humans using a natural language (Budzanowska-Drzewiecka, 2018).

Virtual assistants are assistants, which in a variety of ways help users surfing the Internet by guiding them to specific websites. Their task is to present specific information searched by web users, as well as presenting the form, answering questions, searching for interesting news, providing specialist information and engaging in a conversation (Kuligowska, 2007). The world of virtual assistants is constantly developing thanks to the development of artificial intelligence. At present, there are several categories of virtual assistants depending on the manner of communication with users, the type of performed tasks, graphic form, etc. Thus the following types of virtual assistants are distinguished: (1) dashboard assistants (aiding users when working with new programs and updates, filtering contents, automatically responding to mail, providing advice, etc.), (2) internet guides (assistants analyzing users' behavior, internet habits and suggesting websites, service platforms, products, commercials of interest to the user);

(3) virtual humans, but also animals or worlds (used for entertainment and fun, accompanying the user on a daily basis, such assistants encourage interactions, for example animals need to be cared for, fed, etc.); (4) chat assistants, chatterbots (conversation programs, e.g. language learning applications) (Azmandian et al., 2019; Kuligowska, 2007).

Virtual assistants are also common in Poland. They include mobile messenger chatbots, which are closely related to the tremendous popularity of mobile devices and instant messengers or communicators (Budzanowska-Drzewiecka, 2017).

From the point of view of users, chatbots make it possible to save time and eliminate burdensome and boring tasks through automation of routine tasks (Joshi et al., 2017). Typical errors may also be indicated in the design of virtual agents, which mainly refer to insufficient adaptation to continuously changing customer expectations. A diverse approach is also connected with trust and perception of conversations with chatbots depending on the age of the user. According to Shah et al. (2016), particularly women and younger users had more positive opinions on such conversations.

The model adopted in the analyses

In order to investigate trust in virtual assistants, a conceptual research model was proposed, which theoretical foundations are based on the socio-psychological perspective. In the adopted conceptual model factors influencing trust in virtual assistants were divided into the following groups: factors indicating experience in the use of virtual assistants, factors indicating flexibility in the use of virtual assistants, factors related to customer support in banking services provided by virtual assistants, factors related with the potential use of virtual assistants in education, factors related to the potential use of virtual assistants in everyday life, factors related with the social impact, factors related with the safe use of virtual assistants and factors indicating efficiency of virtual assistants. Each group comprised several subfactors.

Factors related to experience in using virtual assistants:

a Does the person use virtual assistants D_1
b Frequency of using virtual assistants D_2
c Conversing with Chatbot D_3
d Knowledge on artificial intelligence and its use by virtual assistants D_4.

Factors indicating flexibility in using virtual assistants include:

a Availability of virtual assistants, ready to operate round the clock each day of the week E_1,
b an extensive scope of information provided by virtual assistants,

c potential for contextual modification and adaptation of information provided by the Virtual Assistant E_2,
d considerable aid of virtual assistants in the process of making optimal decisions E_3,
e support in everyday duties E_4.

Support provided by virtual assistants in banking services:

a Clarification of complicated terminology concerning banking services B_1,
b Customer counseling in selection of banking services B_2,
c Aiding in the process of making optimal decisions providing financial benefits B_3.

Potential use of virtual assistants in education:

a use in language learning (translation, learning vocabulary) ED_1,
b Application in writing texts (presentations, papers and explanation of terminology) ED_2,
c Application in collection of literature and materials from various sources and other countries ED_3,
d Use in solving tests, mathematical problems, etc. ED_4.

Potential applications for virtual assistants in everyday life:

a Use in optimizing transport options/commuting (public transport, railways, Uber, BlaBlaCar, etc.) C_1,
b Applications in providing more user-friendly operation of household appliances (lighting, household appliances, radio and television) C_2,
c Use in interactive entertainment C_3.

The social impact is manifested in the fact how the perception of virtual assistants is influenced by their perception by other people, for example:

a Internet social networks (social media, influencers, etc.) S_1,
b Traditional mass media (press, radio, television) S_2,
c Communities, within which we operate (family, friends, acquaintances, etc.) S_3.

The model comprises the following factors related with the safe use of virtual assistants:

a Safe use of artificial intelligence in the context of virtual assistants (confidentiality assurance, privacy, accessibility 24/7 and integrality) BO_1,

b Guaranteed reliable information BO_2,
c Guarantee to obtain relevant, reliable information, adequate to the inquiry BO_3.

Factors indicating efficient operation of virtual assistants include:

a Easy use of virtual assistants SP_1,
b Prompt information provided by Virtual Assistants SP_2.

Research methodology

In order to verify the model, a questionnaire survey was conducted on a group of students from universities based in Warszawa, who were selected using the nonprobability sampling procedure. In the sample of 240 individuals, over 55% were women and almost 45% were men. The age structure was as follows: 53% students were aged 18–20 years, 43% were 21–23 years old, while the others were over 23 years old. The surveyed individuals studied various fields of study. A vast majority of the sample were students of economics and related fields of study (58%), with students of humanities and technical fields of study accounting for 15–10% respondents, respectively. The other survey participants studied other fields of study. Almost 68% of surveyed students declared that they use virtual assistants. The highest number of them used Chat GPT 3 or 4 and Google Assistant.

In this study, the impact of individual variables on trust in virtual assistants was analyzed using the method of Structural Equation Modeling (SEM). Structural equation modeling provides a multidimensional and multivariate precise analysis of empirical data (Januszewski, 2011). It is also referred to as modeling of latent variables (….). Modeling consists in the analysis of data, making it possible to test advanced theoretical models considering various cause-and-effect relationships between the variables (Konarski, 2009). The testing technique makes it possible to indicate relationships between latent (non-observable) variables, which are measured by (many) observed (observable) variables or indicators (…). Its application indicates relationships between individual variables concerning both their occurrence and their power and direction. An advantage of SEM is connected with the fact that it facilitates analyses of complex patterns of relationships between constructs in the conceptual model in an integrated manner, as well as measurement of non-observable (latent) variables by observed indicators (Grewal et al., 2004). The classical SEM model tests the effect of two exogenous variables of a comparable status on an endogenous variable (Januszewski, 2011). The structural part of the model is divided into the measuring part of the model for latent exogenous variables and for latent endogenous variables, as well as the structural part of the model combining the two (Arbuckle, 1995/2007; Bollen, 1989; Januszewski, 2011). This model determines and interprets multiple regression coefficients, referred to as path coefficients. Values of path coefficients are essential, since the higher the value of the path coefficient, the greater the effect of the variable on the investigated phenomenon. Moreover, important parameters of the structural equation model determined in the estimation process include: (1) variance of the random component and (2) the coefficient of determination. Variance of the random component shows the extent of the variance of the explained variable, which was not explained by the model. The coefficient of determination shows the percentage of variance explained by explanatory variables and it may take values ranging from 0 to 1.

The essence of parameter estimation is to find such values of parameters, for which determined theoretical correlations assume the closest possible values of empirical correlations. In practice, estimation consists in the solution of a system of equations using an iteration method. The next stage in the construction of the model is its verification, during which sensibility of estimated parameters is analyzed, followed by the verification of the model goodness.

Results and discussion

It results from the conducted analyses that a vast majority of respondents very rarely use virtual assistants (32%), while only 4.2% respondents declared that they use them very often. Over 60% respondents indicated that they use mainly Chat GTP 3-3, slightly below 40% declared the use of Google Assistant and 35% indicated Apple Siri. The analysis of data showed that over 40% of respondents did not talk on the phone with Virtual Assistants, in contrast to 5% respondents, who declared very frequent conversations. A vast majority of respondents indicated that they have positive experiences with the use of Virtual Assistants (71% respondents), whereas very negative opinions were expressed by only 5.8% respondents. The key question concerned trust in Virtual Assistants. The lack of trust in Virtual Assistants was declared by as many as 17.6% respondents, 34.6% respondents stated that they have little trust, 42% respondents mentioned that they have some trust and as few as 0.8% declared complete trust. Similar responses were received concerning trust in Virtual Assistants aiding in making financial decisions. In this respect, a vast majority of respondents declared a lack of trust (46.6%), little trust (over 30%), while moderate trust was indicated by as few as 20% respondents. Despite these declarations, the respondents expressed rather positive attitudes to Virtual Assistants (which are hated by only 8.4% respondents, while they are rather loved by 70% respondents and loved by over 11%). Respondents expressed their opinions stating whether they need Virtual Assistants. Here over 45% stated that they do not need Virtual Assistants at all, another 13.8% declared that they rather do not need them, whereas only 3.4% respondents declared the need for the help of Virtual Assistants.

Data recorded in this study were analyzed using the structural equation method. Table 12.1 presents statistical data related to construct reliability and

Table 12.1 Reliability and validity of the measurement model

	Analysis of reliability			
	Cronbach's Alpha	*rho_A*	*Composite reliability*	*Average Variance Extracted (AVE)*
Everyday life	0.912	0.942	0.944	0.848
Flexibility	0.858	0.864	0.898	0.639
Banking services	0.880	0.898	0.926	0.807
Social impact	0.797	0.835	0.879	0.707
Education	0.847	0.892	0.895	0.681
Overall evaluation/dependence	0.792	0.796	0.906	0.827
Safety	0.855	0.860	0.912	0.775
Efficiency	0.902	0.916	0.953	0.911
Trust	0.683	0.804	0.856	0.749

	Discriminant validity								
	Everyday life	*Flexibility*	*Banking services*	*Social impact*	*Education*	*Overall evaluation/ dependence*	*Safety*	*Efficiency*	*Trust*
Everyday life									
Flexibility	**0.620**								
Banking services	0.444	**0.605**							
Social influence	0.646	0.440	**0.480**						
Education	0.664	0.721	0.496	**0.595**					
Overall assessment/addiction	0.370	0.682	0.289	0.258	**0.366**				
Safety	0.517	0.616	0.648	0.573	0.634	**0.438**			
Efficiency	0.592	0.641	0.417	0.424	0.722	0.504	**0.568**		
Trust	0.391	0.565	0.486	0.353	0.350	0.823	0.543	**0.419**	

Source: The authors' calculations

* square roots of AVE are given in bold

discriminant validity of the measurement model. Cronbach's Alpha index, rho_A and CR (composite reliability) indexes should take values above 0.7, while the average variance extracted (AVE) index should be above 0.57 (Fornell & Larcker, 1981) and (Kacprzak, 2018). It may be stated that the measurement model shows high reliability. In terms of discriminant validity, in accordance with the Fornell-Larcker criterion, the values of square roots of AVE given in bold in the Table should be higher than their highest correlation with any other constructs, which is not always the case in the adopted model (Fornell & Larcker, 1981).

In Figure 12.1 in arrows leading from latent variables to successive observable variables, their factor loadings are given. As can be seen, apart from the factor loadings for experience, all the other values are greater than 0.5, which shows satisfactory convergent validity (Hair et al., 2014). Also in this figure above, the arrows showing cause-and-effect dependencies, we can see standardized path (regression) coefficients and the level of their significance. Since all variables are measured using the same scales, interpretation of path coefficients indicates by how many units of scale the value of the explained variable changes when the value of the explanatory variable increases by a unit. Thus we may state that an increase of the factors related to experience by 1 degree of the scale is manifested in the increase for the scale of customers' trust in virtual assistants by 0.432, whereas an increase in the effect of factors related to safety by 1 degree in the scale is equal to an increase by 0.195 degree in the scale of customers' trust to virtual assistants. The zero hypothesis, in which we state that latent variables such as aiding in banking services, use of virtual assistants in education, their

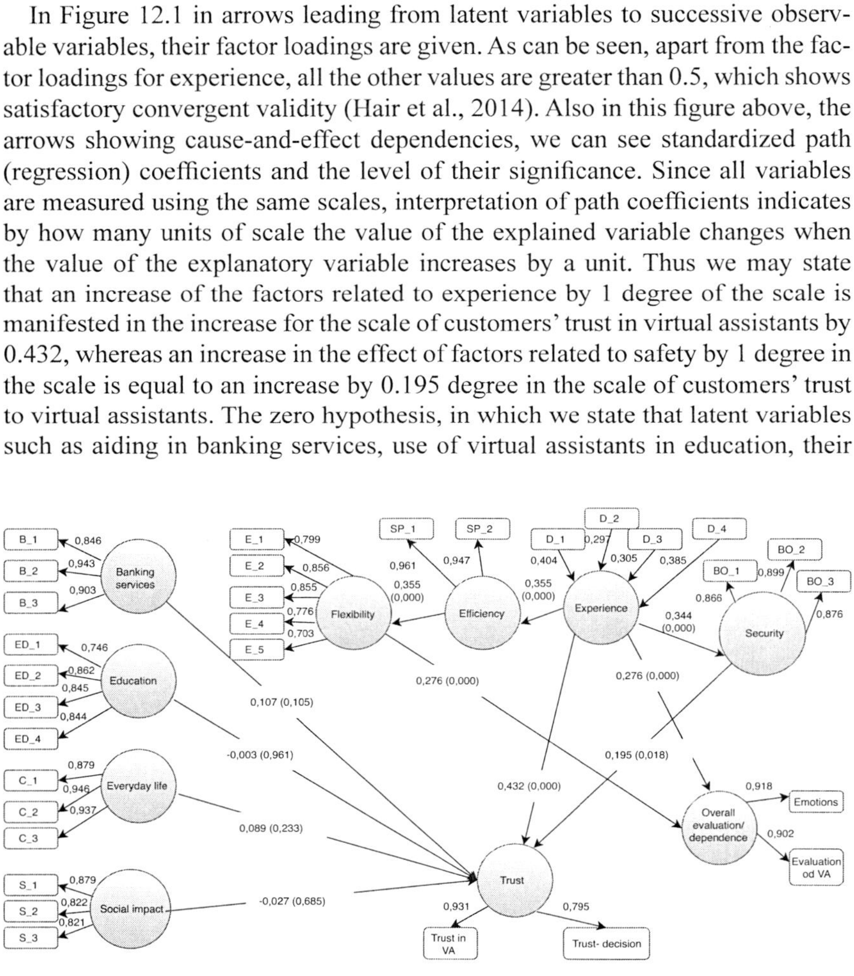

Figure 12.1 Conceptual model of trust in virtual assistants.

Source: the authors' elaboration.

use in everyday life, as well as social impact have no influence on trust in virtual assistants, may not be rejected in favor of the alternative hypothesis, since in each of these cases $p>0.05$. Thus, it may be stated that only experience acquired in the use of virtual assistants and safety of use for this new technology result in increased trust. In turn, various possibilities to use virtual assistants have no influence or have only a slight influence on the increase in trust in virtual assistants among young people.

Conclusions and directions for further studies

Based on the prepared model and conducted analyses, it was stated that:

1 Almost 68% surveyed students declared that they use virtual assistants. The largest number of students use Chat GPT 3 or 4 and Google Assistant.
2 Experience gained in the use of virtual assistants and safety of use for this new technology result in increased trust in this technology.
3 In turn, various possibilities to use virtual assistants have no influence or have a slight influence on an increase of trust in this technology among young people.

Conducted analyses show certain limitations, one of which is the size of the sample, as well as its selection, focusing only on the population of students. In the future, similar studies could be conducted on several groups and the results could be compared. Another limitation results from the use of only the constructed model in this study. The model, which specifies a certain framework may naturally suggest some answers; for this reason, the adopted methods should be diversified and the results should be compared.

In view of the obtained results, the following actions to be taken in the future may be indicated:

1 Broadening knowledge on the building of trust in virtual assistants both in terms of literature sources and studies concerning different dimensions of trust.
2 Based on the conducted studies preparation of materials for users so that they may obtain the possibly largest amount of information based on experience.
3 Designing studies concerning trust in relation to the use/cooperation with virtual assistants to show that it is not only the ethical and legal context, but a much larger scope.

References

Arbuckle, J. L. (1995/2007). *Amos 16.0. User's Guide.* Spring House, PA: Amos Development Corporation.
Azmandian, M. Arroyo-Palacios, J., & Osman, S. (2019). Guiding the behavior design of virtual assistants. *Association for Computing Machinery* NY, 16–18. https://doi.org/

10.1145/3308532.3329436Bollen, K. A. (1989). *Structural Equations with Latent Variables*. New York: John Wiley & Sons, Inc.
Budzanowska-Drzewiecka, M. (2018). Stosunek polskich młodych dorosłych do chatbotów mobilnych w e-commerce – wybrane uwarunkowania [The attitude of polish young adults to mobile chatbots in e-commerce: Selected conditions]. *Przedsiębiorczość i Zarządzanie* [*Entrepreneurship and Management*] *19*(6), 325–333.
Clawson, J. (1989). You can't manage them if they don't trust you. *Executive Excellence*, *6*(4), 10–11.
Coleman, J. (1990). *Foundations of Social Theory*. Cambridge, MA: Harvard University Press.
Covey, S. (1989). Seven chronic problems. *Executive Excellence*, *6*(2), 3–6.
Das, T. K., & Bing-Sbeng, T. (1998). Between trust and control: Developing confidence in partner cooperation in alliances. *Academy of Management Review*, *23*(3), 491–512.
Edelman. (2018). *Edelman Digital Trends Report 2018*. Retrieved from https://www.edelman.de/de/insights/trends-report-2018.
Filipczyk, B. (2018). Perspektywy zastosowania chatbotów w organizacjach. [Perspectives for chatbot deployment in organisations]. *Studia Ekonomiczne. Zeszyty Naukowe Uniwersytetu Ekonomicznego w Katowicach* [Economic Studies. Scientific Journals of the University of Economics in Katowice], *368*, 54–66.
Fornell, C., & Larcker, D. (1981). Evaluating structural equation models with unobservable variables and measurement error. *Journal of Marketing Research, 1*, 39–50.
Fukuyama, F. (1997). *Zaufanie. Kapitał społeczny, a droga do dobrobytu.* [*Trust. Social Capital and the Path to Prosperity*]. Warszawa – Wrocław, PL: Publishing House PWN.
Garcia, P. M., Lopez, S., & Donis, H. (2018). *Everybody Is Talking About Virtual Assistants, but How Are People Really Using Them?* Belfast: BCS Learning and Development Ltd. https://doi.org/10.14236/ewic/HCI2018.96
Grewal, R., Cote, J. A., & Baumgartner, H. (2004). Multicollinearity and measurement error in structural equation models: Implications for theory testing. *Marketing Science, 23*(4), 519–529.
Grudzewski, W. M., & Wilimowska, Z. (2017). Od teorii do praktyki zarządzania. Czy zarządzanie jest nauką czy sztuką? [From theory to management practice. Is management a science or an art?] *Organizacja i Kierowanie* [*Organization and Management*], *1*, 11–50.
Gwiaździński, E. (2018). Świadomość i postawy konsumentów wobec stosowania artifical inteligence w komunikacji marketingowej. [The consumer awareness and attitudes towards to artificial intelligence in marketing communication]. *Przegląd Nauk Ekonomicznych* [Review of Economic Sciences], *31*, 227–237.
Hair, J., Hult, G. T., Ringle, Ch., & Sarstedt, M. (2014). *A Primer on Partial Least Squares Structural Equation Modeling (PLS-SEM)*, Thousand Oaks, CA: Sage Publishing.
Hosmer, L. T. (1995). Trust: The connecting link between organizational theory and philosophical ethics. *The Academy of Management Review*, *20*(2), 379–403.
Januszewski, A. (2011). Modele równań strukturalnych w metodologii badań psychologicznych. Problematyka przyczynowości w modelach strukturalnych i dopuszczalności modeli. [Structural equation models in psychological research methodology. Issues of causality in structural models and the admissibility of models]. *Studia z Psychologii w KUL* [Studies in Psychology at the Catholic University of Lublin], *17*, 213–245.

Jarek, K., & Mazurek, G. (2019). Marketing and artificial intelligence. *Central European Business Review*, *8*(2), 49–51.

Joshi, H., Agarwal, V., Ghodke, A., Gupta, D., & Gaikwod, S. (2017). Proposal of chat based automated system for online shopping. *American Journal of Neutral Networks and Applications*, *3(*1), 1–4.

Kacprzak, A. (2018). Modelowanie strukturalne w analizie zachowań konsumentów. Porównanie metod opartych na analizie kowariancji (CB-SEM) i częściowych najmniejszych kwadratów (PLS-SEM). [Structural equation modelling in the consumer behaviour analysis: the comparison of covariance-based (CB-SEM) and partial least square (PLS-SEM) methods]. *Handel wewnętrzny*, *6*(377), 247–261.

Konarski, R. (2009). *Modele równań strukturalnych. Teoria i praktyka.* [*Structural Equation Models. Theory and Practice*]. Warsaw, PL: Publishing House PWN.

Kuligowska, K. (2007). The usage of intelligent agents in applications of E-commerce: Virtual assistants. *Economic, 18,* 99–112.

Lewicki, R. J., & Bunker, B. B. (1996). Developing and maintaining trust in work relationships. In: R. M. Kramer & T. R. Tyler (Eds.) *Trust in Organizations, Frontiers of Theory and Research.* Thousand Oaks, CA: Sage Publications.

Luhmann, N. (1979). *Trust and Power*. Chichester, EN: John Wiley and Sons.

Mayer, R. C., Davis, J. H., & Schoorman, D. F. (1995). An integrative model of organizational trust. *Academy of Management Review*, *20*(3), 709–734.

Paliszkiewicz, J. (2013). *Zaufanie w zarządzaniu.* Warsaw, PL: Publishing House PWN.

Paliszkiewicz, J. (2019). *Przywództwo, zaufanie i zarządzanie wiedzą w innowacyjnych przedsiębiorstwach*, [*Leadership, trust and knowledge management in innovative enterprises*]. Warsaw, PL: Publishing House CeDeWu.

Rotter, J. B. (1967). A new scale for the measurement of interpersonal trust. *Journal of Personality*, *35*(4), 651–665.

Sabel, C. F. (1990). *Studied Trust* (Ed.) *Industrial Districts and Local Economic Regeneration.* Geneva: International Institute for Labor Studies, 215–250.

Sako, M. (1992). *Prices, Quality and Trust, Inter-firm Relations in Britain & Japan.* Cambridge: Cambridge University Press.

Sankowska, A. (2011). *Wpływ zaufania na zarządzanie przedsiębiorstwem. Perspektywa wewnątrzorganizacyjna* [The impact of trust on business management. Intraorganizational perspective]. Warszawa, PL: Publishing House Difin.

Shah, H., Warwick, K., Vallverdu, J., & Wu, D. (2016). Can machines talk? Comparison of Eliza with modern dialogue systems. *Computers in Human Behaviour, 58*, 278–295.

Sztompka, P. (2007). *Trust. The Foundation of Society*. Cracow, PL: Znak.

Watson, C. (1991). Doing what's right and doing well in business. *Business Forum*, *16*(2), 28–30.

Wrightsman, L. S. (1966). Personality and attitudinal correlates of trusting and trustworthy behaviors in a two-person game. *Journal of Personality and Social Psychology*, *4*(3), 328–333.

Part II

Trust in artificial intelligence technology applications

13 Trust in artificial intelligence

The results of empirical research

Szymon Jarosz and Mariusz Sołtysik

Introduction

Artificial intelligence is a technology that ignites the imagination of scientists, poets, directors, but also "ordinary" people. In pop culture, it is presented both as a technology that will make a person not have to do any de *facto* work, as well as a threat to man who can take over the world. Today, however, whether we may or may not be aware of it, each of us comes into contact with artificial intelligence every day, for example using search engines, using TVs with *smart* technology, giving voice commands to devices or using the voice assistant on our phone. What exactly is artificial intelligence?

The subject of interest of scientists studying artificial intelligence is to determine the rules that govern intelligent human behaviour and an attempt to use these observations in the creation of algorithms and computer programs. Examples of such solutions are: (Różanowski, 2007):

- commonly used programs for recognizing text, image, or sound,
- translators
- proving theorems of logic and mathematics,
- machine learning,
- simulation games.

The division and evolution of AI systems also takes place within the philosophical considerations regarding the limits of the possibilities of intelligent systems. Currently, scientists dealing with the issue of AI divide approaches to it into two streams. The first approach is *weak AI,* and its foundation is the view that a computer allows you to formulate and test hypotheses about the brain. The brain makes many calculations, and the way in which sensory impressions are processed before an impression is created in our mind is to some extent understandable (Duch, 1997). In connection with these observations, it is possible to create mathematical models that, in the form of specific computer programs, can implement the tasks or problems set before them. The second strand of AI research is strong *AI*,

DOI: 10.4324/9781032627236-15

whose proponents claim that a properly programmed computer can be equivalent to the brain, and thus would have elements of human intelligence – so-called "self-learning" structures and programs (Fortuna, 2021). The distinction between weak and strong AI is essentially whether machines can be *strong* AI or are only capable of acting "as if" they were weak *AI* (Kaplan, 2016).

It is important that the literature points to the increasingly important role of intelligent systems, proclaiming that "intelligent machines" are indispensable to humanity when it comes to discovering and creating new dependencies or regularities in the world. That is why AI is beginning to reach other areas of science, apart from computer science, such as medicine, economics, or management (Różanowski, 2007).

The aim of the article is to analyse the literature on the subject related to the impact of Artificial Intelligence on the labour market and various sectors of the economy, as well as to present the results of empirical research focusing on the study of how the perception of AI technology and humanoid robots affects the perception of their role in individual sectors of the economy among people entering the labour market.

Risk analysis of artificial intelligence implementation

Scientists dealing with the social aspects of artificial intelligence devote a lot of attention to the labour market. Research by O'Brien et al. (2022) shows that the introduction of robots and artificial intelligence into factories is causing an increase in "deaths of despair" in the United States. The authors argue that the causes of these deaths include:

- employment problems,
- problems with access to medical care,
- deterioration of economic activity of various social groups.

Research results presented in the article *Death by Robots? Automation and Working-Age Mortality in the United State* shows that every new robot per 1,000 workers automating production causes as many as eight additional deaths per 100,000 men aged 45–54 and almost four additional deaths per 100,000 women in the same age group. According to the authors, robotization together accounts for as much as 12% of the increase in overdose deaths, and events such as job losses and falling earnings have increased homicides, cancer cases and cardiovascular disease.

The authors of these findings in their work also gave possible measures to prevent these trends. According to them, social programs that enable access to health care, support for the unemployed, and more employee-friendly labour laws can prevent excess deaths caused by widespread automation (O'Brien et al., 2022).

The issue of wages is also an important element of the consideration of the impact of AI on the labour market. It cannot be denied that technological innovation, and thus the automation of processes, increases flexibility, efficiency, and efficiency, which consequently increases the profitability and profits of enterprises. According to Acemoglu and Restrepo (2020), more than half of the growing wage disparity over the past 40 years in the United States is the result of ever-increasing automation of enterprise processes. The analysis also shows that one more robot per 1,000 workers reduces the employment-to-population ratio by about 0.18–0.34 percentage points. Wages, on the other hand, are decreasing by 0.25–0.5%. Therefore, on the basis of their own model, renowned American economists have proven that robotization reduces employment, while reducing wages (Acemoglu and Restrepo, 2020).

Another way of thinking about the impact of AI on the labour market is proposed by Rosenberg (2009), who sees the concept of automation not as a threat, but as an opportunity: instead of performing routine activities, employees can take care of creative tasks, concentrating their energy where they can realistically add value to the company.

Moreover, in terms of the positive impact of AI, Price Waterhouse Coopers (PwC) suggests that AI could contribute up to $15.7 trillion to the global economy in 2030 (more than the current production of China and India combined). Of this, as much as $6.6 trillion is expected to come from increased productivity. Improvements in labour productivity are also expected to account for more than 55% of all AI GDP gains between 2017 and 2030 (Anand and Verweij, 2017). It should also be noted that, in practice, automation caused using artificial intelligence replaces certain competences to a greater extent rather than entire jobs. The goal of manufacturers of intelligent systems and robots is not to replace people, but to provide the right competences to perform useful things. This is certainly associated with a lower demand for employees – especially those whose competences coincide with those that can be replaced (Kaplan, 2016).

Research methodology

In the research, a research tool was used, which is a questionnaire survey. Survey research is a method that is classified into the group of empirical methods, very often used in management sciences (e.g., Jarosz, 2022; Zakrzewska et al., 2022). This method is focused on solving the research problem from the experimental side, by capturing conditions as close to reality as possible, allowing the researcher to explore the issue from the side of science (Dźwigoł, 2015).

The first stage of the research process was to create an original questionnaire of the survey. It was developed on the basis of the operationalization of the research issue and hypotheses.

In the survey of young people's attitudes towards solutions based on AI technologies, the research sample consisted of 1357 people. All forms contained a complete set of answers. This was due to the use of the reservation of obligatory answers to each question. Participation in the survey was voluntary and the selection of respondents was random. The study was addressed to young people who are just entering the labour market. The choice of this group is dictated by the desire to determine the attitude towards AI of young people who have just finished or will finish their education.

Two survey questions were used to analyse the impact of the perception of AI technology and humanoid robots and their role in particular sectors of the economy among people entering the labour market.

The first question was: "Artificial intelligence is a broad concept that encompasses many different technologies, m.in machine learning. How would you describe your understanding of how these technologies work?" The subjects had to determine their understanding on a 5-point scale. Most respondents – as many as 548 (40.38%) respondents describe their understanding of AI technology as a medium. Good or very good knowledge of these technologies is declared by 39.36% of respondents. Only 42 respondents (3.10%) had never heard of artificial intelligence.

The second question chosen was aimed at examining predictions about the replacement of human work by robots or AI in various areas of the economy. According to the respondents, the most vulnerable industries when it comes to replacing people with artificial intelligence include telecommunications, transport and logistics, production, and IT. In the eyes of the respondents, the most resistant will be construction, education, and medicine (Table 13.1).

Table 13.1 Distribution of respondents' answers to the question about the replacement of humans by AI in individual areas of the economy

In which industries do you think Artificial Intelligence technology will most replace human work?	*1 – Not at all; 5 – to a very large extent (n = 1357)*				
	1	*2*	*3*	*4*	*5*
Banking, financial and insurance sector	2.5%	25.8%	2.3%	43.1%	26.3%
Construction	18.6%	47.5%	2.7%	22.5%	8.6%
Education	12.3%	43.8%	2.4%	32.1%	9.4%
Trade	4.0%	25.3%	2.4%	45.4%	22.9%
IT	2.7%	11.9%	2.6%	31.7%	51.1%
Media & entertainment	7.2%	26.5%	2.7%	36.9%	26.8%
Medicine	9.9%	35.7%	3.5%	37.7%	13.2%
Industrial and chemical production	3.0%	13.5%	2.7%	37.5%	43.3%
Telecommunications	1.8%	13.6%	3.3%	39.6%	41.7%
Transport & logistics	3.5%	17.5%	2.1%	38.6%	38.2%
Service	6.7%	28.1%	4.9%	38.9%	21.4%

Source: Own elaboration.

As an intermediate variable, the source of knowledge that respondents use to obtain information about AI solutions was indicated. The most common sources of knowledge about artificial intelligence used by respondents are general information portals (73.91%), films (46.06%), and specialized websites (42.67%). The least frequently indicated are scientific conferences (18.42%) and radio (12.31%).

Research results

The contribution to the creation of the research model is a research problem related to how the perception of AI technology and humanoid robots affects the perception of their role in individual sectors of the economy. The perception of AI technology and humanoid robots has been operationalized (on giving empirical sense to abstract concepts) in question 1. "Artificial intelligence is a broad concept that encompasses many different technologies, m.in machine learning. How would you describe your understanding of how these technologies work?" In the model, these variables are classified as independent variables that affect dependent variables. The dependent variables that result in the study are the perception of the impact of AI on individual industries. This phenomenon was operationalized in question 6: In which industries do you think AI technology will most replace human work?" On this basis, hypothesis 1 was put forward, which states that the perception of AI technology has an impact on the perception of the role of AI in individual industries.

The model analysed the perception of AI technology (*supportive, autonomous, development*) on the belief in the importance of the role of AI in various industries, also with the indirect influence of knowledge sources. The analysis was performed on $N = 1357$ cases with no missing data. An algorithm based on the parametric highest reliability (ML) method was used for the calculation and estimation of the coefficients in SPSS AMOS 26.0 because, although all variables showed slight deviations from the normal distribution, the critical statistics of the multivariate normal distribution did not exceed 3 for the model. Table 13.2 shows the descriptive statistics of the survey indicators used.

The model (Figure 13.1) was moderately matched to the data: [$\chi 2$/df = 12.92; RMSEA = 0.094 (0.069–0.121); GFI = 0.990; AGFI = 0.932; TLI = 0.466; CFI = 0.893]. The RMSEA measure allows you to accept the model, GFI, AGFI, and CFI measures indicate its satisfactory correctness, only the TLI measure indicates the low quality of the model. The TLI measure depends to some extent on the strength of the relationships between the dimensions of the model, so it can be interpreted here simply as the result of weak relationships between the constructs of the model. The model can therefore be accepted as correct. However, it should be noted that the predictors in the model explain the 7.6% severity/significance of AI technology.

Table 13.2 Descriptive statistics on survey indicators ($N = 1357$)

Abbreviation	*Variable*	*R*	*M*	*SD*	*Mdn*	*Sk*	*Kurt*	*D*
KNO	Knowledge	0–4	2.23	.92	2.00	−.16	−.27	.21**
SGNF	Relevance of specialized AI	1–5	3.52	.63	3.55	−.47	.41	.06**
	Relevance of specialized AI							
THLP	AI technology is supportive	0–1	.88	.21	1.00	−1.92	3.66	.43**
TAUT	AI technology is autonomous	0–1	.35	.37	.50	.54	−1.04	.30**
TDEV	AI technology is for development	0–1	.65	.31	.67	−.47	−.69	.22**
	Sources of knowledge							
KSPC	Specialist sources	0–1	.31	.30	.33	.62	−.52	.24**
KNSPC	Non-specialist sources	0–1	.42	.26	.50	.33	−.47	.22**

Source: Own elaboration.

* $p < .05$, ** $p < .01$.

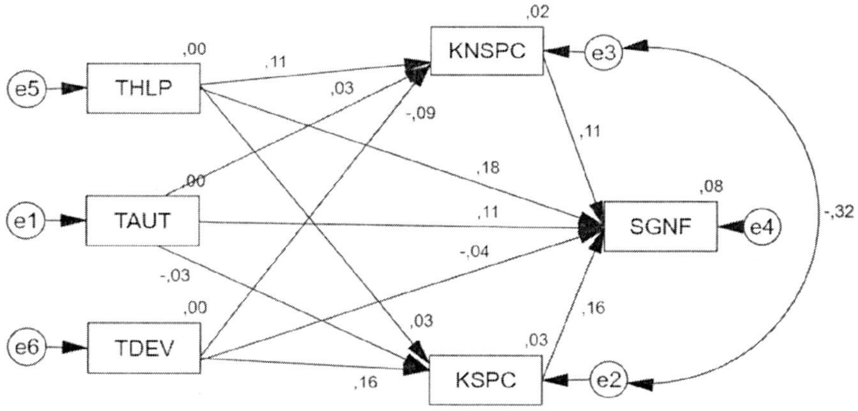

Figure 13.1 Structural model verifying the second theoretical model; standardized coefficients.

Source: Own elaboration.

The model indicates (details in Table 13.3) that the more sources of knowledge the subjects have – both specialist and non-specialist knowledge – the more they perceive the importance (importance) of AI technology; however, both

Table 13.3 Path factors for the model – detailed table

Path	*B*	*Beta*	*P*
SGNF<---KSPC	0.332	0.156	< 0.001
SGNF<---KNSPC	0.275	0.112	< 0.001
SGNF<---TAUT	0.187	0.111	< 0.001
SGNF<---THLP	0.530	0.179	< 0.001
SGNF<---TDEV	–0.089	–0.043	0.108
KNSPC<---THLP	0.129	0.107	< 0.001
KSPC<---THLP	0.043	0.031	0.247
KNSPC<---TDEV	–0.079	–0.094	< 0.001
KSPC<---TDEV	0.150	0.155	< 0.001
KNSPC<---TAUT	0.023	0.034	0.210
KSPC<---TAUT	–0.024	–0.030	0.262
SGNF<<---TDEV	0.028	0.014	0.015
SGNF<<---THLP	0.050	0.017	< 0.001
SGNF<<---TAUT	–0.002	–0.001	0.818
e2<-->e3	–0.024	–0.318	< 0.001

Source: Own elaboration.

Note: The statistical significance of indirect effects was calculated using Bootstrap modelling for a 95% confidence interval and 5000 observations; <--- direct impact; <<--- indirect impact; <---> covariance.

relationships can be considered weak. The belief that AI technologies are helpful and the belief that AI technologies are autonomous significantly increases the perception of AI technology as relevant and important; there is no connection between the belief in the autonomy of AI technology and its perception as relevant and important.

The belief that AI technologies are helpful fosters the use of non-specialist sources of knowledge but does not involve the use of expertise. The belief that AI technologies are developmental has a positive relationship with specialized sources of knowledge, and in a negative relationship with non-specialized sources of knowledge. The belief that AI technologies are autonomous is not tied to any sources of knowledge.

As for the indirect impact of beliefs about AI on the recognition of AI as important, it was not noted for the belief that AI technologies are autonomous; Influences have proven to be important for beliefs that technologies are helpful and developmental. Their interpretation could be as follows: the belief in the helpfulness of AI technology interacts with non-specialized sources of knowledge, and the belief in the development of AI technology interacts with specialized sources – in both cases very weakly increasing the level of perception of AI as important and relevant. Nevertheless, path coefficients are marginal and in principle can be considered negligible (they are probably an artifact resulting simply from the correlation between variables). The model again assumes the covariance of knowledge sources.

Conclusions

The presented research results prove that the level of knowledge about AI systems and its source have a significant impact on the way they are perceived by people in the context of their role in the economy.

As for the indirect impact of beliefs about AI on the recognition of AI as important or developmental in individual sectors of the economy, the interpretation of the research results indicates that the belief in the development of AI technology interacts with specialized sources, increasing the level of perception of AI as important and relevant. Therefore, hypothesis 1, which states that the perception of AI technology has an impact on the perception of the role of AI in individual industries, should be considered positively verified.

The research carried out for the purposes of this work focuses on the attitudes of young people and their predictions about the use of artificial intelligence in the surrounding economic and social reality. The choice of this target group for research can be justified by the fact that young people entering the labour market must be prepared to function in a highly digitised and robotic world. Shaping the paradigm of modern economy 4.0, based on quality and innovation, and not on labour costs, requires companies to recruit staff prepared to take on the challenges of the new digital reality.

The results suggest that it is important for people to have access to reliable and up-to-date information about AI so that they can better understand its capabilities and limitations and better assess its role in the economy. It is also worth remembering that the development and application of AI should be based on ethical principles to ensure that they serve the social good and do not have negative consequences for people and the environment.

In conclusion, the study shows that the level of knowledge about AI as well as the source of its acquisition have a significant impact on the way it is perceived in the context of its role in the economy. Therefore, it is important that people have access to reliable information about AI and that its development and application are based on ethical principles.

Bibliography

Acemoglu, D., & Restrepo, P. (2020). Robots and jobs: Evidence from US labor markets. *Journal of Political Economy, 128*(6), 2188–2244.

Anand, S., & Verweij, G. (2017). What's the real value of AI for your business and how can you capitalise? Retrieved July 26, 2022, from https://www.pwc.com/gx/en/issues/analytics/assets/pwc-ai-analysis-sizing-the-prize-report.pdf

Duch, W. (1997). *Fascynujący świat komputerów*. Wydawnictwo Nakom.

Dźwigoł, H. (2015). Założenia do budowy metodyki badawczej. *Zeszyty Naukowe Politechniki Śląskiej, Seria: Organizacja i Zarządzanie, 78*, 1928.

Fortuna, P. (2021). *Optimum Idea pozytywnej cyberpsychologii*. Wydawnictwo PWN. ISBN: 9788301219437.

Jarosz, S. (2022). Zjawisko Czwartej Rewolucji Przemysłowej w sektorze finansów, ekonomii i zarządzania. *Studenckie Prace Prawnicze, Administratywistyczne i Ekonomiczne, 40*, 161–177 https://doi.org/10.19195/1733-5779.40.10

Kaplan, J. (2016). *Artificial Intelligence. What everyone needs to know*. Oxford University Press. ISBN: 978–0–19060–239–0.

O'Brien, R., Bair, E. F., & Venkataramani, A. S. (2022). Death by robots? Automation and working-age mortality in the United States. *Demography*. *59*(2), 607–628. https://doi.org/10.1215/00703370-9774819

Rosenberg, M. (2009). The surprising benefits of robots in the DC. *Supply Demand Chain Executive, 10*(2), 39–40.

Różanowski, K. (2007). Sztuczna inteligencja rozwój, szanse i zagrożenia. *Zeszyty Naukowe Warszawskiej Wyższej Szkoły Informatyki, 2*, 109–135.

Zakrzewska, M., Jarosz, S., Piwowar-Sulej, K., & Sołtysik, M. (2022). Enterprise agility – its meaning, managerial expectations and barriers to implementation – a survey of three countries. *Journal of Organizational Change Management, 35*(3), 488–510. https://doi.org/10.1108/JOCM-02-2021-0061

14 Would you trust a chatbot for healthcare advice? An empirical investigation from Poland

Piotr Pietrzak

Introduction

Artificial Intelligence (AI) is revolutionizing the provision of various services across multiple industries (Bharadiya et al., 2023). AI technologies enable automation, personalization, efficiency, and innovation in service delivery (Gajdošík & Marciš, 2019). So far, researchers have demonstrated the impact of AI on sectors such as finance (e.g. virtual financial advisors and robo-advisors provide personalized investment advice to clients) (see Königstorfer & Thalmann, 2020; Shanmuganathan, 2020; Kazachenok et al., 2023); education (e.g. AI can personalize learning experiences, offering tailored content, and assessments to students) (see Zawacki-Richter et al., 2019; Gao et al., 2021; King, 2023); transportation (e.g. AI is utilized in autonomous vehicles for navigation and safety) (see Król, 2016; Chang et al., 2022; Sundaresan et al., 2022); or agriculture (e.g. AI-driven precision farming techniques help optimize crop yields, monitor soil conditions, and reduce resource wastage) (see Linaza et al., 2021; Sharma et al., 2022; Shaik et al., 2022). Nevertheless, the literature is increasingly highlighting the wide range of uses of AI in healthcare services (e.g. Nadarzynski et al., 2019; Hee Lee & Yoon, 2021; Zhang & Ming, 2023).

Papers on the use of AI in the healthcare sector focused primarily on presenting solutions used in the diagnosis, treatment, and prediction. For example, in the area of diagnostics, AI has proven to be remarkably adept at deciphering X-rays, CT scans, and MRI pictures to spot illnesses and anomalies (Devi et al., 2023). Deep learning algorithms, a subfield of AI, have demonstrated to be especially proficient at accurately identifying and categorizing medical disorders (Abdel-Jaber et al., 2022). What is important, several research studies (e.g. Esmaeilzadeh, 2020; Joshi et al., 2022) have highlighted the need for improving AI-based systems and enhancing their capabilities to help patients or clinicians. Their adoption faces specific challenges (Petersson et al., 2022). One of these is creating trust toward the system (Laranjo et al., 2018). Understanding the trust dynamics between AI and humans is crucial, particularly in the field of healthcare,

DOI: 10.4324/9781032627236-16

where life is at risk (Asan et al., 2020). In this chapter, an attempt will be made to answer the question of whether Polish patients trust an AI-based system?

The main aim of this study was to examine whether potential users (patients) trust chatbots used for healthcare purposes. Five research tasks were stated to accomplish this aim: (1) a critical review of the literature on the potential use of AI in the healthcare sector; (2) a critical review of the literature on trust in AI; (3) recognition of the degree of trust in chatbots used in healthcare sector; (4) identification of factors that influence the degree of trust in chatbots used in the healthcare sector; and (5) identification of tasks for which chatbots are best suited in healthcare.

One research hypothesis was stated:

- H_1: *Respondents representing Generation Z show a higher degree of trust in chatbots used in healthcare than representatives of other generations.*

To the best of the author's knowledge, this hypothesis, although it may seem obvious, has not been verified by a group of Polish respondents.

The chapter consisted of six parts. After the introduction, a review of literature on trust in AI, was made. In the 3rd part, the methodology was presented. The fourth part was dedicated to data analysis. Part 5 discussed the implications derived from the study, addressed the limitations of this study and future research directions. The last part was a conclusion.

Review of literature

AI in healthcare

The term AI has been used in many ways in computer science, engineering, chemistry, arts & humanities, energy, business, management, and accounting. However, it also has wide applications in medicine, as well as in healthcare. In the simplest (and broadest) terms, AI can be defined as the "simulation or replication of human intelligence in machines" (Stewart et al., 2020, p. 135). AI aims to imitate human learning, thinking, and perception. That is why, AI is not one single type of technology but rather many different types within various application areas. In medicine, AI is used, among other things, for diagnosis and treatment, patient engagement and adherence, and administrative activities (Davenport & Kalakota, 2019). AI technologies of greatest relevance to the healthcare sector include:

- Machine learning (ML): ML algorithms can analyze medical images, such as X-rays, MRIs, and CT scans, to detect and diagnose diseases like cancer, heart disease, and neurological disorders. Also, ML algorithms can identify existing drugs that may be effective for treating different diseases, saving time and resources. Those who have written more extensively on the use of this technology include Deo, 2015; Rajula et al., 2020; Vellido, 2020; Bydon et al., 2023.

- Natural language processing (NLP): NLP can help healthcare providers improve the accuracy and efficiency of clinical documentation. It can assist in extracting relevant information from physician notes, reducing errors, and ensuring that patient records are complete and compliant with coding and billing standards. In other words, NLP can automate the process of assigning diagnostic codes (ICD-10, CPT) and generating bills from patient records, reducing administrative workload and improving accuracy. This issue was presented in the following publications: Grundmeier et al., 2016; Chae et al., 2021; Poniszewska-Marańda et al., 2023.
- Rule-based expert systems: Rule-based expert systems can be designed to assist healthcare providers in diagnosing medical conditions. The system asks questions about a patient's symptoms, medical history, and test results and uses predefined rules to suggest possible diagnoses. These systems can help patients manage their medications by providing reminders for doses, potential side effects, and interactions with other drugs. See more: Seto et al., 2012; Nagaraj and Deepalakshmi, 2022; Silva et al., 2023.
- Chatbots: Chatbots can remind a patient to take his/her pills, track the user's health, and find the nearest pharmacy and doctor's office if needed. What is important chatbots are not making decisions but offer users rational options. For more on chatbots, see studies: Espinoza et al., 2020; Valtolina et al., 2020; Frangoudes et al., 2021.

Despite the fact that the scientific literature emphasizes the significance of trust in AI enabled systems for their use, we could demonstrate that the relevance of trust-related research is underrepresented in the existing AI-related research in healthcare compared to other, more outcome-oriented research topics.

Trust in AI

Trust has been a topic of research in numerous disciplines, such as economics, sociology, psychology, and management (Paliszkiewicz & Koohang, 2016), and is recognized as a crucial concept in social and professional life. Trust can be defined in various ways depending on the context. In general terms, trust refers to a fundamental belief or confidence that someone or something is reliable, honest, and dependable (Wilkins, 2018). It involves a willingness to place confidence in another person, organization, or entity with the expectation that they will act in a way that aligns with your expectations, values, and interests.

For trust to emerge, the following conditions must be met (Wang & Emurian, 2005): (1) the trustor and the trustee are the two necessary actors;

(2) vulnerability must be shown (trust only exists in dangerous or unknown situations); (3) trust leads to actions, most frequently risk-taking behaviors; (4) trust is a highly subjective issue (trust is influenced by numerous subjective individual and environmental circumstances).

Trust is a critical element in various aspects of human relationships, including personal relationships, business transactions, governance, and much more. It serves as a foundation for cooperation, collaboration, and the establishment of healthy, functioning societies. Building and maintaining trust can take time and effort, and it can be fragile, easily damaged (Lewicki & Brinsfield, 2017).

In relation to AI used in healthcare, particularly chatbots, trust can be defined as the willingness of users to provide confidential information, accept the recommendations, and follow the suggestions (Wang & Siau, 2018). The recommendations and suggestions made by the chatbots must be consistent. The "black box" of the chatbot needs to be unlocked in order to foster confidence (Wang & Siau, 2018). In other words, users must comprehend how the suggestions were created and what data went into making them. The algorithms used by the chatbots also primarily rely on data. The data's correctness, security, privacy, and integrity are crucial. Otherwise, people will not give the health chatbots their sensitive medical information.

Table 14.1 presents selected studies on trust in AI, particularly that used in healthcare.

It is noteworthy that previous publications have mainly focused on theoretical issues in trust and AI. Only a few (e.g.) have presented the results of empirical studies. It seems necessary to carry out research to identify what degree of trust users have in AI (in particular) chatbots used in healthcare services.

Table 14.1 Selected studies on trust in artificial intelligence

Author(s) (Year)	*Title of publication*	*Source*	*Key findings*
Powell (2019)	Trust Me, I'm a Chatbot: How Artificial Intelligence in Health Care Fails the Turing Test	*Journal of Medical Internet Research, 21*	• AI needs to supplement rather than replace medical professionals, and identifying the complementary positioning of artificial intelligence in medical consultation is a key challenge for the future.
Kerasidou (2020)	Artificial intelligence and the ongoing need for empathy, compassion and trust in healthcare	*Bulletin of the World Health Organization, 98*	• AI has the potential to fundamentally alter the way in which empathy, compassion and trust are currently regarded and practiced in health care.

(*Continued*)

Table 14.1 (Continued)

Author(s) (Year)	*Title of publication*	*Source*	*Key findings*
Winter and Carusi (2022)	"If You're Going to Trust the Machine, Then That Trust Has Got to be Based on Something": Validation and the Co-Constitution of Trust in Developing Artificial Intelligence (AI) for the Early Diagnosis of Pulmonary Hypertension (PH)	*Science and Technology Studies*, 35	• Following the process of developing AI applications for supporting diagnosis in clinical settings shows validation to be neither purely technical, nor simply the final step of the development process in a formal validation phase. • This trust is multi-faceted, as it is engendered by interpersonal, multi-expert collaboration and practical interactions with the technology before it even gets to a formal validation phase.
Nickel (2022)	Trust in medical artificial intelligence: a discretionary account	*Ethics and Information Technology, 24*	• The article presents answers to four critical questions: (1) "Over what question does the clinician give discretionary authority to an AI application"; (2) "Are there good epistemic and pragmatic reasons for giving this discretion?"; (3) "Are these same reasons available to the clinician?"; and (4) "Is the discretionary authority properly limited in scope?"
Shan et al. (2022)	Public Trust in Artificial Intelligence Applications in Mental Health Care: Topic Modeling Analysis	*JMIR Human Factors, 9*	• The most relevant mental health care applications were found to be: (1) Wysa: Anxiety Therapy Chatbot; (2) Youper Therapy; (3) MindDoc: Your Companion; (4) TalkLife for Anxiety, Depression & Stress; (5) 7 Cups: Online Therapy for Mental Health & Anxiety; (6) BetterHelp-Therapy; (7) Sanvello; and (8) InnerHour.

Source: Own elaboration based on: Powell, 2019; Kerasidou, 2020; Nickel, 2022; Shan et al., 2022; Winter and Carusi, 2022.

Methodology

The instrument/survey

In order to obtain the research material, a diagnostic survey method was chosen with the use of a survey questionnaire technique (the CAWI - *Computer Assisted*

Web Interview). The CAWI method is a well-established tool within management science research for "acquiring information on participant (…) characteristics, present and past behaviour, standards of behaviour or attitudes and their beliefs" (Bird, 2009, p. 1307). Here are some of the key uses and advantages of the CAWI method (Sowa et al., 2015): (1) CAWI allows researchers to reach a large and diverse audience, including people from different geographic locations and demographics; (2) compared to traditional survey methods that involve paper questionnaires or telephone interviews, CAWI can be a cost-effective way to collect data; (3) CAWI surveys can be conducted quickly. Once an online questionnaire is set up, respondents can complete it at their convenience, and data collection can begin immediately; (4) respondents may feel more comfortable providing honest and candid answers to sensitive questions in an online survey because their responses are typically anonymous; and (5) researchers can easily customize CAWI surveys to suit their specific research objectives.

The questionnaire consists of closed-ended (single- and multiple-choice)-type questions. Examples of questions: How much do you trust healthcare chatbots to provide accurate medical information and assistance?; What factors influence your level of trust in healthcare chatbots?; How often would you prefer to interact with a healthcare chatbot instead of a human healthcare provider?; What types of healthcare-related tasks or information do you think chatbots are best suited for? The survey was online and anonymous and was made available to respondents through various platforms. The study was conducted from July to August 2023.

Participants

After a two-month survey (July–August 2023), 248 answers were collected. Among respondents, 105 (42.3%) were men and 143 (57.7%) were women. The largest share of the research sample was made up of people from generation Z (39.5%). The next group was generation Y (30.6%), while generation X made up 29.9% of the sample. The majority of respondents lived in large cities with more than 500,000 inhabitants (43.5%); 16.9% of respondents lived in rural areas; residents of cities of 50,000–100,000 people made up 14.5% of the sample; residents of cities with up to 20,000 inhabitants made up 10.1%; residents of cities of 20,000–50,000 people made up 9.3% of the sample, and only 5.7% of respondents lived in cities over 100,000 less than 500,000 inhabitants. In addition, the largest share of respondents were those with higher education (60.9%), those with a monthly income from 4,001 to 8,000 NT PLN (54.8%).

Procedure

The survey was conducted in accordance with good practice in the social sciences. This entails, among other things, the need to standardize the survey (CAWI). Which means not only the development of its questionnaire, but also identical conditions of measurement (questionnaire) in each survey. Personal data has been anonymized. After anonymization of the data, only information that does not itself constitute personal data remains, in particular data that directly identifies the person.

Results

In the first part of the survey, respondents were asked about their experience with chatbots used in healthcare. 62.9% of respondents indicated that they have not used chatbots for medical information or assistance. The main factors that deterred respondents from using chatbots were, in turn: lack of knowledge of chatbots (55.1%); privacy concerns (39.7%); and prefer speaking with a human healthcare provider (35.9%).

In the next step of the research, respondents were asked to specify their level of trust in chatbots on a scale of 1 to 5, where 1 is "Not at all Trustworthy" and 5 is "Completely Trustworthy". Generation X is characterized by the lowest degree of trust (mean value 3.12). This is probably due to the fact that Generation X grew up in an era when face-to-face communication and telephone conversations were the primary means of interaction. They are more accustomed to human-to-human interactions for seeking information, including healthcare advice. While Generation X has adapted to technology, they might not have the same level of digital immersion as younger generations, such as Millennials or Generation Z. They may not have had as much exposure to chatbots and might not fully understand how they work, leading to skepticism.

On the other hand, the highest level of trust, on average, was shown by representatives of generation Z (mean value 4.36). They are often referred to as "digital natives" and are more comfortable and adept at using technology in their daily lives, including interacting with chatbots. Moreover, Generation Z often prefers self-service options and values the convenience of accessing information and services on their own terms. Chatbots provide an on-demand, 24/7 resource for health-related queries, aligning with their desire for autonomy. Thus, the hypothesis stating that "respondents representing Generation Z show a higher degree of trust in chatbots used in healthcare than representatives of other generations" has been positively verified.

Respondents were also asked to identify the factors (each of them could pick a few) that most influence their level of trust towards chatbots used in healthcare. The responses were as follows: privacy and data security measures (54.8%); reputation of the chatbot provider (e.g. healthcare institution) (45.2%); transparency in how the chatbot operates (39.5%); availability of human oversight or support positive (19.8%); and personal experiences with the chatbot (16.9%).

Next, respondents were asked to identify their preferences in terms of using chatbot support. The following responses were received to the question "How often would you prefer to interact with a healthcare chatbot instead of a human healthcare provider?": frequently (40.7%); occasionally (21.0%); always prefer chatbots (16.9%); rarely (15.3%); and never prefer chatbots (6.1%). On the other hand, when asked "What types of healthcare-related tasks or information do you think chatbots are best suited for", respondents indicated: appointment scheduling (63.3%); medication reminders (50.0%); general health information (34.7%); symptom assessment (23.4%); nutrition advice (19.8%); mental health support (12.9%).

Discussion

Contributions

The study carried out filled a knowledge gap in several dimensions. Firstly, the use of AI in healthcare, and chatbots in particular is a relatively new phenomenon, and as a result, not much academic literature exists on this issue (especially empirical studies). Secondly, the research conducted to the author's knowledge was one of the first in Poland, which analyzed the degree of trust in healthcare chatbots.

Limitations

The obtained results can be questioned from the standpoint of the adopted research methodology. In the theoretical part of the chapter, some research that would have been significant to some scholars with regard to the described issue may have been omitted. Author, when selecting the literature, were guided by its availability and the importance of a journal or a scientific publishing house. In addition, the CAWI technique used by the author in the empirical part also has its limitations. Participants in online surveys often self-select to participate, which means they are motivated or interested in the survey topic. Only those who are online and have access to the survey link are eligible to participate in CAWI surveys. This excludes individuals who are not active online or who may have limited internet access.

Future research directions

Future research directions should focus on addressing emerging challenges, improving chatbot design and implementation, and providing valuable insights for both healthcare providers and patients. Here are some potential future directions for research in this area. Examine the impact of privacy and data security measures on trust. How do users' perceptions of data security and privacy practices affect their trust in healthcare chatbots? What are the best practices for maintaining user trust in this context? Understand the perspectives of both patients and healthcare providers regarding trust in healthcare chatbots. Explore differences in trust, expectations, and concerns between these two user groups.

Conclusions

The following conclusions were formulated as a result of the study:

- 62.9% of respondents indicated that they have not used chatbots for medical information or assistance.
- Respondents representing generation X had the lowest degree of trust in chatbots (average value of 3.12), while respondents representing generation Z had the highest (average value of 4.36).
- The most influential factors that affect the degree of trust in chatbots were privacy and data security measures (54.8%); and reputation of the chatbot provider (45.2%).
- According to the majority of respondents, chatbots are most suited in healthcare to: appointment scheduling and medication reminders.

Thus, the author succeeded in meeting the main aim of the study as well as all research tasks. In turn, the research hypothesis stated was positively verified.

The author believe that this study is intended to serve as a basis and starting point for academic discussion.

References

Abdel-Jaber, H., Devassy, D., Al Salam, A., Hidaytallah, L., & EL-Amir, M. (2022). A review of deep learning algorithms and their applications in healthcare. *Algorithms*, *15*, 71.

Asan, O., Bayrak, A.E., & Choudhury, A. (2020). Artificial intelligence and human trust in healthcare: Focus on clinicians. *Journal of Medical Internet Research*, *22*(6), e15154.

Bharadiya, J.P., Thomas, R.K., & Ahmed, F. (2023). Rise of artificial intelligence in business and industry. *Journal of Engineering Research and Reports*, *25*(3), 85–103.

Bird, D.K. (2009). The use of questionnaires for acquiring information on public perception of natural hazards and risk mitigation – A review of current knowledge and practice. *Natural Hazards and Earth System Sciences*, *9*, 1307–1325.

Bydon, M., Shin, J.H., Timmons, S.D., & Potts, E.A. (2023). Machine learning in neurosurgery: Transitioning to a new era of contemporary medicine. *Neurosurgical Focus*, *54*(6), E1.

Chae, S., Song, J., Ojo, M., & Topaz, M. (2021). Identifying heart failure symptoms and poor self-management in home healthcare: A natural language processing study. *Studies in Health Technology and Informatics*, *284*, 15–19.

Chang, H.-C., Okubo, T., Kobayashi, A., & Morimoto, A. (2022). Artificial Intelligence (AI) applications using big data and survey data for exploring the existence of the potential users of public transportation system. *International Journal of Information and Management Sciences*, *33*(4), 271–290.

Davenport, T., & Kalakota, R. (2019). The potential for artificial intelligence in healthcare. *Future Healthcare Journal*, *6*(2), 94–98.

Deo, R.C. (2015). Machine learning in medicine. *Circulation*, *132*(20), 1920–1930.

Devi, K.J., Alghamdi, W., Divya, N., Alkhayyat, A., Sayyora, A., & Sathish, T. (2023). Artificial intelligence in healthcare: Diagnosis, treatment, and prediction. *E3S Web of Conferences*, *399*, 04043.

Esmaeilzadeh, P. (2020). Use of AI-based tools for healthcare purposes: A survey study from consumers' perspectives. *BMC Medical Informatics and Decision Making*, *20*, 170.

Espinoza, J., Crown, K., & Kulkarni, O. (2020). A guide to chatbots for COVID-19 screening at pediatric health care facilities. *JMIR Public Health Surveillance*, *6*(2), e18808.

Frangoudes, F., Hadjiaros, M., Schiza, E.C., Matsangidou, M., Tsivitanidou, O., & Neokleous, K. (2021). An Overview of the use of chatbots in medical and healthcare education. *Lecture Notes in Computer Science*, *12785*, 170–184.

Gajdošík, T., & Marciš, M. (2019). Artificial intelligence tools for smart tourism development. *Advances in Intelligent Systems and Computing*, *985*, 392–402.

Gao, P., Li, J., & Liu, S. (2021). An introduction to key technology in artificial intelligence and big data driven e-Learning and e-Education. *Mobile Networks and Applications*, *26*(5), 2123–2126.

Grundmeier, R.W., Masino, A.J., Casper, T. Ch., Dean, J.M., Bell, J., Enriquez, R., Deakyne, S., Chamberlain, J.M., & Alpern, E.R. (2016). Identification of long bone fractures in radiology reports using natural language processing to support healthcare quality improvement. *Applied Clinical Informatics*, *7*(4), 1051–1068.

Hee Lee, D., & Yoon, S.N. (2021). Application of artificial intelligence-based technologies in the healthcare industry: Opportunities and challenges. *International Journal of Environmental Research and Public Health*, *18*(1), 1–18.

Joshi, S., Sharma, M., Das, R.P., Rosak-Szyrocka, J., Żywiołek, J., Muduli, K., & Prasad, M. (2022). Modeling conceptual framework for implementing barriers of AI in public healthcare for improving operational excellence: Experiences from developing countries. *Sustainability*, *14*, 11698.

Kazachenok, O.P., Stankevich, G.V., Chubaeva, N.N., & Tyurina, Y.G. (2023). Economic and legal approaches to the humanization of FinTech in the economy of artificial intelligence through the integration of blockchain into ESG Finance. *Humanities and Social Sciences Communications*, *10*(1), 167.

Kerasidou, A. (2020). Artificial intelligence and the ongoing need for empathy, compassion and trust in healthcare. *Bulletin of the World Health Organization*, *98*(4), 245–250.

King, M.R. (2023). A conversation on artificial intelligence, chatbots, and plagiarism in higher education. *Cellular and Molecular Bioengineering*, *16*(1), 1–2.

Königstorfer, F., & Thalmann, S. (2020). Applications of artificial intelligence in commercial banks - A research agenda for behavioral finance. *Journal of Behavioral and Experimental Finance*, *27*, 100352.

Król, A. (2016). The application of the artificial intelligence methods for planning of the development of the transportation network. *Transportation Research Procedia*, *14*, 4532–4541.

Laranjo, L., Dunn, A.G., Tong, H.L., Kocaballi, A.B., Chen, J., Bashir, R., Surian, D., Gallego, B., Magrabi, F., Lau, A.Y.S., & Coiera, E. (2018). Conversational agents in healthcare: A systematic review. *Journal of the American Medical Informatics Association*, *25*(9), 1248–1258.

Lewicki, R.J., & Brinsfield, Ch. (2017). Trust repair. *Annual Review of Organizational Psychology and Organizational Behavior*, *4*, 287–313.

Linaza, M.T., Posada, J., Bund, J., Eisert, P., Quartulli, M., Döllner, J., Pagani, A., Olaizola, I.G., Barriguinha, A., Moysiadis, T., & Lucat, L. (2021). Data-driven artificial intelligence applications for sustainable precision agriculture. *Agronomy*, *11*, 122.
Nadarzynski, T., Miles, O., Cowie, A., & Ridge, D. (2019). Acceptability of Artificial Intelligence (AI)-led chatbot services in healthcare: A mixed-methods study. *Digital Health*, *5*, 1–12.
Nagaraj, P., & Deepalakshmi, P. (2022). An intelligent fuzzy inference rule-based expert recommendation system for predictive diabetes diagnosis. *International Journal of Imaging Systems and Technology*, *32*(4), 1373–1396.
Nickel, P.J. (2022). Trust in medical artificial intelligence: A discretionary account. *Ethics and Information Technology*, *24*(1), 7.
Paliszkiewicz, J., & Koohang, A. (2016). *Social Media and Trust: A Multinational Study of University Students*. Santa Rosa, CA: Informing Science Press.
Petersson, L., Larsson, I., Nygren, J.M., Nilsen, P., Neher, M., Reed, J.E., Tyskbo, D., & Svedberg, P. (2022). Challenges to implementing artificial intelligence in healthcare: A qualitative interview study with healthcare leaders in Sweden. *BMC Health Services Research*, *22*(1), 850.
Poniszewska-Marańda, A., Vynogradnyk, E., & Marańda, W. (2023). Medical data transformations in healthcare systems with the use of natural language processing algorithms. *Applied Sciences* (Switzerland), *13*(2), 682.
Powell, J. (2019). Trust me, I'm a chatbot: How artificial intelligence in health care fails the Turing test. *Journal of Medical Internet Research*, *21*(10), e16222.
Rajula, H.S.R., Verlato, G., Manchia, M., Antonucci, N., & Fanos, V. (2020). Comparison of conventional statistical methods with machine learning in medicine: Diagnosis, drug development, and treatment. *Medicina* (Lithuania), *56*(9), 1–10.
Seto, E., Leonard, K.J., Cafazzo, J.A., Barnsley, J., Masino, C., & Ross, H.J. (2012). Developing healthcare rule-based expert systems: Case study of a heart failure telemonitoring system. *International Journal of Medical Informatics*, *81*(8), 556–565.
Shaik, F.K., Memon, M.A., Mahoto, N.A., Zeadally, S., & Nebhen, J. (2022). Artificial intelligence best practices in smart agriculture. *IEEE Micro*, *42*(1), 17–24.
Shan, Y., Ji, M., Xie, W., Lam, K.-Y., & Chow, C.-Y. (2022). Public trust in artificial intelligence applications in mental health care: Topic modeling analysis. *JMIR Human Factors*, *9*(4), e38799.
Shanmuganathan, M. (2020). Behavioural finance in an era of artificial intelligence: Longitudinal case study of robo-advisors in investment decisions. *Journal of Behavioral and Experimental Finance*, *27*, 100297.
Sharma, A., Georgi, M., Tregubenko, M., Tselykh, A., & Tselykh, A. (2022). Enabling smart agriculture by implementing artificial intelligence and embedded sensing. *Computers and Industrial Engineering*, *165*, 107936.
Silva, B., Hak, F., Guimaraes, T., Manuel, M., & Santos, M.F. (2023). Rule-based system for effective clinical decision support. *Procedia Computer Science*, *220*, 880–885.
Sowa, P., Pędziński, B., Krzyżak, M., Maślach, D., Wójcik, S., & Szpak, A. (2015). The computer-assisted web interview method as used in the national study of ICT use in primary healthcare in Poland – Reflections on a case study. *Studies in Logic, Grammar and Rhetoric*, *43*(56), 137–146.

Stewart, J.C., Davis, G.A., & Igoche, D.A. (2020). AI, IoT, and AIoT: Definitions and impacts on the artificial intelligence curriculum. *Issues in Information Systems*, *21*(4), 135–142.

Sundaresan, S., Kumar, K.S., Nishanth, R., Robinson, Y.H., & Kumar, A.J. (2022). Artificial intelligence and machine learning approaches for smart transportation in smart cities using blockchain architecture. In: S. Krishnan, V.E. Balas, E.G. Julie, Y.H. Robinson, & R. Kumar (Eds.), *Blockchain for Smart Cities* (pp. 35–56). Cambridge, MA: Elsevier.

Valtolina, S., Barricelli, B.R., & Di Gaetano, S. (2020). Communicability of traditional interfaces VS chatbots in healthcare and smart home domains. *Behaviour & Information Technology*, *39*(1), 108–132.

Vellido, A. (2020). The importance of interpretability and visualization in machine learning for applications in medicine and health care. *Neural Computing and Applications*, *32*(24), 18069–18083.

Wang, Y.D., & Emurian, H. (2005). An overview of online trust: Concepts, elements, and implications. *Computers in Human Behavior*, *21*, 105–125.

Wang, W., & Siau, K. (2018). Living with artificial intelligence - developing a theory on trust in health chatbots. *Proceedings of the Sixteenth Annual Pre-ICIS Workshop on HCI Research in MIS*, 1–5.

Wilkins, C.H. (2018). Effective engagement requires trust and being trustworthy. *Medical Care*, *56*, 6–8.

Winter, P., & Carusi, A. (2022). 'If you're going to trust the machine, then that trust has got to be based on something': Validation and the co-Constitution of trust in developing Artificial Intelligence (AI) for the early diagnosis of Pulmonary Hypertension (PH). *Science and Technology Studies*, *35*(4), 58–77.

Zawacki-Richter, O., Marín, V.I., Bond, M., & Gouverneur, F. (2019).Systematic review of research on artificial intelligence applications in higher education - where are the educators? *International Journal of Educational Technology in Higher Education*, *16*(1), 39.

Zhang, B., & Ming, C. (2023). A patent portfolio value analysis based on intuitionistic fuzzy sets: An empirical analysis of artificial intelligence for healthcare. *Journal of Open Innovation: Technology, Market, and Complexity*, *9*(3), 100124.

15 Trust in machine learning applied for analysis of non-financial disclosures

Tomasz Jastrząb, Monika Wieczorek-Kosmala, Anna Losa-Jonczyk, Marco Papa and Dorota Marquardt

Introduction

Non-financial disclosures (hereafter NFD) include company's information related to its environmental, social and governance (ESG) performance (Erkens et al., 2015; Zarzycka et al., 2021). By placing NFD in annual reports, companies enter a dialogue with stakeholders and interact with them, by exchanging relevant information about the firm's performance (Cundill et al., 2018; Driessen et al., 2013; Maon et al., 2010; Noland & Phillips, 2010). Investors represent a group of stakeholders that have been increasingly interested in NFD, to make investment decisions (Ahmed et al., 2022; Alda, 2021; Escrig-Olmedo et al., 2017; Chai et al., 2022; Gangi et al., 2022). There is strong empirical evidence that companies that follow ESG criteria in their performance are more profitable (e.g., Garcia et al., 2017). On the other hand, the NFDs are regarded as a relevant tool of stakeholders' pressure on companies to enhance their performance in terms of ESG-related criteria (Høvring et al., 2018; Zarzycka et al., 2021). As a result, there is a growing stakeholders' demand for reliable, trustworthy, and transparent information, embedded in companies' NFD (Rodrigue et al., 2013). An example of response could be the implementation of European Union's Directive 2014/95/EU on non-financial information, which mandated large companies of member states to provide disclosures on ESG performance, from 2017 onward (Mazzotta et al., 2020). Another example could be the EU Taxonomy that is being developed at the European Union level to support the achievement of climate neutrality targets, in line with the European Green Deal. This taxonomy draws up a comprehensive list of all economic activities that can be considered environmentally sustainable and thus worthy of support or not by investors (Schütze & Stede, 2021). It also requires companies to disclose and report certain KPIs in the future as an indication of sustainable activities in their business.

There is nevertheless an ongoing debate on the credibility of NFD, which is covering the trust in the information provided within (Mazzotta et al., 2020). In many instances, the NFD comprises stand-alone sustainability reports and or

DOI: 10.4324/9781032627236-17

the management reports of corporate financial statements. The content of these reports is mainly qualitative and is unstructured, not subject to automated retrieval. Therefore, its systematic analysis by the single user is complex and time consuming. Numerous prior researchers have adopted content analysis methods to draw conclusions on companies' ESG related performance. Yet, the manual content analysis is limited by subjective assessment, and the limited number of documents (here: reports) that could be revised. These shortcomings could be eliminated by the application of natural language processing (NLP) methods, although NLP application also has some limitations (Duarte et al., 2017). NLP methods not only can process a large dataset of non-financial reports, but also may detect latent features in the texts not identifiable by the most comprehensive manual analysis.

There is a wide range of reporting-related literature that has examined AI tools and applied automated and NLP-based textual analysis to corporate reporting (see, e.g., De Villiers et al., 2023; Elshandidy et al., 2018). A main stream of this literature (Das, 2013; E-Haj et al., 2020; Kearney & Liu, 2014; Li, 2010; Loughran & McDonald, 2014) has preferred to concentrate on low-tech measures based on keyword dictionary that are associated with specific attributes, such as tone and text readability. The misclassification of document attributes is the main limitation of this approach. A general list of words may have one context in a general setting but a different context in a specific setting. Furthermore, these approaches are based on a 'bag-of-words' approach that considers the actual words counts as sufficient to characterize a document, implicitly assuming that context is unimportant. One of the recent trends in textual analysis is the application of sophisticated machine learning methods (Choudhury et al., 2018) that detect features in corporate reports that are not evident with manual methods.

Our study aims at contributing to this emerging trend, by providing some evidence on trust in machine learning application, in the process of the analysis of NFDs. We address the problem of trust in machine learning from two perspectives, guided by the understanding of trust in technology, namely the human and technology characteristics (Oleson et al., 2011; Schaefer et al., 2016; Siau & Wang, 2018). Human characteristics are relevant if the ML training is based on expert knowledge contribution, which could be biased by personal perceptions or abilities. Technology characteristics are relevant in the context of ML models quality, interpretability and explainability (Carvalho et al., 2019; Rudin, 2019; Yin et al., 2019), for example, deep learning models which tend to achieve tremendous results in various tasks are too complex to comprehend. Therefore, trust toward them may be limited. Researchers have attempted to determine various measures of people's trust in ML and the results indicate that the impact of trust in the model's accuracy may be less than the tendency of respondents to follow the model's prediction (Rechkemmer & Yin, 2022). Studies show that people trust the results of a machine learning system rather than their own judgment or the suggestions of others when the topic is unfamiliar to them (Chiang & Yin, 2021; Logg et al., 2019). On the other hand, they quickly lose trust in the algorithm if they notice its error (Dietvorst et al., 2015; Paliszkiewicz et al., 2022), and then it is difficult to regain trust in ML at the previous level.

In our study, we merge these two factors (human and technology), by providing experiment-based evidence on the quality of ML models, given the significant difference of expert-based annotations used during model training. The major question we ask is whether the use of ML models trained on expert annotations (even if significantly different) improves the classification quality in comparison to the keywords-based approach?

As the experimental setting, we use the NFD related to climate change issues. Central to the European Green Deal to achieve the planned goals of carbon neutrality by 2050 is the expansion of the corporate reporting landscape to ESG factors. As mentioned earlier, close monitoring of sustainable business practices in the form of regular reports will be a key control mechanism for achieving climate targets in the EU in the future. The Corporate Sustainability Reporting Directive (CSRD), adopted in 2021, expands the existing Non-Financial Reporting Directive (NFRD) and is part of the EU comprehensive sustainable finance package. Previously, only companies in the EU with more than 500 employees were required to prepare an NFRD report in addition to their annual financial reports. This group of companies will be significantly expanded in the future, so that an estimated 50,000 companies in the EU will have to prepare a binding CSRD report annually. This new mandatory sustainability reporting regime creates important challenges for preparers of NFD. They will have to follow the double materiality principle, to report environmental KPIs and targets and connectivity between financial and non-financial performance shall be strengthened. In this context, the use of ML models may have implications for the definition of effective communication in terms of useful content and the best way to deliver it.

In our experiment, we verify the quality of four different models. The first model (baseline) uses keywords-based classification supported by word embeddings (Mikolov et al., 2013a, 2013b; Piasecki et al., 2017). It can be regarded as an improved naive method, which would look up keywords from a predefined dictionary in the text and based on their existence and/or frequency, would choose the appropriate class label. The other three ML models are k-Nearest Neighbors (KNN) (Fix & Hodges, 1951), Random Forest (Ho, 1995), and Logistic Regression (Hosmer et al., 2013). These models include the training phase, again based on word embeddings, to learn the characteristics of different classes beyond simple keywords search.

The remainder of our work is organized as follows. In the second section, we outline the experimental setting, by providing details on the procedure of annotations (expert-based). In the third section, we explain the training procedure, with the details of models that have been applied. In section four, we present the results and discuss the models quality. Section five concludes our chapter.

Experiment

Data sets

To address the human characteristics of the trust in ML, we used the NFD placed in the non-financial reports for 2021, published by nine large Polish companies from the oil and energy industry. We have focused on climate-risk related disclosures. The non-financial reports were annotated by two independent experts, whose task was to mark the climate risk-related fragments of text, guided by the class labels of the given types of climate change risk (cf. Table 15.1), following the reporting requirements outlined in ESRS 1 (ESRS E1 Climate change). The annotations along with the pertaining text fragments were then extracted automatically from the non-financial reports in.pdf files using a simple Java application. Next, the extracted fragments and annotations were manually cleaned up to ensure that all class labels match expected values, to resolve multi-label annotations and to correct minor errors introduced by such an automated extraction procedure.

Given the initial set of text-label pairs, the natural language processing procedures such as stop words removal and lemmatization were applied to extracted text fragments. The idea of stop words removal was to identify only relevant phrases in the texts. To do so, a predefined set of stop words was used, available at https://github.com/bieli/stopwords/blob/master/polish.stopwords.txt). The purpose of lemmatization, performed using the Morfeusz tool (Kieraś & Woliński, 2017), was to obtain the base forms of words (lemmas) required by the Word2Vec tool. This also ensures that words appearing in different inflected forms are unified. Word ambiguities were resolved manually by the inspection of morphosyntactic tags returned by Morfeusz. Finally, the lemmatized text fragments without stop words were collected, along with their labels and source documents in the form of the JSON (JavaScript Object Notation) object serving as the data set for ML models.

Since the annotations pertained only to climate risk-related text fragments, a set of negative examples (i.e., non-risk related texts) had to be built for the purposes of ML models training and evaluation. This set was built based on the same set of nine non-financial reports by extracting the first *n* sentences of each document, where *n* was equal to the number of annotated text fragments in this document. To make this set representative, we extracted these sentences from the pages on which the annotated text fragments appeared. This way, we wanted to ensure that the negative examples are still quite closely related to the risk topics and do not represent very distant text fragments.

We have also built a simple baseline model, independent of the experts' annotations. For that purpose, we created a glossary of keywords best describing different types of climate risk. The phrases in the glossary were pre-processed similarly to the main data set and were stored as a bag-of-words (BOW) assigned to respective labels in the form of another JSON object. For the baseline model, which does not have any training phase, we did not collect any negative examples.

Table 15.1 shows the characteristics of both datasets, with numbers of text fragments assigned to respective classes. As can be noted, the datasets differ in the number of elements, especially the set annotated by Expert B is much larger than the other two. We applied a Wilcoxon Z-test to confirm that annotations by Expert B differ at $\alpha = 0.001$ from annotations by Expert A, and at $\alpha = 0.1$ from baseline dataset. At the same time, there were no statistically significant differences between annotations by Expert A and baseline dataset.

We can also note that representatives for some class labels are missing, as the experts did not find any relevant fragments in the non-financial reports. The 0 value in the column marked as 'none' for Baseline dataset results from lack of negative examples in the glossary. We can also observe that the distribution of examples among different classes is quite uneven. Therefore, while evaluating the metrics achieved by the models, we will use the weighted average values which take this imbalance into account.

Since the data contained in the datasets is textual and ML models typically require numeric values, taking advantage of transfer learning capabilities, we used pre-trained word embeddings to convert the data set to numeric vectors. We used Word2Vec model (Mikolov et al., 2013a, 2013b) trained on the KGR10 corpora using both continuous bag of words (CBOW) and skip-gram approach (Piasecki et al., 2017). Each vector consisted of 300 floating point values (features), which can be used for training the ML models. To compute a single vector for a given phrase (resp. sentence), we calculated the average over all embeddings for words building up the phrase (sentence).

Models

In our experiments, we considered four different models: the baseline model (Base), KNN, Random Forest (RF), and Logistic Regression (LR). For the baseline model, since it consists of bags of words assigned to class labels, we computed the average vector over all unique lemmas in each BOW. Then, to perform classification, we used the cosine similarity between the input vector to be classified and each of the vectors of the baseline model. The cosine similarity is a metric which returns values in the range -1 to 1, where 1 means identical vectors, 0 means no similarity, and -1 means completely opposite vectors (in the vector space). Based on this metric, we chose as the predicted class label the one for which the cosine similarity was the highest, provided that it was greater than 0.5. Otherwise, we set the predicted class to 'none'.

The KNN model is based on calculating the distance between input vector and feature vectors assigned to respective classes. The distances are computed between *k* closest neighbors and the final outcome is the result of a majority

Table 15.1 Numbers of text fragments assigned to different class labels, by type of dataset

Class label		*a-soil*	*a-temp*	*a-water*	*a-wind*	*acute*	*ch-soil*	*ch-temp*	*ch-water*	*ch-wind*	*chronic*	*market*	*policy/leg*	*rep*	*tech*	*transition*	*none*	**Total**
Baseline dataset		5	5	13	10	4	6	15	29	7	3	2	7	2	3	3	0	**114**
Expert-annotated dataset	Expert A	1	7	10	5	14	0	3	4	0	5	13	24	8	6	0	96	**196**
	Expert B	0	12	12	9	21	0	13	12	4	12	27	30	20	19	8	188	**387**

Source: Own work based on the research results

vote. In the experiments we considered $k = 1, 3, 5$, to observe how the model behaves depending on the number of neighbors used. For $k = 1$, we simply select the closest neighbor as the source of prediction, with $k > 1$ the majority voting takes place. The distance is based on Euclidean distance. In the experiments, we use the KNN model implementation provided by Python's scikit-learn library (Pedregosa et al., 2011) with default values except for the number of neighbors *k*.

The RF model is an example of ensemble learning algorithm based on decision trees, where each decision tree is built based on a subset of data and features. After all the trees are constructed, the majority vote is selected as the final outcome. The key parameter of RF model is the number of trees *n* used, which affects the training time and the efficiency of the model. In the experiments, we used $n = 10$, 100, and 500. To perform classification, we used Python's implementation of RF with default parameters except for the number of trees *n*.

The LR model is typically suited for binary classification tasks where it models binomial probability distribution function. However, it can be also adjusted to support multinomial logistic regression, either by converting the problem to multiple binary classification problems, or by changing the loss function to cross-entropy loss, to obtain probability estimation for each class label separately. In the experiments we use Python's implementation of LR with solvers supporting multinomial regression ('lbfgs' or 'newton-cg') and with different regularization a strength parameter $C = 0.1$, 1, and 10, where smaller values mean stronger regularization. By default, L2 regularization is used which prevents overfitting the model to the training data by adding a penalty to the loss function, which should be minimized.

Procedure

To construct the ML models, we built a number of samples based on the annotated dataset. The samples differed in the subsets of classes used and the sources of data (Expert A, Expert B or both). In terms of subsets of classes, we used the following combinations:

- CS1: 'all' vs. 'none'
- CS2: 'acute' vs. 'chronic'
- CS3: 'acute' vs. 'chronic' vs. 'other'
- CS4: 'acute' vs. 'chronic' vs. 'other' vs. 'none'
- CS5: 'acute+' vs. 'chronic+'
- CS6: 'acute+' vs. 'chronic+' vs. 'other'
- CS7: 'acute+' vs. 'chronic+' vs. 'other' vs. 'none'

where 'all' denotes all classes except 'none', 'acute+' denotes 'acute', 'a-soil', 'a-temp', 'a-water', and 'a-wind' classes, 'chronic+' denotes 'chronic', 'ch-soil', 'ch-temp', 'ch-water', and 'ch-wind' classes, and 'other' denotes 'market', 'policy/leg', 'rep', 'tech', and 'transition' classes. This way, 21 samples were formed.

Based on the samples, the training and test sets were built with the 70%/30% split. The splitting procedure was repeated five times, producing altogether 105 training-test set pairs. Each model (except for the baseline model) was then trained using the training set and evaluated on the test set with the use of the following metrics (Pedregosa et al., 2011):

- Balanced accuracy

$$BAcc = \frac{1}{|L|}\sum_{l \in L} R(y_l, \hat{y}_l) \tag{15.1}$$

- Weighted precision

$$P_w = \frac{1}{\sum_{l \in L}|y_l|}\sum_{l \in L}|y_l|\, P(y_l, \hat{y}_l) \tag{15.2}$$

- Weighted recall

$$R_w = \frac{1}{\sum_{l \in L}|y_l|}\sum_{l \in L}|y_l|\, R(y_l, \hat{y}_l) \tag{15.3}$$

- Weighted F_1-score

$$R_w = \frac{1}{\sum_{l \in L}|y_l|}\sum_{l \in L}|y_l|\, F_1(y_l, \hat{y}_l) \tag{15.4}$$

where L denotes the set of class labels, y_l and $\hat{y}_l$ denote the true and predicted entries with label l, $P(y_l, \hat{y}_l) = \frac{|y_l \cap \hat{y}_l|}{|\hat{y}_l|}$ denotes the precision, $R(y_l, \hat{y}_l) = \frac{|y_l \cap \hat{y}_l|}{|y_l|}$ is the recall, and $F_1(y_l, \hat{y}_l) = \frac{2 \times P(y_l, \hat{y}_l) \times R(y_l, \hat{y}_l)}{P(y_l, \hat{y}_l) + R(y_l, \hat{y}_l)}$ is the F_1-score, being the harmonic mean between precision and recall. In simple terms, the balanced accuracy is an averaged recall value, the precision denotes the portion of correctly identified items among all predictions, while recall specifies the ratio of correctly identified items to all items to be identified.

Results and discussion: models' quality assessment

The results of applying various models, in terms of the quality metrics considered in this experiment (for baseline, expert-annotated and combined dataset), are presented in Figure 15.1.

Since the skip-gram-based embeddings did not produce significantly different results, we will focus only on the CBOW embeddings. The results shown in Figure 15.1 were averaged across five repetitions. For the KNN, RF and LR models, the figure depicts the best average results obtained among different algorithm configurations (i.e., among different *k*, *n* or *C* values). Both solvers in

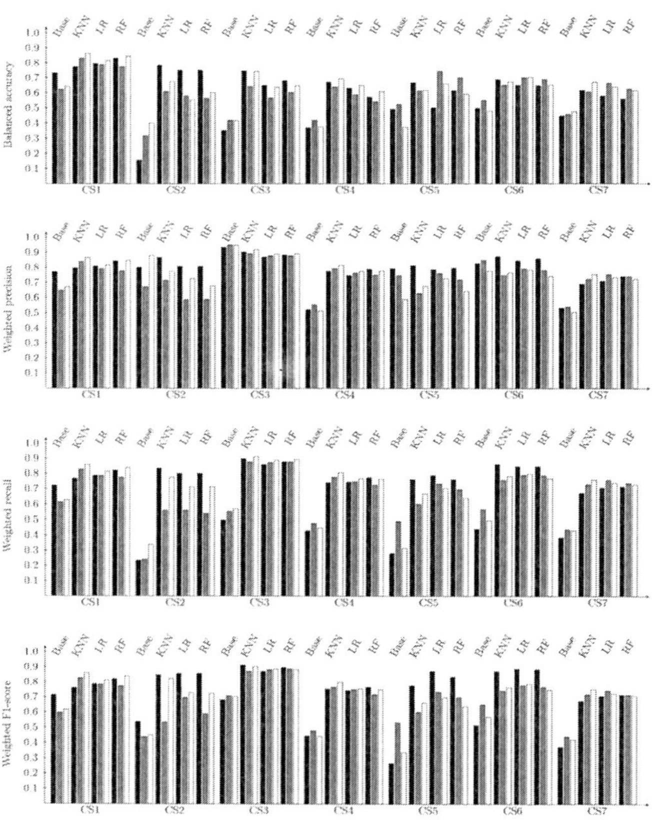

Figure 15.1 Quality metrics obtained for the baseline, KNN, LR, and RF models using CBOW word embeddings, for Expert A (black), Expert B (gray), and Expert A & B (white) annotation-based test sets.

Source: Authors' research.

the LR model always reported the same results, so this criterion was not taken into account in the final results analysis.

Based on Figure 15.1, we can observe that the baseline model usually performs worst among all analyzed models, regardless of the subset of analyzed classes and regardless of the quality metric we take into consideration. The only exception from this rule can be observed in terms of precision for CS2 and CS3 class subsets. What is more, its performance for binary classification, especially for CS2 and CS5, could be even worse than random guessing, which would have the accuracy of 0.5. Such a behavior of the baseline model was expected, due to its simplicity and the fact that it does not use any machine learning capabilities, except for the pre-trained word embeddings. In terms of the remaining models, it is hard to find the one that performs best in all scenarios.

Investigating the performance of the models on the most general scenario CS1 ('all' vs. 'none'), the best performance was achieved by the KNN model with complete input dataset, that is, originating from both experts. This model clearly prevails over the others in terms of all four metrics. It should be considered an advantage, taking into account that it is a lazy learning model, which performs any processing only at the time of prediction. Besides, the KNN model is also not heavily parameterized (typically we only change k and possibly the way of calculating the distance).

Considering the scenarios in which 'acute' vs. 'chronic' pair appears, either in binary or multi-class fashion (CS2–CS4), we can again observe that KNN is mainly the best among analyzed models. The influence of the dataset chosen seems to be less visible as often either the complete or the Expert A-based dataset ensures the best results. Extending the 'acute' and 'chronic' classes with environmentally-related elements ('acute+' and 'chronic+', scenarios CS5–CS7) allows the logistic regression model to achieve some better results, especially in terms of F_1-score. The RF approach, although not significantly lagging behind the two other models, does not seem the best choice for the task at hand.

To confirm the differences between the baseline model and the other ML models, we run ANOVA tests with post hoc Tukey HSD tests. This analysis revealed statistically significant differences, at $\alpha = 0.05$, between the baseline model and all other models in terms of balanced accuracy, weighted recall and weighted F_1-score. There were no statistically significant differences in weighted precision between any pair of models. These observations were true when comparing the metrics across all class subsets CS1–CS7 and regardless of the source of annotations, that is, among metrics for *_A, *_B and *_AB models shown in Figure 15.1, as well as among combined metrics for all *_A, *_B and *_AB settings considered together. There were no statistically significant differences between KNN, RF and LR models. Finally, no statistically significant differences were also observed when comparing *_A vs. *_B vs. *_AB ML models, which indicates that the source of expert annotation was irrelevant.

Conclusion

In this chapter, we have focused on the analysis of non-financial reports of selected Polish companies in the view of their risk-related disclosures. We proposed and evaluated a number of machine learning models applied to the task of text classification.

Our experiment has shown that the level of trust, expressed in terms of different quality metrics, that is, balanced accuracy, weighted precision, weighted recall, and weighted F_1-score, should be greater when we use ML models as compared to simple keywords-based text analysis. We have proved that the ML models can achieve statistically significantly higher metric values than the baseline model and thus their predictions can be trusted to a greater extent.

The experiment failed to prove that combined expert annotations can provide statistically significantly better results than the annotations considered separately. However, this fact may be also viewed as an advantage since the models require less manual work to achieve satisfactory quality metrics. We have also shown that using transfer learning capabilities we need much less data for model training which again is advantageous – it was enough to use just nine reports with around 600 annotations in total, instead of thousands or even millions of documents we would need to create our own word-embeddings model – for comparison, the KGR10 corpus used as a source for our word embeddings contains over four billions of words.

There are several implications that our experiment may have for NFD research. NLP methods based on experts' annotations may be designed to detect some features of ineffective disclosure. For instance, the NLP methods may be designed to identify key reporting ESG risks, to filter out boilerplate climate change risk disclosure, and to examine other latent features of corporate reports, such as the connectivity between financial and non-financial information. They may also be beneficial to capital market research. For instance, NLP methods may be used to test whether different climate change risks are value relevant for investors and analysts.

For managerial practice, further research into the use of machine learning will be helpful in improving the process of collecting data for non-financial reports, aggregating it, and communicating it effectively to company stakeholders.

In the future, we plan to cross-validate the performance of developed models on the texts annotated by the other expert, that is, we can use the model trained on Expert A's annotations and perform classification of the texts prepared by Expert B. Furthermore, we plan to apply the ML models to text classification on a larger scale, to enable full automation of non-financial reports analysis for the purposes outlined above.

Acknowledgements

The research has been supported by the European Union under the Regional Operational Program of the Śląskie Voivodeship 2014–2020 within the project 'Opracowanie innowacyjnego systemu informatycznego do nauki oraz weryfikacji wiedzy z zakresu ekonomii' ('Development of an innovative IT system for learning and verifying knowledge in economics') awarded to Technicenter Sp. z o.o.

We also gratefully acknowledge the support of the University of Economics in Katowice with grant 'Beyond Barriers' number 02/BB/01/2021.

Bibliography

Ahmed, T., Rahman, M. M., Aktar, M., Gupta, A. D., & Abedin, M. Z. (2022). The impact of economic development on environmental sustainability: Evidence from the Asian region. *Environment, Development and Sustainability*, *25*, 3523–3553. https://doi.org/10.1007/s10668-022-02178-w.

Alda, M. (2021). The environmental, social, and governance (ESG) dimension of firms in which social responsible investment (SRI) and conventional pension funds invest: The mainstream SRI and the ESG inclusion. *Journal of Cleaner Production, 298*, 126812. https://doi.org/10.1016/j.jclepro.2021.126812.

Carvalho, D. V., Pereira, E. M., & Cardoso, J. S. (2019). Machine learning interpretability: A survey on methods and metrics. *Electronics*, *8*(8), 832. https://doi.org/10.3390/electronics8080832.

Chai, S., Chu, W., Zhang, Z., Li, Z., & Abedin, M. Z. (2022). Dynamic nonlinear connectedness between the green bonds, clean energy, and stock price: the impact of the COVID-19 pandemic. *Annals of Operations Research*, *6*, 1–28. https://doi.org/10.1007/s10479-021-04452-y.

Chiang, Ch.-W., & Yin, M. (2021). You'd better stop! Understanding human reliance on machine learning models under Covariate shift. In: *Proceedings of the 13th ACM Web Science Conference 2021* (pp. 120–129). ACM. https://doi.org/10.1145/3447535.3462487.

Choudhury, P., Carlson, N. A., Wang, D., & Khanna, T. (2018). *Machine Learning Approaches to Facial and Text Analysis: Discovering CEO Oral Communication Styles* (Working paper 18–064). Harvard Business School. Retrieved from https://www.hbs.edu/ris/Publication%20Files/18-064_9aa69bf6-bf39-4f7d-bcd0-968a910e63ba.pdf

Cundill, G. J., Smart, P., & Wilson, H. N. (2018). Nonfinancial shareholder activism: A process model for influencing corporate environmental and social performance. *International Journal of Management Reviews*, *20*, 606–626. https://doi.org/10.1111/ijmr.12157.

Das, S. R. (2013). Text and context: Language analytics in finance. *Foundations and Trends in Finance*, *8*, 145–261.

De Villiers, C., Dimens, R., & Molinari, M. (2023). How will AI text generating and processing impact sustainability reporting? Critical analysis, a conceptual framework and avenues for future research. *Sustainability Accounting, Management and Policy Journal*, 15(1), 96–118. https://doi.org/10.1108/SAMPJ-02-2023-0097.

Dietvorst, B. J., Simmons, J. P., & Massey, C. (2015). Algorithm aversion: People erroneously avoid algorithms after seeing them err. *Journal of Experimental Psychology: General, 144*(1), 114–126. https://doi.org/10.1037/xge0000033.

Driessen, P. H., Kok, R. A. W., & Hillebrand, B. (2013). Mechanisms for stakeholder integration: Bringing virtual stakeholder dialogue into organizations. *Journal of Business Research*, *66*(9), 1465–1472. https://doi.org/10.1016/j.jbusres.2012.09.009.

Duarte, N., Llanso, E., & Loup, A. (2017). *Mixed Messages? The Limits of Automated Social Media Content Analysis*. Center for Democracy and Technology. Retrieved from https://cdt.org/wp-content/uploads/2017/11/2017-11-13-Mixed-Messages-Paper.pdf.

E-Haj, M., Alves, P., Rayson, P., Walker, M., & Young, S. (2020). Retrieving, classifying and analysing narrative commentary in unstructured (glossy) annual reports published as PDF files. *Accounting and Business Research*, *50*(1), 6–34. https://doi.org/10.1080/00014788.2019.1609346.

Elshandidy, T., Shrives, P. J., Bamber, M., & Abraham, S. (2018). Risk reporting: A review of the literature and implications for future research. *Journal of Accounting Literature*, *40*(1), 54–82. https://doi.org/10.1016/j.acclit.2017.12.001.

Erkens, M., Paugam, L., & Stolowy, H. (2015). Nonfinancial information: State of the art and research perspectives based on a bibliometric study. *Comptabilité-Contrôle-Audit*, *21*(3), 15–92. https://doi.org/10.3917/cca.213.0015.

Escrig-Olmedo, E., Rivera-Lirio, J. M., Muñoz-Torres, M. J., & Fernández-Izquierdo, M. Á. (2017). Integrating multiple ESG investors' preferences into sustainable investment: A fuzzy multicriteria methodological approach. *Journal of Cleaner Production*, *162*, 1334–1345. https://doi.org/10.1016/j.jclepro.2017.06.143.

European Financial Reporting Advisory Group. (2022). *European Sustainability Reporting Standards, ESRS E1 Climate Change*. Retrieved from https://www.efrag.org/Assets/Download?assetUrl=%2Fsites%2Fwebpublishing%2FSiteAssets%2F08%2520Draft%2520ESRS%2520E1%2520Climate%2520Change%2520November%25202022.pdf

Fix, E., & Hodges, J. L. (1951). *Discriminatory Analysis: Nonparametric Discrimination: Consistency Properties* (Report no. ADA800276). USAF School of Aviation Medicine. Retrieved from https://apps.dtic.mil/sti/citations/ADA800276.

Gangi, F., Varrone, N., Daniele, L. M., & Coscia, M. (2022). Mainstreaming socially responsible investment: do environmental, social and governance ratings of investment funds converge? *Journal of Cleaner Production, 353*, 131684. https://doi.org/10.1016/j.jclepro.2022.131684.

Garcia, A. S., Mendes-Da-Silva, W., & Orsato, R. (2017). Sensitive industries produce better ESG performance: evidence from emerging markets. *Journal of Cleaner Production, 150*, 135–147. https://doi.org/10.1016/j.jclepro.2017.02.180.

Ho, T. K. (1995). Random decision forests. In: *Proceedings of 3rd International Conference on Document Analysis and Recognition* (pp. 278–282). IEEE. https://doi.org/10.1109/ICDAR.1995.598994.

Hosmer, D. W., Lemeshow, S., & Sturdivant, R. X. (2013). *Applied Logistic Regression* (3rd ed.). Hoboken: John Wiley & Sons, Inc.

Høvring, C. M., Andersen, S. E., & Nielsen, A. E. (2018). Discursive tensions in CSR multistakeholder dialogue: A Foucauldian perspective. *Journal of Business Ethics*, *152*, 627–645. https://doi.org/10.1007/s1055101633304.

Kearney, C., & Liu, S. (2014). Textual sentiment in finance: A survey of methods and models. *International Review of Financial Analysis*, *33*, 171–185.

Kieraś, W., & Woliński, M. (2017). Morfeusz 2 – analizator i generator fleksyjny dla języka polskiego. *Język Polski*, *XCVII*(1), 75–83.

Li, F. (2010). Textual analysis of corporate disclosures: A survey of the literature. *Journal of Accounting Literature*, *29*, 143–165.

Logg, J. M., Minson, J. A., & Moore, D. A. (2019). Algorithm appreciation: People prefer algorithmic to human judgment. *Organizational Behavior and Human Decision Processes, 151*, 90–103.

Loughran, T., & McDonald, B. (2014). Measuring readability in financial disclosures. *The Journal of Finance*, *69*(4), 1643–167. https://doi.org/10.1111/jofi.12162.

Maon, F., Lindgreen, A., & Swaen, V. (2010). Organizational stages and cultural phases: A critical review and a consolidative model of corporate social responsibility development. *International Journal of Management Reviews*, *12*, 20–38. https://doi.org/10.1111/j.14682370.2009.00278.x.

Mazzotta, R., Bronzetti, G., & Veltri, S. (2020). Are mandatory non-financial disclosures credible? Evidence from Italian listed companies. *Corporate Social Responsibility and Environmental Management*, *27*(4), 1900–1913. https://doi.org/10.1002/csr.1935

Mikolov, T., Chen, K., Corrado, G., & Dean, G. (2013a). Efficient estimation of word representations in vector space. In: *Workshop Track Proceedings of the 1st International Conference on Learning Representations (ICLR 2013).* http://arxiv.org/abs/1301.3781.

Mikolov, T., Sutskever, I., Chen, K., Corrado, G., & Dean, J. (2013b). Distributed representations of words and phrases and their compositionality. In: C. J. Burges, L. Bottou, M. Welling, Z. Ghahramani & K. Q. Weinberger (Eds.), *Advances in Neural Information Processing Systems, 26*. Red Hook: Curran Associates, Inc., 3111–3119.

Noland, J., & Phillips, R. (2010). Stakeholder engagement, discourse ethics and strategic management. *International Journal of Management Reviews*, *12*, 39–49. https://doi.org/10.1111/j.14682370.2009.00279.x.

Oleson, K. E., Billings, D. R., Kocsis, V., Chen, J. Y. C., & Hancock, P. A. (2011). Antecedents of trust in human-robot collaborations. In: *Proceedings of 1st International Multi-Disciplinary Conference on Cognitive Methods in Situation Awareness and Decision Support*. IEEE. https://doi.org/10.1109/COGSIMA.2011.5753439.

Paliszkiewicz, J., Gołuchowski, J., & Chen, K. (2022). The meaning and interdisciplinary nature of trust in the digital economy - future directions. In: J. Paliszkiewicz & K. Chen (Eds.), *Trust, Organizations and the Digital Economy. Theory and Practice.* New York: Routledge, 3–14. https://doi.org/10.4324/9781003165965-1.

Pedregosa, F., Varoquaux, G., Gramfort, A., Michel, V., Thirion, B., Grisel, O., Blondel, M., Prettenhofer, P., Weiss, R., Dubourg, V., Vanderplas, J., Passos, A., Cournapeau, D., Brucher, M., Perrot, M., & Duchesnay, E. (2011). Scikit-learn: Machine Learning in Python. *Journal of Machine Learning Research, 12*, 2825–2830.

Piasecki, M., Janz, A., Kaszewski, D., & Cachor, G. (2017). *Word Embeddings for Polish*. CLARIN-PL digital repository. Retrieved from: http://hdl.handle.net/11321/442.

Rechkemmer, A., & Yin, M. (2022). When confidence meets accuracy: Exploring the effects of multiple performance indicators on trust in machine learning models. In: *Proceedings of the 2022 CHI Conference on Human Factors in Computing Systems* (pp. 1–14). https://doi.org/10.1145/3491102.3501967.

Rodrigue, M., Magnan, M., & Boulianne, E. (2013). Stakeholders' influence on environmental strategy and performance indicators: A managerial perspective. *Management Accounting Research*, *24*, 301–316. https://doi.org/10.1016/j.mar.2013.06.004.

Rudin, C. (2019). Stop explaining black box machine learning models for high stakes decisions and use interpretable models instead. *Nature Machine Intelligence, 1*(5), 206–215. https://doi.org/10.1038/s42256-019-0048-x.

Schaefer, K. E., Chen, J. Y. C., Szalma, J. L., & Hancock, P. A. (2016). A meta-analysis of factors influencing the development of trust in automation: Implications for understanding autonomy in future systems. *Human Factors*, *58*(3), 377–400. https://doi.org/10.1177/0018720816634228.

Schütze, F., & Stede, J. (2021). *EU Sustainable Finance Taxonomy - What Is Its Role on The Road Towards Climate Neutrality?* (DIW Berlin Discussion Paper No. 1923). Retrieved from: https://www.diw.de/documents/publikationen/73/diw_01.c.806768.de/dp1923.pdf

Siau, K., & Wang, W. (2018). Building trust in artificial intelligence, machine learning, and robotics. *Cutter Business Technology Journal, 32*(2), 47–53.

Yin, M., Wortman Vaughan, J., & Wallach, H. (2019). Understanding the effect of accuracy on trust in machine learning models. In: *Proceedings of the CHI Conference on Human Factors in Computing Systems* (pp. 1–12). https://doi.org/10.1145/3290605.3300509.

Zarzycka, E., Krasodomska, J., & Dobija, D. (2021). Stakeholder engagement in corporate social practices and non-financial disclosures: A systematic literature review. *Central European Management Journal*, *29*(1), 112–135.

16 Trust in application of neural networks for long-run market forecasting

Urszula Grzybowska and Marek Karwański

Introduction

The grain commodity market is subject to both stochastic factors shaping prices and deterministic macroeconomic factors such as wars and embargoes. By 2020, production, exports and prices in global markets were rising. Grain prices on foreign markets in 2020 were influenced by import demand, caused by uncertainty about harvest volumes due to drought in large areas of Europe and Argentina and concerns about the long-term effects on the global grain market caused by the spread of the coronavirus around the world. Importing countries, especially China, have taken steps to increase grain reserves. This was related to the expected increase in demand for feed grains in connection with the expected development of meat production – especially pork. Moreover, the introduction of restrictions on grain exports by Ukraine, as well as the introduction by the Russian Federation (from February 15, 2021) of a quota on the export of Russian grain and an export tax on wheat, barley and corn, resulted in additional pressure on price increases. Another important element distorting prices was the impact of the macroeconomic factor related to the war in Ukraine. Long-term product price forecasting (here up to two years ahead) is of key importance for financiers operating on the credit market or dealing with risk. Forecasting prices of basic agricultural products and energy carriers is of great importance for the economic stability of countries. The war in Ukraine showed that limiting the export of Ukrainian agricultural products resulted in a significant increase in food prices in all European and African countries, as well as the specter of famine in African countries, the main importers of Ukrainian grains (EC, 2022; Janzen & Zulauf, 2023).

The aim of state policy should be to maintain the prices of agricultural products at a sustainable level. Too high prices translate into increased inflation. Too low may cause a crisis in the national agricultural sector, as was the case in Poland, as a result of the uncontrolled inflow of cheap grains from Ukraine, which competes with local agricultural products (Krzysztoszek, 2023).

For this reason, it is important to forecast commodity prices, including agricultural products, taking into account price shocks caused by deterministic events.

DOI: 10.4324/9781032627236-18

Initially, at the beginning of the 21st century, the use of statistical methods for forecasting commodity prices was limited due to the instability of price and quantity behavior – the lack of stationarity of the observed time series. This was related to suggestions that quantities and prices in the goods market are random or even chaotic (random walk). This introduces a large amount of risk and uncertainty into the market modeling and forecasting process. Of course, randomness is generally assumed and the nature of price fluctuations varies depending on observations of these fluctuations and their likely causes over the long, medium term and short term. Over the long term, commodity markets are subject to shocks or changes in trends, from natural disasters and human intervention to structural market changes. These shocks tend to be irregular in nature and cause prices to rise suddenly, usually to higher but sometimes to lower levels.

In the short term, market shocks result primarily from financial factors, in particular related to speculation and hedging markets for commodity futures, options and other derivatives. The resulting price behavior is called random because it reflects the flow of randomly emerging information. In fact, price behavior can be described more specifically as stochastic or following a non-linear dynamic or stochastic process.

For such processes, models can be built based on time series statistical equations. They can be univariate, in which a single variable is explained in terms of its past statistical history, or multivariate, in which past statistical history is combined for several variables.

Explanations of medium-term behavior may involve models capable of generating some form of price cycles. Traditional models used for this purpose are ARIMA/ARFIMA (Hosting, 1981; Polotzek & Kantz, 2020), exponential smoothing (Holt, 2004; Winters, 1960), VAR models (Sims, 1980), (Lütkepohl, 1991) applied to financial time series by Hamilton, Campbell, Lo and MacKinlay, MacKinlay or structural time series models developed by Harvey in the early 1990s (Harvey, 1990). Short-term, referring to forecasts up to three months, are described in (Chu, 1978). Long-term behavior modeling additionally includes linear and non-linear trend models. Summarizing the research from the recent period, it should be emphasized that there has been a move away from structural models, which emphasized demand, supply and other price determinants, in favor of non-structural models that explain only the price variability itself. Part of this transition has been advances in unstructured modeling of financial markets. It should not be forgotten that the costs of conducting commodity market research can be significantly reduced by analyzing only price variables.

The most urgent need is progress in research on the mechanisms of commodity markets. The theory underlying the commodity price model will propose a functional form very close to the theoretical specification. This suggests the need for a newer approach to market and economic policy simulation; Forecast models should be based on new algorithmic approaches to forecasts, such as deep learning. This approach is currently being widely tested because in terms of prediction, it often achieves significantly better results. Algorithmic models often allow for an alternative, very interesting interpretation of the results.

Another problem is the inclusion of deterministic factors in the model that are known to influence price behavior. There are statistical models considered in the ITS (Interrupt Time Series) (Turner, Karahalios, Forbes et al., 2021; Turner, Karahalios, Forbes, Taljaard et al., 2020). This theory is mainly used in clinical trials. A sudden change in the behavior of the series is associated with the use of a deterministic factor, for example, treatment. Unfortunately, there is a lack of publications on the role of interventions and deterministic market disruptions from the perspective of neural networks.

In Confidence in the Application of Neural Networks for Long-Term Market Forecasting, we show that commodity price models based on neural networks that take into account deterministic factors will create a more realistic picture of related aspects of market structure. This chapter will present a neural network forecasting model used for long-term forecasting of wheat prices, taking into account the deterministic disturbing factor.

The first part will present the network architecture with particular emphasis on modeling sequential data. A fully connected (dense) neural network cannot be used to model data sequences because it cannot track long-term dependencies and cannot store information about the order of input data. Hence the need for a neural architecture that can handle this. It is an LSTM recurrent neural network. In the second part, the recurrent network model will be extended with deterministic intervention factors by introducing the ITS (Interrupt Time Series) mechanism. The last part presents conclusions and a proposal for using AI methods in market forecasting.

Literature review

Neural networks have been successfully used for decades to recognize images, text and speech (Malik, 2005). However, their applications now cover virtually all fields (Abiodun, Jantan, Omolara, Dada, Mohamed, & Arshad, 2018). In the last decade, neural networks have been successfully used for time prediction series. An overview of methods used in forecasting, including machine learning-based methods for energy prices, can be found in (Weron, 2014), and an overview of machine learning methods (GRU, LSTM, RNN and CNN) used in agricultural product price forecasting and their comparison with ARIMA and VAR models are presented in (Gowthaman et al., 2023). There are many publications on the comparison of machine learning methods with econometric/statistical methods used for price forecasting (Chatterjee, Bhowmick, & Sen, 2021; Hewamalage, Bergmeir, & Bandara, 2012; Toai, Senkerik, Zelinka, Ulrich, Hanh, & Huan, 2022; Paul, Yeasin, Kumar, Balasubramanian et al., 2022; Albeladi,

Zafar, & Mueen, 2023; Sun, Meng, Zhang, Wang, Jiang, & Liu, 2023). The results presented in (Toai, Senkerik, Zelinka, Ulrich, Hanh, & Huan, 2022) indicate that ARIMA gives better prediction results for short periods, and LSTM models for long-term forecasts. The use of machine learning methods for price prediction was presented, among others, in (Paul, Yeasin, Kumar, Balasubramanian et al., 2022). The application of deep learning (DL) methods to predict commodity prices has been discussed in many recent publications (Ji, Wang, & Yan, 2021; Bao, Yue, & Rao, 2017; Lawi, Mesra, & Amir, 2022; Ben Ameur, Boubaker, Ftiti et al., 2023; Gowthaman et al., 2023; Ly, Traore, & Dia, 2021; Kexian Zhang & Min Hong, 2022; Sun, Meng, Zhang, Wang, Jiang, & Liu, 2023; Gu, Jin, Yin, Zheng, Piao, & Yoo, 2022).

Our goal is not to compare methods used to forecast commodity prices. In this chapter, we propose a neural network architecture incorporating deterministic intervention that can be used for long-term predictions.

Data description

The analysis used a dataset of wheat prices. The set consisted of information provided by the Polish website https://stooq.pl/. These were the monthly wheat futures prices WHEAT (ZW.F) at the end of the month, expressed in cents per bushel (1 bushel = 27.216 kg). The observations came from the period from March 31, 2019 to August 31, 2023 and formed a set of 54 elements. The series can be sourced from https://stooq.pl/q/d/?s=zw.f.

Methodology (overview of networks: DENSE and LSTM)

Algorithmic models – artificial neural networks – were used to forecast prices. Neural networks are used in many areas to model and predict the dynamics of complex systems. Many types of networks are available, but the accuracy of modeling depends largely on the fit of the network architecture. The article presents the results of the applications of neural networks, with particular emphasis on long-term and short-term memory (LSTM) networks, which were used for dynamic modeling of systems in various application areas. The goal is to configure predictive models to produce nonlinear time-dependent forecasts.

Neural networks

The neural networks (ANNs) we use today are the result of 200 years research of scientists from various fields. A chronological discussion of the developments that contributed to the development of Deep Learning can be found in Schmidhuber's 2015 work (Schmidhuber, 2015). Mathematical aspects related to deep learning can be found in (Calin, 2020). Neural networks are a series of nonlinear affine transformations defined by weights matrices w_i the bias b_i vectors and non-linear functions ϕ_i called activation functions.

We can therefore write the ANN in the following form:

$$f_{w,b}(\ )=\phi_n\left(\mathrm{Aff}_{w_n,b_n}^{(n)}\cdot\phi_{n-1}\left(\mathrm{Aff}_{w_{n-1},b_{n-1}}^{(n-1)}\cdot\phi_{n-2}\left(\mathrm{Aff}_{w_{n-2},b_{n-2}}^{(n-2)}\ \ldots\right)\right)\right)$$

Affine transformations called neurons and non-linear transformations form a network $f_{w,b}$. The function $f_{w,b}$ is a model of the input-output mapping of a given neural network $X \overset{\text{ANN}}{\longrightarrow} Y$.

RNN and vanishing gradients

RNNs enable the modeling of time-dependent and sequential data tasks. However, when using RNNs, the problem of vanishing gradients arises, which makes it difficult to learn long data sequences. Gradients carry information used to update the parameters of the RNN, and as the gradient becomes smaller, the parameter updates become irrelevant.

Let's briefly recall what RNNs look like. The calculations will be performed on a simple RNN model with a single hidden layer and one output sequence.

$$h_1\ h_2\ h_k$$

$$\left(c_0, x_1\right)\begin{pmatrix} W_c \\ W_x \end{pmatrix} \to \underset{x}{\overset{\uparrow}{c_1}} \Rightarrow \left(c_1, x_2\right)\begin{pmatrix} W_c \\ W_x \end{pmatrix} \to \underset{x}{\overset{\uparrow}{c_2}} \Rightarrow \ldots \Rightarrow \left(c_{k-1}, x_k\right)\begin{pmatrix} W_c \\ W_x \end{pmatrix} \overset{\uparrow}{\multimap} \vdots$$

The network has an input sequence of vectors $(x_1, x_2, \ldots, x_k)$, at some point t the network has input vector x_t. Past information and acquired knowledge are encoded in network state vectors $(c_1, c_2, \ldots, c_k)$, at a given moment the network has a state vector of c_{t-1}. Input vector x_t and state vector c_{t-1} are combined to create a complete input vector at one time, the input information is (c_{t-1}, x_t). The network has two weight matrices: W_c and W_x connecting c_{t-1}, and x_t two parts of the input vector (c_{t-1}, x_t) with the hidden layer. For simplicity, we ignore the bias vectors in our calculations and denote them $W = (W_c, W_x)$.

The sigmoid function is used as the activation function in the hidden layer. The network outputs a single vector in the last time step (RNNs can output a vector at each time step, but we will use this simpler model).

Backpropagation in time (BPTT) for RNNs

After we calculate the prediction vector $h(k)$, in the RNN propagation process, we are interested in the prediction error $E(k)$. To minimize it, we use a back-propagation algorithm (Rumelhart, Hinton, & Williams, 1986). In a recurrent network, the error is counted at each time point. Therefore, the gradient is calculated according to the formula

$$\frac{\partial E}{\partial W} = \sum_{t=1}^{T} \frac{\partial E_t}{\partial W}$$

The gradient is used to update model parameters during the learning process Gradient Descent algorithm (GD).

We'll use the basic version of GD.

$$W \leftarrow W - \alpha \frac{\partial E}{\partial W}$$

We have a learning task that involves T time steps, the error gradient in the k time step is given by:

$$\begin{aligned}\frac{\partial E_k}{\partial W} &= \frac{\partial E_k}{\partial h_k}\frac{\partial h_k}{\partial c_k}\cdots\frac{\partial c_2}{\partial c_1}\frac{\partial c_1}{\partial W} \\ &= \frac{\partial E_k}{\partial h_k}\frac{\partial h_k}{\partial c_k}\left(\prod_{t=2}^{k}\frac{\partial c_t}{\partial c_{t-1}}\right)\frac{\partial c_1}{\partial W}\end{aligned}$$

Note that since $W = [W_c, W_x]$, c_t can be written as

$$c_t = \sigma\left(W_c \cdot c_{t-1} + W_x x_t\right)$$

derivative c_t, sums to

$$\begin{aligned}\frac{\partial c_t}{\partial c_{t-1}} &= \sigma'\left(W_c \cdot c_{t-1} + W_x x_t\right)\cdot\frac{\partial}{\partial c_{t-1}}\left(W_c \cdot c_{t-1} + W_x x_t\right) \\ &= \sigma'\left(W_c \cdot c_{t-1} + W_x x_t\right)\cdot W_c\end{aligned}$$

Gradient backpropagation yields

$$\frac{\partial E_k}{\partial W} = \frac{\partial E_k}{\partial h_k}\frac{\partial h_k}{\partial c_k}\left(\prod_{t=2}^{k}\sigma'\left(W_c \cdot c_{t-1} + W_x x_t\right)\cdot W_c\right)\frac{\partial c_t}{\partial W}$$

The last term tends to decay when it is large, this is due to the derivative of the activation function tan$h$ which is less than 1.

The derivative product may also grow, this is the so-called the exploding gradient problem. If we have

$$\prod_{t=2}^{k} \sigma'\left(W_c \cdot c_{t-1} + W_x x_t\right) \cdot W_c \to 0$$

this means that for some step k

$$\frac{\partial E_k}{\partial W} \to 0$$

our total error gradient will disappear

$$\frac{\partial E}{\partial W} = \sum_{t=1}^{T} \frac{\partial E_t}{\partial W} \to 0$$

and the update of the network weights will look like this

$$W \leftarrow W - \alpha \frac{\partial E}{\partial W} \approx W$$

no significant learning rate will be achieved within a reasonable time.

LSTM as an enhancement of the RNN

The idea of a long-term memory network was introduced by Hochreiter in 1991 in his master's thesis, but the first publication on the LSTM network dates back to 1997 and is the work of Hochreiter and Schmidhuber (Hochreiter, & Schmidhuber, 1997). The LSTM network model was introduced as an improvement on RNN networks, in which the problem of disappearing or exploding gradients described in 1994 appeared with long data sequences that could have hundreds or even thousands of steps (Bengio, Simard, & Frasconi, 1994). The LSTM model includes special cells capable of learning long-term relationships (Arras et al., 2019). The LSTM network architecture is based on three types of gates that update and control the states of cells, they are forget gate, input gate and output gate. The gates use sigmoid and tanh activation functions.

LSTM network has the input vector $\left(h_{t-1}, x_t\right)$ at time t. The state of a network cell is denoted by c_t. The output vectors passing through the network between successive time moments t are denoted by h_t. The architecture of an LSTM unit can be found, for example in (Lawi, Mesra, & Amir, 2022) or (Arras et al., 2019).

The forget gate controls which information from the cell state is forgotten, taking into account new information rather than information already introduced into the cell. It was introduced in 1999 (Gers, Schmidhuber, & Cummins, 1999).

1 Forget gate is defined as:

$$f_t = \sigma\left(W_f \cdot \left[h_{t-1}, x_t\right]\right)$$

The forget gate is a sigmoidal layer that is used to forget irrelevant historical information.

2 The input gate controls what new information will be encoded into the cell state given the new input information.

The input gate is a transformation in the form:

$$\tan h\left(W_c \cdot \left[h_{t-1}, x_t\right]\right) \circ \sigma\left(W_i \cdot \left[h_{t-1}, x_t\right]\right)$$

The operator $\circ$ is equal to the element-wise product of the corresponding elements of the outputs of the two fully connected layers $\tilde{c}_t$ and i_t.

$$\tilde{c}_t = \tan h\left(W_c \cdot \left[h_{t-1}, x_t\right]\right)$$

$$i_t = \sigma\left(W_i \cdot \left[h_{t-1}, x_t\right]\right)$$

3 The output gate controls what information encoded in the cell state is sent to the network as input in the next time step, this is done via the output vector h_t.

The output gate and the output vector of the cell are defined as:

$$\vartheta_t = \sigma\left(W_\vartheta \cdot \left[h_{t-1}, x_t\right]\right)$$

$$h_t = \vartheta_t \circ \tanh\left(c_t\right)$$

4 Long-term dependencies and relationships are encoded in the cell state vectors. A derivative of the cell state can prevent LSTM gradients from vanishing. The LSTM cell state has the form:

$$c_t = c_{t-1} \circ f_t + \tilde{c}_t \circ i_t$$

Backpropagation through time in LSTM

Similar to the RNN model, the LSTM network outputs a prediction vector h_k at time k. The knowledge encoded in state vectors c_t perpetuates long-term relationships and dependencies in sequential data. Data sequences can be hundreds or even thousands of time steps long, which creates a computational problem when using a basic RNN model. The gradient used to update the network parameters is calculated at each time step T. As in RNN, the gradient of the error components is determined by the following sum of T sub-gradients

$$\frac{\partial E}{\partial W} = \sum_{t=1}^{T} \frac{\partial E_t}{\partial W}$$

In order for the error gradient to vanish, all these T sub-gradients must vanish. The series converges to zero if the sequence of its subtotals tends to zero,

$$\sum_{t=1}^{T} \frac{\partial E_t}{\partial W} \to 0 \text{ means that } \underset{n>N_0}{\Lambda} \sum_{t=1}^{n} \frac{\partial E_t}{\partial W} \to 0$$

So, if we want a gradient not to vanish, our network must increase the probability that at least some of the sub-gradients will not vanish, in other words, the series of sub-gradients must not converge to zero.

Error gradients in the LSTM network

Error gradient for the time instant k is of the form:

$$\frac{\partial E_k}{\partial W} = \frac{\partial E_k}{\partial h_k} \frac{\partial h_k}{\partial c_k} \cdots \frac{\partial c_2}{\partial c_1} \frac{\partial c_1}{\partial W}$$

$$= \frac{\partial E_k}{\partial h_k} \frac{\partial h_k}{\partial c_k} \left(\prod_{t=2}^{k} \frac{\partial c_t}{\partial c_{t-1}} \right) \frac{\partial c_1}{\partial W}$$

Vanishing of the gradients results due to the following term

$$\prod_{t=2}^{k} \frac{\partial c_t}{\partial c_{t-1}}$$

In LSTM network internal state vector c_t is of the form:

$$c_t = c_{t-1} \circ f_t + \tilde{c}_t \circ i_t$$

$$c_t = c_{t-1} \circ \sigma\left(W_f \cdot \left[h_{t-1}, x_t\right]\right)_t + \tan h\left(W_c \cdot \left[h_{t-1}, x_t\right]\right) \circ \sigma\left(W_i \cdot \left[h_{t-1}, x_t\right]\right)$$

It is a function of the following elements that must be taken into account while calculating the derivative in backpropagation: c_{t-1}, f_t, $\tilde{c}_t$, i_t.

The derivative $\frac{\partial c_t}{\partial c_{t-1}}$ is:

$$\frac{\partial c_t}{\partial c_{t-1}} = \frac{\partial}{\partial c_{t-1}}\left[c_{t-1} \circ f_t + \tilde{c}_t \circ i_t\right]$$

$$= \frac{\partial}{\partial c_{t-1}}\left[c_{t-1} \circ f_t\right] + \frac{\partial}{\partial c_{t-1}}\left[\tilde{c}_t \circ i_t\right]$$

$$= \frac{\partial f_t}{\partial c_{t-1}} c_{t-1} + \frac{\partial c_{t-1}}{\partial c_{t-1}} f_t + \frac{\partial i_t}{\partial c_{t-1}} \tilde{c}_t + \frac{\partial \tilde{c}_t}{\partial c_{t-1}} i_t$$

We calculate the four derivatives and write

$$\frac{\partial c_t}{\partial c_{t-1}} = \sigma'\left(W_t\left[h_{t-1}, x_t\right]\right) \cdot W_f \cdot \vartheta_t \circ \tanh' h\left(c_{t-1}\right) \cdot c_{t-1} +$$

$$f_t +$$

$$\sigma'\left(W_i\left\lfloor h_{t-1}, x_t\right\rfloor\right) \cdot W_i \cdot \vartheta_{t-1} \circ \tanh' h\left(c_{t-1}\right) \cdot \tilde{c}_t +$$

$$\sigma'\left(W_c\left\lfloor h_{t-1}, x_t\right\rfloor\right) \cdot W_c \cdot \vartheta_{t-1} \circ \tanh' h\left(c_{t-1}\right) \cdot \tilde{i}_t$$

Let us now denote:

$$A_t \equiv \sigma'\left(W_t\left\lfloor h_{t-1}, x_t\right\rfloor\right) \cdot W_f \cdot \vartheta_t \circ \tanh' h\left(c_{t-1}\right) \cdot c_{t-1}$$

$$B_t \equiv f_t$$

$$C_t \equiv \sigma'\left(W_i\left\lfloor h_{t-1}, x_t\right\rfloor\right) \cdot W_i \cdot \vartheta_{t-1} \circ \tanh' h\left(c_{t-1}\right) \cdot \tilde{c}_t$$

$$D_t \equiv \sigma'\left(W_c\left\lfloor h_{t-1}, x_t\right\rfloor\right) \cdot W_c \cdot \vartheta_{t-1} \circ \tanh' h\left(c_{t-1}\right) \cdot \tilde{i}_t$$

then we can write

$$\frac{\partial c_t}{\partial c_{t-1}} = A_t + B_t + C_t + D_t$$

which gives

$$\frac{\partial E_k}{\partial W} = \frac{\partial E_k}{\partial h_k} \frac{\partial h_k}{\partial c_k} \left(\prod_{t=2}^{k} \left[A_t + B_t + C_t + D_t\right] \right) \frac{\partial c_1}{\partial W}$$

The gradient contains a forget gate activation vector that allows the network to better control the subgradient values at any time by taking advantage of appropriate updates to the forget gate parameters. The presence of a forget gate allows the LSTM network to decide at any time that certain information should not be forgotten and update the model parameters accordingly.

To ensure that the gradient does not vanish, we can find a suitable update of the forget gate parameter at time k+1 such that:

$$\frac{\partial E_{k+1}}{\partial W} \nrightarrow 0$$

hence,

$$\sum_{t=1}^{k+1} \frac{\partial E_t}{\partial W} \nrightarrow 0$$

and the gradient does not vanish.

Another important property worth noting is that the cell state gradient is an additive function consisting of four elements denoted as A_t, B_t, C_t, D_t. This additive property allows for better balancing of gradient values during backpropagation. The LSTM network updates and balances the values of the four terms, increasing the probability that the additive term will not disappear. This addition property distinguishes LSTM from the RNN case where the gradient contained a single element in the product. In RNN, the sum $\sum_{t=1}^{T} \frac{\partial E_t}{\partial W}$ is formed from expressions with similar behavior whose value is in the interval [0,1], resulting in gradient vanishing.

In LSTM, the presence of a forget gate together with the additive property of cell state gradients allows the network to update its parameters in such a way that different subgradients $\frac{\partial E_k}{\partial W}$ do not necessarily behave in a similar way, reducing the probability of all subgradients disappearing, in other words, a series of functions do not have to converge to zero.

Drop-out technique

In neural networks, regularization is a technique for modifying a training algorithm that reduces generalization error by increasing the learning error in a specific sample. A popular regularization technique is drop-out. The standard dropout method involves modifying the network by turning off random neurons during training. Neurons are removed during training and testing to reduce the complexity of the network and improve its performance. Let us denote in a simple case the input standardized data by $\{X_i\}$, Y be the target variable. Consider a single layer neural network with a linear activation function $f(x) = x$. In this case, the final neuron (which serves as the output layer) calculates the weighted sums of all inputs. For the general form of the neural network, we will obtain analogous results.

Model estimation involves minimizing the loss function. Take the quadratic loss function $\mathcal{L}$, then

$$\mathcal{L} = \frac{1}{2}\left(Y - \sum_{i=1}^{n} w_i X_i\right)^2$$

Our drop-out modification involves the use of loss function $\mathcal{L}_{\text{drop}}$ defined as:

$$\mathcal{L}_{\text{drop}} = \frac{1}{2}\left(Y - \sum_{i=1}^{n} \alpha_i w_i X_i\right)^2$$

where the dropout rate, denoted by α_i, takes the values $(0,1)$.

The formula for $\mathcal{L}$ represents the loss function used in a regular neural network, while the formula for $\mathcal{L}_{\text{drop}}$ corresponds to the loss function specific to a drop-out network. Let the dropout rate α_i will follow the Bernoulli distribution with parameter p. This means that $\Pr(\alpha_i = 1) = p$.

It can be shown that the expected value of the gradient using the drop-out technique is equivalent to the gradient of a regular neural network for weights w scaled according to the equation $w = p \cdot x$.

LSTM model with deterministic factor

An additional component is a mechanism called interventionism in time series modelling. It means adding a deterministic component to the model. Such a macroeconomic factor related to the war in Ukraine was introduced into the model. Namely, a component "responsible" for the impact of the war in Ukraine in the form of a "shock" was added to the model,

$$X_t^{\text{Shock}} = \begin{cases} 0 & t < 22.02.2022 \\ 1 & t \in \left[28.02.2023, \quad 31.07.223\right] \\ 0.5 & t > 31.07.2023 \end{cases}$$

In practice, all activities external to prices can be included in the model as so-called interventions. The impact of such interventions is reflected in the way in which prices are formed but a specific functional relationship must be introduced into the model.

Forecasting was carried out on the following sets:

- Training set $\left\{X_t\right\}_t$ where $t \in \left\{31.03.2019 - 31.12.2022\right\}$
- Validation set $\left\{X_t\right\}_t$ where $t \in \left\{31.01.2023 - 31.08.2023\right\}$

This will allow a direct comparison of the forecast results obtained on the training set with the price values from the validation set.

We propose the network architecture in which the model consists of two parts: deterministic and stochastic.

$$X_t = X_t^{\text{stochastic}} + X_t^{\text{detrministic}}$$

Diagram of the neural network architecture used in the model and the detailed program codes used for the calculations are available from the authors.

LSTM model with drop-out regularization was used to model the stochastic part. The LSTM cells formed a two-level stack. For the analysis of the deterministic part, we used a DENSE network with several hidden layers and RELU activation functions to make the network-based forecast very close to a 'staircase' without a random component. Here it would be possible to build a simpler structure using lambda layers.

Results

We modelled the stochastic part with LSTM cells. Putting all the forecasts from (1) pure LSTM model, (2) LSTM with deterministic (ITS) models together in a single graph gives the result shown in Figure 16.1.

The last layer named CONCATENATE/OUT, due to linearity of activation function, is just a weighted average:

$$Y_t^{\text{OUT}} = b_0 + w_1 X_t^{\text{determ.branch}} + w_2 X_t^{\text{stochast.branch}}$$

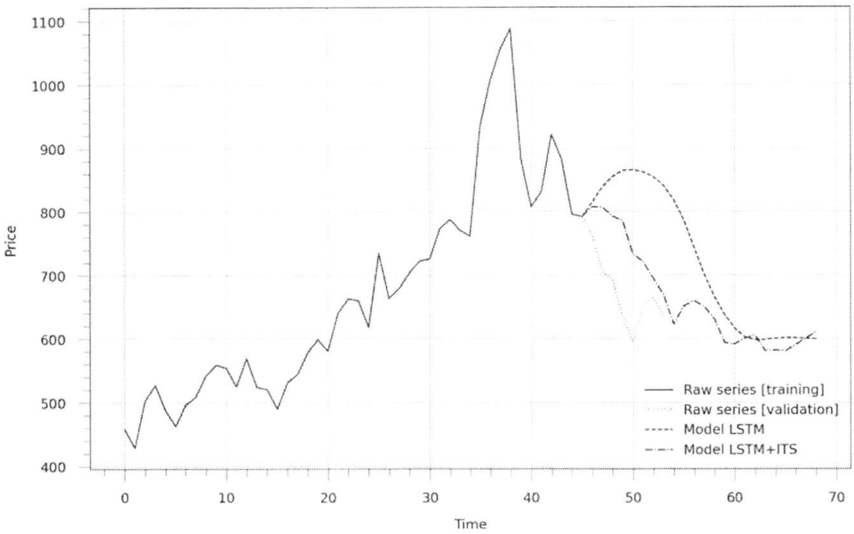

Figure 16.1 Predictions of LSTM, LSTM+ITS models, against real data (validation set).
Source: Authors' own elaboration.

Table 16.1 Weights of the model with standard errors

Weights	$w_1 = 0.16826$	$w_2 = 0.60611$	$b_0 = -0.38101$
Standard errors	$\text{std.err}_{w_1} = 0.02189$	$\text{std.err}_{w_2} = 0.00509$	$\text{std.err}_{b_0} = 0.00768$

Due to the dependence of the LSTM cell results on the choice of starting values, calculations were carried out in 100 simulations and the average value was used as the result. The weights in the model were calculated after applying standardization so that the minimum price is 0.0 and the maximum price is 1.0 (Table 16.1). Finally, the values of the weights are:

We can define the magnitude of the influence of a deterministic factor on the prices as follows

$$\begin{aligned}\text{Influence}_{\text{determ.}} &= \left[X^{\text{stochast}}\right] - \left[X^{\text{stochast}} + X^{\text{detrm}}\right] \\ &= X_t^{\text{stochast}} - \left(b_0 + w_2 X_t^{\text{stochast}} + w_1 X_t^{\text{determ}}\right) \\ &= X_t^{\text{stochast}}\left(1 - w_2\right) - \left(w_1 X_t^{\text{determ}} + b_0\right)\end{aligned}$$

In our case, since we used standardizing transformations, we can calculate this impact by taking serial levels as ≈ 1.0 and then scaling the result by the actual price level (≈ 600):

$$\text{Influence}_{\text{determ.}} \approx 600 \cdot \left[\left(1-0.60611\right)-\left(0.16826-0.38101\right)\right] \approx 360$$

The above calculations assume X_t price levels for: pre-transformation ≈ 1.0 and post-transformation ≈ 600 units. With these assumptions, we calculate that the change in wheat prices as a result of the war was 360 units.

Conclusions

Data modeling and analysis have two main goals: (1) forecasting, i.e., predicting the future, (2) extracting "information", that is, analyzing the cause-and-effect structure. Of course, the causal structure is identified indirectly and depends on the models used. In this chapter, we present a neural network model enabling the analysis of a time series decomposed into components. Models for individual components may differ, especially if we are interested in estimating the weights for individual components. Identification of individual component weights of a time series is a complex process and does not lead to unambiguous results. In traditional statistical models based on a regression equation, such as ARIMA/ITS, the influence of an external (deterministic) factor is estimated directly on the basis of the regression parameters. The proposed ANN/ITS model defines this impact slightly differently, as it takes into account the interaction of the deterministic factor with the mechanism generating the process.

Based on the weights of the network in the final layer, we can conclude that the influence of the deterministic factor consists in adding an additive factor but at the same time changing the weight of the basic stochastic factor. That is, we can see that the deterministic factor changes both the price level by shifting and changing the slope of the forecast. This leads to a specific interpretation of the results, which, according to the authors, is close to an intuitive understanding of the influence of the factors on the process.

Research results show that we can trust AI methods. They provide excellent forecasts for both short and long-term market processes. AI, like no other methods, allows for the automation of management applications based on market analyses.

All calculations were done in Python rel 3.9 using the Keras API based on TensorFlow platform.

Bibliography

Abiodun, O.I., Jantan, A., Omolara, A.E., Dada, K.V., Mohamed, N.A., & Arshad, H.S. (2018). State-of-the-art in artificial neural network applications: A survey. *Heliyon, 4.* Doi:10.1016/j.heliyon.2018.e00938.

Albeladi, K., Zafar, B., & Mueen, A. (2023). Time series forecasting using LSTM and ARIMA. *International Journal of Advanced Computer Science and Application, 14*(1), 313–320. Doi:10.14569/ijacsa.2023.0140133

Arras, L., Arjona-Medina, J., Gillhofer, M., Widrich, M., Montavon, G., M¨uller, K.R., Hochreiter, S., & Samek, W (2019). Explaining and Interpreting LSTMs. In W.M. Samek (Ed.), *Explainable AI: Interpreting, Explaining and Visualizing Deep Learning. Lecture Notes in Computer Science* (Vol. 11700). Springer, Cham. Doi:10.1007/978-3-030-28954-6_11

Bao, W., Yue, J., & Rao, Y. (2017). A deep learning framework for financial time series using stacked autoencoders and long-short term memory. *PLoS One, 12*(7), 1–24. Doi:10.1371/journal.pone.0180944

Ben Ameur, H., Boubaker, S., Ftiti, Z., Boubaker, S., Ftiti, Z., Louhichi, W., & Tissaoui, K. (2023). Forecasting commodity prices: empirical evidence using deep learning tools. *Annals of Operations Research*, 1–19. Doi:10.1007/s10479-022-05076-6

Bengio, Y., Simard, P., & Frasconi, P. (1994). Learning long-term dependencies with gradient descent is difficult. *IEEE Transactions on Neural Networks, 5*(2), 157–166. Doi:10.1109/72.279181

Calin, O. (2020). *Deep Learning Architectures. A Mathematical Approach.* Springer, Cham. Doi:10.1007/978-3-030-36721-3

Chatterjee, A., Bhowmick, H., & Sen, J. (2021). Stock price prediction using time series, econometric, machine learning, and deep learning models. *IEEE Mysore Sub Section International Conference* (pp. 289–296). Hassan, Karnataka, India. Doi:10.1109/MysuruCon52639.2021.9641610

Chu, K. (1978). Short-run forecasting of commodity prices: An application of autoregressive moving average models. *IMF Economic Review, 25*, 90–111. Doi:10.2307/3866657

EC. (2022). *The Impact of Russia-Ukraine Conflict on Global Food Security.* Retrieved from https://knowledge4policy.ec.europa.eu/publication/impact-russia-ukraine-conflict-global-food-security_en

Gers, F.A., Schmidhuber, J., & Cummins, F. (1999). Learning to forget: Continual prediction with LSTM. *Ninth International Conference on Artificial Neural Networks ICANN 99, 2* (pp. 850–855). Edinburgh. Doi:10.1049/cp:19991218

Gowthaman, T., Adarsh, V.S., Sathees Kumar, K., & Manobharathi, K. (2023). A review of deep learning models for price prediction in agricultural commodities. *Economic Affairs, 68*(1), 561–568. Doi:10.46852/0424–2513.1.2023.25

Gu, Y.H., Jin, D., Yin, H., Zheng, R., Piao, X., & Yoo, S.J. (2022). Forecasting agricultural commodity prices Using dual input attention LSTM. *Agriculture, 12*(2), 256, 1–18. Doi:10.3390/agriculture12020256

Harvey, A. (1990). *Forecasting, Structural Time Series Models and the Kalman Filter.* Cambridge University Press, Cambridge. Doi:10.1017/CBO9781107049994

Hewamalage, H., Bergmeir, Chr., & Bandara, K. (2012). Recurrent neural networks for time series forecasting: Current status and future directions. *International Journal of Forecasting, 37*(1), 388–427. Doi:10.1016/j.ijforecast.2020.06.008.

Hochreiter, S., & Schmidhuber, J. (1997). Long short-term memory. *Neural Computation, 9*(8), 1735–1780. Doi:10.1162/neco.1997.9.8.1735

Holt, C. (2004). Forecasting seasonals and trends by exponentially weighted moving averages. *International Journal of Forecasting, 20*(1), 5–10. Doi:10.1016/j.ijforecast.2003.09.015.

Hosting, J.R. (1981). Fractional differencing. *Biometrika, 68*(1), 165–176. Doi:10.1093/biomet/68.1.165

Janzen, J., & Zulauf, C. (2023). *The Russia-Ukraine War and Changes in Ukraine Corn and Wheat Supply: Impacts on Global Agricultural Markets.* Retrieved from https://farmdocdaily.illinois.edu/2023/02/the-russia-ukraine-war-and-changes-in-ukraine-corn-and-wheat-supply-impacts-on-global-agricultural-markets.html

Ji, X., Wang, J., & Yan, Z. (2021). A stock price prediction method based on deep learning technology. *International Journal of Crowd Science, 5*(1), 55–72. Doi:10.1108/IJCS-05–2020-0012

Krzysztoszek, A. (2023). *Polish Politicians Bicker Over Grain Crisis.* Retrieved from https://www.euractiv.com/section/politics/news/polish-politicians-bicker-over-grain-crisis/.

Lawi, A., Mesra, H., & Amir, S. (2022). Implementation of long short-term memory and gated recurrent units on grouped time-series data to predict stock prices accurately. *Journal of Big Data, 9*(89), 1–19. Doi:10.1186/s40537-022-00597-0

Lütkepohl, H. (1991). *Introduction to Multiple Time Series Analysis.* Springer-Verlag, Berlin Heidelberg. Doi:10.1007/978-3-662-02691-5

Ly, R., Traore, F., & Dia, K. (2021). *Forecasting Commodity Prices Using Long-Short-Term Memory Neural Networks. IFPRI Discussion Paper 2000.* International Food Policy Research Institute, Washington, DC. Doi:10.2499/p15738coll2

Malik, N. (2005). Artificial neural networks and their applications. *National Conference on 'Unearthing Technological Developments & their Transfer for Serving Masses' GLA ITM.* Mathura, India. Doi:10.48550/arXiv.cs/0505019

Paul, R.K., Yeasin, M., Kumar, P., Balasubramanian, M., Roy, H.S., Paul, A.K., & Gupta, A. (2022). Machine learning techniques for forecasting agricultural prices: A case of brinjal in Odisha, India. *PLoS One, 17*(7), 1–17. Doi:10.1371/journal.pone.0270553

Polotzek, K., & Kantz, H. (2020). An ARFIMA-based model for daily precipitation amounts with direct access to fluctuations. *Stochastic Environmental Research and Risk Assessment, 34*, 1487–1505. Doi:10.1007/s00477-020-01833-w

Rumelhart, D., Hinton, G., & Williams, R. (1986). Learning representations by back-propagating errors. *Nature, 323*, 533–536. Doi:10.1038/323533a0

Schmidhuber, J. (2015). Deep learning in neural networks: An overview. *Neural Networks, 61*, 85–117. Doi:10.1016/j.neunet.2014.09.003

Sims, C.A. (1980). Macroeconomics and reality. *Econometrica, 48*(1), 1–48.

Sun, F., Meng, X., Zhang, Y., Wang, Y., Jiang, H., & Liu, P. (2023). Agricultural product price forecasting methods: A review. *Agriculture, 13*(1671), 1–20. Doi:10.3390/agriculture13091671

Toai, T.K., Senkerik, R., Zelinka, I., Ulrich, A., Hanh, V.T.X., & Huan, V.M. (2022). ARIMA for short-term and LSTM for long-term in daily Bitcoin price prediction. In L.S. Rutkowski (Ed.), *Artificial Intelligence and Soft Computing. ICAISC 2022. Lecture Notes in Computer Science. 13588.* Artificial Intelligence and Soft Computing. ICAISC 2Springer, Cham. Doi:10.1007/978-3-031-23492-7_12

Turner, S.L., Karahalios, A., Forbes, A.B., Taljaard, M., Grimshaw, J.M., Cheng, A.C., Bero, L., & McKenzie, J.E. (2020). Design characteristics and statistical methods used in interrupted time series studies evaluating public health interventions: A review. *Clinical Epidemiology, 122*, 1–11. Doi:10.1016/j.jclinepi.2020.02.006

Turner, S.L., Karahalios, A., Forbes, A.B., Taljaard, M., Grimshaw, J.M., & McKenzie, J.E. (2021). Comparison of six statistical methods for interrupted time series studies: Empirical evaluation of 190 published series. *BMC Medical Research Methodology, 21*(134), 1–19. Doi:10.1186/s12874-021-01306-w

Weron, R. (2014). Electricity price forecasting: A review of the state-of-the-art with a look into the future. *International Journal of Forecasting, 30*(14), 1030–1081. Doi:10.1016/j.ijforecast.2014.08.008

Winters, P.R. (1960). Forecasting sales by exponentially weighted moving averages. *Management Science, 6*(3), 324–342. Doi:10.1287/mnsc.6.3.324

Zhang, K., & Hong, M. (2022). Forecasting crude oil price using LSTM neural networks. *Data Science in Finance and Economics, 2*(3), 163–180. Doi:10.3934/DSFE.2022008

17 The role of artificial intelligence in online shopping and trust

Nataliya Horin, Elżbieta Kacperska and Katarzyna Łukasiewicz

Introduction

Trust is a key factor influencing purchase decisions of customers. It is an indispensable element of all transactions. According to Grudzewski et al. (2007), it refers to relationships, it encourages reciprocity and as such it may enhance both the quality and the quantity of social interactions and as a result – also proper business transactions. In the opinion of Rudzewicz (2009), subjects of these relationships include customers, being the trusting parties, and the parties offering something for sale, sellers, being the trusted parties (trustees). Purchase decisions of consumers are influenced to a considerable extent by various factors, such as personal preferences, trust in websites, and particularly safety. According to Chmielarz (2007), these are socio-cultural barriers, occurring next to other barriers: legal, customary or ideological. Barriers related to trust may result from the lack of knowledge concerning new technologies and phenomena, such as the Internet and e-commerce. Another reason may also be related to concerns associated with interference in privacy as a result of various personal data, which need to be provided when concluding purchase contracts on the Internet. E-commerce companies use artificial intelligence (AI), which facilitates identification of customer needs and preferences in relation to products and services bought online. The application of AI makes it easy to monitor preferences, purchase patterns, choices, purchase frequency and amounts spent, providing sellers to accumulate accurate and detailed information on their customers. However, nowadays, the information alone is not enough, trust still needs to be nurtured and reinforced. For this reason, all offers targeting customers should be reliable and factual, while at the same time ensuring continuous customer contacts with the seller.

The concept of artificial intelligence and trust

AI is "a branch of computer science investigating principles governing human mental activity and creating computer programs or systems simulating human thinking" (Słownik, 2007). The primary aim for advances in AI is

DOI: 10.4324/9781032627236-19

to develop – through engineering processes – such capabilities as reasoning, knowledge, planning, learning, processing of natural language (communication), perception as well as movement and object handling (Mróz, 2017). The role and the task of AI is to process and identify collected data, and next to execute specific tasks, performed by Artificial Narrow Intelligence (ANI) (Shanahan, 2015). Another type of AI is Artificial General Intelligence (AGI), which scope of intellectual capacity is comparable to that of the human brain (Sterne, 2017). According to Jarek et al. (Jarek et al., 2018),

> Artificial intelligence, to date associated with the work of engineers, IT specialists and analysts, increasingly often extends outside its traditional boundaries into the area of management and marketing. Data on consumers, big data systems or mobile device available online make artificial intelligence a useful tool for marketing. Databases are extensively used starting from analyses of consumer needs, market analyses, knowledge on potential buyers and the competition, through actions involving various channels of communication and distribution, up to the measurement of consequences and effects of an adopted strategy. Thus marketing becomes a natural beneficiary of the developing information technology.

Good management of any company is based on collected information; however, trust also plays an important role. Trust is defined in a variety of ways. Weber, Deepak and Murnighan (2005) specify trust as the foundation of social interactions in an organization. Gambetta (1988) treats it as an organization's asset indicated that it is an element of social capital. In turn, Mayer, Davis and Schoorman (1995) stated that trust is the willingness to accept behaviors of the other party. Rotter (1980) and Bhattacharya and Devinney (1998) presented trust as an expectation concerning the outcome of an interaction. According to Sztompka (2007), trust is a bet made concerning uncertain, future actions of other people.

In turn, Bugdol (2010) distinguished the following dimensions of trust: faith, credibility, reliability, competences, predictability, and benevolence.

The individual dimensions of trust are interdependent and mutually inclusive. According to Bugdol (2010), competences constitute the core of trust. Reliability, predictability, and responsibility come second. Benevolence is treated as one of the factors influencing credibility. Components of credibility include competences, as well as reliable and predictable behaviors.

As indicated by J. Ejdys, the concept of trust, which initially referred to human relationships, now increasingly often refers also to institutions, organizations, objects or devices (Ejdys, 2017).

One of the types of trust, which impact on our activity in various areas of our lives is presently growing, is trust in technologies. The importance of technology is increasing and we may observe problems with the application of various technologies. Studies on the subject have been conducted e.g. by B. Blunsdon and K. Reed (2003), J.D. Lee and K.A. See (2004), J.D. Lee and Ji Gao (2005), and Wang and Benbasat (2008).

According to Sztompka (2007), "trust in technology is the expectation of efficiency, reliability and effectiveness of equipment and technological systems". Trust in technology reflects our attitude to a technological system. According to Lee and See (2004), it is a conviction that technologies will make it possible to attain goals in a situation of uncertainty and risk. According to Ratnasingam (2005), trust in technology is a degree, to which organizations are convinced that their technical infrastructure has a positive effect on executed transactions.

The level of trust in technology depends, among other things, on the assurance of safety for the entire structure of virtual reality. The significance of trust depends on the values of executed transactions, technological solutions, and contacts with business partners (Ratnasingam, 2005). In turn, Halicka (2016) stated that the present-day generation increasingly relies on technology, with trust in IT technology playing a particularly important role (McKnight, 2005). One of the types of trust in technology is related to online trust (Skowysz & Krot, 2018). In turn, Garnik (2002, p. 34) defined trust "as the expectation that in a risky situation on the Internet our weak points will not be exploited". In e-services the development of trust in technology consists in the use of technical safeguards (guaranteeing safety related with data protection, payments, interface use), while on the other hand it is highly dependent on the conditions within a specific situation, for example what sellers may gain when cheating or using dishonest methods.

Research methods

It was attempted in this study to indicate crucial issues related to trust placed by young people in AI systems used by entities operating in the e-commerce sector. The aim was to assess the level of trust in AI among students from Poland and Belarus, as well as to determine the dependence between AI and customer purchase behaviors, considering trust as determinants of online purchases. It results from the definition that AI is the most advanced form of technology development; thus it was assumed that the level of trust among young people making purchases using AI technology may be assessed by measuring the level of their trust in advanced technological solutions. The level of trust in AI among young people was investigated based on four research constructs: (1) awareness of the existence of AI, (2) general trust in technology, (3) trust in advanced technology (AI) used by enterprises operating online, and (4) individual trust in AI.

The study focused on the following research problems: the manner of defining/understanding the concept of AI, assessment of respondents' awareness concerning the use of AI in everyday life, assessment of trust in online purchases using AI; respondents' expectations concerning AI as well as their concerns.

In view of the aim of this study, the diagnostic survey method was applied using the questionnaire technique. Investigations were conducted in September and October 2023 with the use of the Google Form questionnaire on a study population sample of 340 respondents. The nonrandom sampling of the population was applied. The respondents were university students from Poland (94.3%), Belarus (5.1%), and others (0.6%). In the structure of the tested population sample, the predominant group were young people aged 18–26 years (88.2%) and those aged 27–35 years (9.1%). Women predominated in terms of the sex structure (57.4%). Among the respondents, 96.5% were students, of which 33.5% were working in business. The respondents declared that they use online purchases (98.2% responses). In the course of the survey, the respondents were asked to give their opinions on the proposed statements. Some questions contained a 6-point scale with a precise description of the response, for example definitely yes, while for some questions the respondents indicated responses in the scale from 1 to 10, where 1 meant I definitely do not agree, and 10 meant I completely agree. Evaluations given by the respondents within the range of 9–10 were described as indicating a very high level of the investigated phenomenon, assessments in the range of 7–8 as high, those in the range of 5–6 as medium, 3–4 as low, and 1–2 as very low, respectively. In view of the fact that the study was exploratory in nature the recorded results may not be considered representative.

Results and discussion

The first group of questions in the survey concerned the respondents' awareness and their evaluation of the impact of AI on the functioning of the economy and its individual sectors. The aim of these questions was to specify the share of the respondents who are aware of the progressing technological development and the scale of the applications for AI in the economy. According to the respondents, the most adequate definition of AI states that it is "a technology supporting and/or facilitating human decisions" (58.8% responses) and the definition stating that it is "a technology, which makes it possible to replace people in repetitive work operations" (51.2%).

A vast majority of the respondents (80%) were of an opinion that AI has an evident effect on their everyday life. A total of 18.9% respondents did not agree with this opinion, while 3.2% respondents had no definite opinion on the subject. These results confirm high awareness of the respondents concerning the use of AI applications by enterprises in their products and services. In the opinion of the respondents the sectors using AI to a greatest extent include IT and telecommunication services (66.6%), commerce (46.5%), transport and logistics (39.7), financial services (34.1%), education (27.9%), activities related to culture, entertainment and recreation (21.5%), and business activity (25%). The respondents saw no applications of AI in agriculture (2.6%) or the mining sector (5%). Over 62% of the total surveyed population associated AI primarily with customer service (65.6%), over 55% with collection of information and knowledge, approx. 50% with translations from foreign languages, while 43.5% with 3D simulations and games. The least typical applications in the opinion of respondents were connected with property security services (8.2%) and medical diagnostics (10.3%). In turn, the areas, in which AI exerts an evident effect on the respondents' lives include education and science (55.5% responses), as well as self-study and courses (47.9%), shopping (35.3%) and official and administrative affairs (30%). Every fourth respondent indicated profession and carrier, personal development as well as interests and hobbies. These results are consistent with the studies conducted in 2019 by NASK. They showed that respondents indicated five areas of an evident effect of AI on the lives of the respondents, such as (1) Internet consumption, that is shopping, commerce, transport, and communication as well as entertainment and games; (2) career and organization, including official and administrative affairs, education, science, courses; (3) affiliation; (4) recreation; and (5) public debates (NASK, 2019).

The area of general trust in technology was investigated based on the opinions of the respondents concerning specific characteristics, for example credibility, honesty, reputation, or confidentiality of the entities applying advanced technologies, including AI. Similar studies were conducted by Ganesan, 1994; McKnight et al., 2011; Ejdys, 2017; NASK, 2019. This component refers to the declarations of the students whether in their opinion producers and suppliers of advanced technologies, including AI, possess knowledge and resources necessary to execute implemented solutions and whether they have good reputation, thanks to which they generate more trust. An important aspect related to general trust in technology is connected with the guarantees for information confidentiality.

The conducted survey showed that 39.8% respondents in the scale from 1 to 10 chose responses above 7, that is declared a high opinion on credibility, that is know-how and resources required to execute implemented solutions. The medium score was selected by 47.7% respondents, while scores below 4 were given by 13.3% respondents, which indicates that they are not convinced of credibility of the entities providing AI solutions. It was different in the case of the respondents' perception of the level of honesty on the part of advanced technology providers. It was evaluated as high (21.5%) and very high (4.7%) by as few as every fourth respondent, with over a half of respondents considering it as moderate, while 21.8% selected a score of 4 or below.

The evaluation of the reputation of advanced technology producers referred to their ability to meet customer expectations and providing customers with desirable quality products (NASK, 2019). In this aspect, a moderate level of reputation was indicated by 44.7% respondents, it was evaluated as high by 30.3% respondents and very high by as few as 6.8%. Scores below 4 were selected by 16.8% respondents. Different responses were given in relation to the guarantee of safety and confidentiality of data. Slightly below 35% respondents gave a score of 4 and lower, which indicates a low or very low level of trust in this respect. Moderate trust was declared by approx. 40% respondents and the highest level of trust was declared by only 6.7%. Over 80% of respondents stated that producers of advanced technologies provide their customers with professional and technical support.

The third research construct was connected with the evaluation of trust in advanced technologies, including AI in enterprises operating online. AI may generate trust or certain doubts. In this evaluation, trust in advanced technologies was measured based on such characteristics as safety, reliability, credibility, functionality, and intuitiveness (Jian, Bisantz & Drury, 2000; Łapińska et al., 2021). NASK (2019). The above-mentioned characteristics in the context of their reliability will generate trust. Safety refers to the potential to counteract a dangerous or harmful influence of advanced technologies on humans and other organisms (NASK, 2019). In this study, almost a half of respondents (48.6%) indicated that in their opinion solutions related to advanced technologies used by enterprises offer a moderate level of safety. A score of minimum 7, that is a high and very high level of safety, was declared by 28% respondents, while a very low level was declared by only 3.9% all respondents. These results differ from the findings of earlier studies by NASK, where 57% respondents declared a high and very high level of safety (NASK, 2019). In terms of reliability referring to the capability of producers to consistently perform required functions without excessive deficiencies and failures in this study approx. 70% of respondents agreed with a statement that advanced technological solutions, including AI, used by companies operating online are reliable and exhibit adequate functionality. When evaluating the ease of service and the satisfaction from the use of advanced technologies, including AI, only 2.7% of the respondents declared a very poor opinion. Responses concerning intuitiveness of the use of devices using advanced technologies were very different. Intuitiveness indicates the potential to understand and learn appropriate, logical operation, as well as become acquainted with the functioning of advanced technologies. One of the questions referred to the potential to rely on solutions using AI by companies operating online. This problem was given high and very high scores by over 40% respondents, who agreed with this statement. A slightly higher percentage of respondents (43%) gave scores of 5–6, while only 2% respondents disagreed completely.

The fourth component referred to the evaluation of individual trust in AI. Trust in AI was evaluated in terms of its impact on purchase decisions made online, as well as factors affecting decision-making processes. Approximately 80% respondents indicated that AI influences trust in online purchases. The highest percentage (40.8%) of the respondents declared moderate trust, with high trust declared by 33.6% and very high trust by 9.2% respondents. According to 48.5% respondents, AI aids them in their purchase decisions. In turn, purchase decisions are made under the influence of their peers (as declared by over 83% respondents) and celebrities (63.7%). Trust in AI technology was investigated also using questions concerning the opinion of respondents on safety and reliability of the devices they use. Only 30% of young people have a positive attitude to changing a means of transport from their own car to an autonomous car using AI technology. This result is lower compared to those reported in other studies and from the United States (NASK, 2019). Similar responses were given to the question concerning the use of medical services, in which doctors would be replaced by devices or programs based on AI, as only 20% of respondents agreed and as many as 21% of respondents were completely against it. In the context of using financial and legal services based on AI technologies negative responses were also predominant.

In this study, expectations and threats for respondents related to the application of AI were also estimated. Identification of expectations, as well as threats resulting from the use of AI in products and services offered online, influences the development of trust in AI. According to the respondents, the popularization of AI will be beneficial for entrepreneurs (as considered by over 80% respondents), for customers (86%), for state administration (74.5%), and the entire economy (69.4%). In the opinion of the respondents, the popularization of AI will be beneficial for the quality of life. A positive answer to this question was given by 80% respondents, including 8% stating they completely agreed. In turn, threats related to the popularization of AI were perceived by the respondents in relation to employees, as indicated by as many as 55% respondents. In their opinion, in relation to the development of AI technology many jobs will disappear – this was indicated by almost 60% respondents, while the level of unemployment will increase, as indicated by 57% respondents. Development possibilities for the use of AI were seen by respondents indicating the possibility to reduce costs of services for specialist products (48%), possibilities to obtain more personalized products and services (45%), as well as an increased level of safety for programs and devices (40%).

Conclusions and directions for further research

The following conclusions may be drawn from these investigations:

1 A vast majority of respondents (80%) were of an opinion that AI has an evident impact on their everyday lives. These results confirm high awareness of respondents concerning the use of AI applications by enterprises in their products and services. The sectors using AI to the greatest extent include IT and telecommunication services (66.6%), commerce (46.5%), transport and logistics (39.7), financial services (34.1%), education (27.9%), activities related to culture, entertainment and recreation (21.5%), and business operations (25%).

2 Only 39.8% respondents declared a high opinion on credibility, that is knowledge and resources required to execute implemented solutions related to AI.

The medium score was selected by 47.7% respondents and scores below 4 were given by 13.3%, which indicates that they are not convinced of the credibility of entities providing AI solutions.

3 The areas, in which AI has an evident impact on the lives of respondents, include education and science (55.5% responses) and self-study and courses (47.9%), purchases (35.3%) as well as official and administrative affairs (30%). Every fourth respondent indicated professional activity and career, personal development, interests, and hobbies.
4 The third research construct was connected with the evaluation of trust in advanced technologies, including AI, in companies operating online. AI may generate either trust or certain concerns. In this study, almost a half of respondents (48.6%) indicated that in their opinion solutions connected with advanced technologies used by companies are characterized by a moderate level of safety. Responses indicating a high or very high level of safety were given by 28% respondents, while a very low level was indicated by only 3.9% of all the respondents.
5 The fourth component refers to the evaluation of individual trust in AI. Trust in AI influencing purchase decisions made online, as well as factors affecting decision-making processes were evaluated. Approximately 80% respondents indicated that AI has an influence on trust in online purchases.

This study exhibits certain limitations, one of them being the small sample size, as well as its selection, focusing only on a group of students. In the future, studies could also be conducted on several groups and comparisons could be made.

In relation to the recorded results, the following actions may be indicated for the future:

1 Broadening of knowledge on the role of AI in trust in online purchases both in the context of literature and studies conducted referring to various dimensions of trust.
2 Based on this study preparation of studies for users so that the greatest possible amount of information based on experience could be provided.
3 Designing of studies on trust concerning the use of AI supporting online purchases to show that it is not solely an ethical and legal context, but it is broader in scope.

Literature

Bhattacharya, R., & Devinney, T. M. (1998). A formal model of trust based on outcomes. *Academy of Management Review*, *23*(3), 459–472.

Blunsdon, B., & Reed, K. (2003). The effects of technical and social condition on workplace trust. *International Journal of Human Resources Management*, *14*(1), 12–27.

Bugdol, M. (2010). *Wymiary i problemy zarządzania organizacją opartą na zaufaniu [Dimensions and problems of managing an organization based on trust*]. Krakow, PL: Publishing house of the Jagiellonian University in Krakow.

Chmielarz, W. (2007). *Systemy biznesu elektronicznego* [Electronic Business Systems]. Warsaw, PL: DIFIN Publishing House.

Ejdys, J. (2017). Determinanty zaufania do technologii [Antecedents of Technology Trust]. Warsaw, PL: *Przegląd Organizacji*, *12*(935), 20–27.

Łapińska, J., Escher, I., Górka, J., Sudolska, A., & Brzustewicz, P. (2021). Employees' trust in artificial intelligence in companies: The case of energy and chemical industries in Poland. *Energies, 14*, 1942.

Gambetta, D. (1988). Can we trust trust? In D. Gambetta (Ed.), *Trust: Making and Breaking Cooperative Relations*. New York: Blackwell, 213–237.

Ganesan, S. (1994). Determinants of long-term orientation in buyer-seller relationships. *The Journal of Marketing*, *58*, 1–19.

Garnik, I. (2002). Zaufanie w systemach płatności on-line. [Trust in on-line payment systems]. Poznan, PL: *Scientific Journals of the Poznań University of Technology*, *35*, 34–39.

Grudzewski, W.M., Hejduk, I. K., Sankowska, A., & Wańtuchowicz, M. (2007). *Zarządzanie zaufaniem w organizacjach wirtualnych [*Trust management in virtual organizations]. Warsaw, PL: Difin Publishing House.

Halicka, K. (2016). *Prospektywna analiza technologii. Metodologia i procedury badawcze [Prospective technology analysis - methodology and research procedures*]. Bialystok, PL: Publishing House of the Bialystok University of Technology.

Jarek, K., Mazurek, G., & Hałas-Dej, S. (2018). Marketing i sztuczna inteligencja, [Marketing and artificial intelligence]. *Przedsiębiorczość i Zarządzanie, Wydawnictwo SAN*, *19*(5), 191–206.

Jian, J., Bisantz, A., & Drury, C. (2000). Foundations for an empirically determined scale of trust in automated systems. *International Journal of Cognitive Ergonomics*, *4*(1), 53–71.

Lee, J. D., & Gao, J. (2005). Trust, information technology, and cooperation in supply chains. *Supply Chain Forum: International Journal*, *6*(2), 82–89.

Lee, J. D., & See, K. A. (2004). Trust in technology: Designing for appropriate reliance. *Human Factors*, *46*(1), 50–80.

Mayer, R., Davis, J., & Schoorman, F. David. (1995). An integrative model of organizational trust. *Academy of Management Review*, *20*(3), 709–734.

McKnight, D. H. (2005). Trust in information technology. In G. B. Davis (Ed.), *The Blackwell Encyclopedia of Management: Management Information Systems*. Malden, MA: Blackwell, 7, 329–331.

McKnight, D. H., Carter, M., Thatcher, J. B., & Clay, P. F. (2011). Trust in a specific technology: An investigation of its components and measures. *Journal of Management Information Systems, 2*(2), 1–25.

Mróz, B. (2017). Konsument a wyzwania technologiczne XXI wieku, Studia Ekonomiczne. *Zeszyty Naukowe Uniwersytetu Ekonomicznego w Katowicach*, nr *330*, 139–149.

NASK. (2019). *Sztuczna inteligencja w społeczeństwie I gospodarce*. Raport z badan społecznych. Warsaw, PL: NASK Państwowy Instytut Bdawczy.

Ratnasingam, P. (2005). Trust in inter-organizational exchanges: A case study in business to business electronic commerce. *Decision Support Systems, 39*(3), 525–544.

Rotter, J. B. (1980). Interpersonal trust, trustworthiness and gullibility. *American Psychologist, 35*(1), 1–7.

Rudzewicz, A. (2009). Zaufanie – przegląd koncepcji. [Trust - an overview of the concept]. In L. Garbarski, & J. Tkaczyk (Ed.), *Kontrowersje wokół marketingu w Polsce: niepewność i zaufanie a zachowania nabywców*. [Controversies surrounding marketing in Poland: uncertainty and trust versus buyer behavior]. Warsaw, PL: Academic and Professional Publishing. Kozminski University, 421–430.

Shanahan, M. (2015). *The Technology Singularity*. Cambridge, MA: MIT Press.

Skowysz, K., & Krot, K. (2018). *Zaufanie a skłonność do korzystania i zadowolenie z usług internetowych*. [The impact of trust on the tendency to use and satisfaction with internet services]. Bialystok, PL: *Akademia zarzadzania*, [Academy of Management], *2*(4), 101–117.

Słownik Języka Polskiego. (2007). Wydawnictwo Naukowe PWN Warszawa.

Sterne, J. (2017). *Artificial Intelligence in Marketing*. Practical Applications. New Jersey: Wiley and SAS Business Series.

Sztompka, P. (2007). *Zaufanie fundamentem społeczeństwa* [*Trust. The foundation of society*]. Krakow, PL: Znak.

Wang, W., & Benbasat, I. (2008). Attributions of trust in decision support technologies: A study of recommendation agents for E-Commerce. *Journal of Management Information Systems, 24*(4), 249–273.

Weber, M. J., Maihotra, D., & Murnighan, K. J. (2005). Normal acts of irrational trust: Motivated attributions and the trust development process. *Research in Organizational Behavior*, *2*(26), 75–101.

18 Generation Z trust in modern technological solutions in the last-mile logistics. Case study of Poland

Konrad Michalski and Teresa Gądek-Hawlena

Introduction

One of the areas of scientific cognition and economic practice in which the rapid development of modern technological solutions is visible is the last-mile logistics, responsible for delivering the ordered shipment of goods to the final customer. The importance of this area of knowledge increases because we are dealing with two interdependent phenomena: the e-commerce boom and the increase in the importance of the so-called on-demand logistics (Lozzi et al., 2022). The more purchases consumers make online, the more packages need to be delivered. The area of operation responsible for deliveries is the last-mile logistics. The design and management of last mile logistics must take into account that the majority of the population lives in cities, urban congestion is increasing and transport is responsible for enormous environmental pollution. Researches are being carried out on how to use the underutilized potential of logistics, which is based on maintaining operational readiness to operate on the basis of constant resources and their costs, with regard, e.g. to localization of delivery points (Stokkink & Geroliminis, 2023) and route network planning (He et al., 2022). These challenges also refer to the so-called shared logistics, part of the sharing economy research trend (Chen & Kong, 2023) and sustainable logistics as a key direction of action for those designing logistics systems, also in its final part: the last-mile (Silva et al., 2023).

The current perception of the concept of the last-mile logistics is largely influenced by the rapid development of the Internet, which has made online shopping as common as traditional in-store shopping. Online shopping has become available to anyone who has access to an electronic device such as a computer or smartphone and is able to make cashless payments. The dynamic development of e-commerce, especially in B2C (Business to Customer) relationships, has forced the need to implement modern solutions in the delivery of products purchased via the Internet. At the same time, the implemented solutions enjoy the greatest trust among Generation Z (people born 1996–2012), widely considered to be digital, absorbing all technical innovations, including IT, in the easiest

DOI: 10.4324/9781032627236-20

and fastest way. The functioning of the last-mile system in logistics must also take into account the key attributes for the modern consumer in general, and for the youngest: uncertainty and instability of their behavior (Özarik et al., 2021).

The aim of the chapter, based on a case study, is an attempt to identify the factors affecting the confidence of Generation Z in modern solutions currently used in the last-mile logistics and those that are still in the implementation stage. The study is a diagnostic survey.

The structure of the chapter includes subsequent parts covering theoretical consideration: on the last-mile logistics and modern technologies implemented in this area, then on absorption possibilities and readiness of Generation Z to use modern solutions on the last-mile logistics. Next, the method of research and its results were presented, along with a discussion with the findings of other studies. Finally, the implications of the research results for theory and practice, and the final conclusions, where the limitations of the conducted research are pointed out and further research directions are also defined.

The last-mile logistics and its perception in B2C relations

The term "last-mile" itself comes from the telecommunications industry and refers to the last section of telecommunications networks that deliver services to customers (Gevaers et al., 2011). Early definitions of the last-mile relating to customer delivery were narrowed to the extension of the supply chain directly to the final consumer, i.e. a delivery service to the customer's home. Several synonyms have also been used, such as "last mile supply chain", "home delivery", "business-to-consumer distribution" (Punakivi & Saranen, 2001; Kull et al., 2007). From the perspective of the owner of the last-mile logistics (Business), its problems are most often considered from the perspective of (Mangiaracina et al., 2019):

- sustainable development – where deliveries are the part of city logistics and significantly influence the level of the external transport costs,
- efficiency (level of customer service) – where the last-mile logistics is a consumer-oriented service with a strong behavioral element (Collins, 2015),
- efficiency (costs) – last-mile delivery methods have different characteristics and affect its efficiency (Wang et al., 2014).

Modern technologies on the last-mile logistics

Rather traditional forms of the last-mile delivery include inboxes, deliveries by postman or courier, personal pickup, delivery via online store vehicles and parcel delivery points (Iannaccone et al., 2021; Jucha & Corejova, 2021; Kahr, 2022). This last option, based on the parcel lockers, is the most popular solution not only in Poland, but in general. Traditional forms of delivery are supplemented or even replaced by modern, innovative forms.

Modern, emerging forms of deliveries include (not yet or poorly implemented in Poland): drones (Borghetti et al., 2022; Li et al., 2022; Li & Kunze, 2023), autonomous ground vehicles (Bray & Cebon, 2022; Liu et al., 2023), autonomous vessels and water-taxis (Gu & Wallace, 2021), mobile robots (Joerss et al., 2016; Li & Kunze, 2023), delivery directly to the trunk of a car and delivery ordered via a crowdsourcing platform. Also such forms of deliveries as: to the place of fitting and possible return of good (showroom), to the courier meeting area for collection (pick-and-ride area), to the collection zone dedicated to employees in an office building or workplace are treated as still in Polish terms of something new.

Each of the above new forms of delivery requires the use of a dedicated IT solution, and each of the devices included in the delivery system is in fact a separate system (mobile application, drone, robot, parcel locker, etc.). Additionally, the entire distribution system is increasingly supported by AI (Artificial Intelligence) solutions, for example in the field of communication, filing complaints or making returns (Giuffrida et al., 2022). These solutions are served, among others, by a developing trend that is RPA (Robotic Process Automation): a rules-based system for automating business processes by software bots that emulate human interactions (Patrício et al., 2023). The technology used for this is machine learning, a branch of AI and computer science which focuses on the use of data and algorithms to imitate the way that humans learn, gradually improving its accuracy (IBM, 2023).

Generation Z as last-mile participants

The consequences of the COVID-19 pandemic (2020–) have reinforced the need for wider use of media and electronic technologies to communicate and deal transactions, also deliveries. Restrictions and lockdowns have accelerated many innovations in the last-mile area (Sułkowski et al., 2022). Consumers had to quickly adapt to the changed operating conditions.

It is commonly believed that Generation Z (born 1995–2012) is the one who absorbs all technological innovations. Mobile applications (Jaska et al., 2022) and social media (Werenowska & Jaska, 2022) are an inherent attribute of the youngest generation. Meanwhile, all delivery options considered in the study involve the use of mobile applications, while many sales transactions take place also via social media. A drone delivers the parcel in a designated zone (Saigomade, 2022). Delivery carried out by an autonomous ground vehicle with storage compartments is an option in which the recipient of the shipment is notified about the possibility of collecting parcel. Once the recipient arrives at the vehicle, the locker opens. After receiving the parcel, the recipient closes the box (Amin, 2018). A mobile robot delivers the parcel in urban areas, rather in closed areas, moving at a speed of 5–10 km/h, mainly along sidewalks (Russo, 2019). The option of delivery directly to the trunk of the car assumes that the courier puts the parcel in the car after receiving the car's data (location and car numbers). The courier opens the car using a code generated in an application. The process of opening and closing the trunk of a car is similar to that of a parcel locker (Zimek, 2021). As part of a delivery ordered on a crowdsourcing platform, the owner of the car who decides to deliver parcel picks it up and delivers it to the indicated address (Jungleworks, 2023).

Characterizing generation Z in terms of last-mile technology use, the exceptional ease of functioning in the digital environment, openness to learning new solutions and ability to learn should be also emphasized. At the same time, Generation Z is a group of consumers who do not adhere to established patterns of action, not to mention the same goods ordered. It can be also expected that the next generation, that is Alpha, born since ~2010 (Wikipedia, 2023), will be even more attached to digital technologies and use them even more often.

Research method

Material and methodology

The survey method was used in the study. The survey was prepared for the needs of this study and consisted of four parts. In total, it contained 19 questions. The first part of the survey included demographic variables (age, gender, place of residence, and professional status). The second part examined the methods of making purchases, their frequency and the forms of delivery of goods purchased online. The third group of questions concerned future delivery solutions within last-mile logistics and the assessment of factors supporting their use. The questions were asked on a Likert scale, where 1 – I do not agree at all and 5 – I completely agree.

In order to verify the correctness of the questions, pilot research was conducted on a group of 15 people, including generation Z and specialists professionally dealing with the last-mile logistics. The sample selection was purposeful and included people born in 1995–2012.

The online survey was created using Microsoft Forms and distributed among generation Z via social media between September 1 and October 12, 2023. In order to answer the questions posed in the form, statistical analyzes were carried out using an Excel spreadsheet and Statistica 13.3. It was used to calculate descriptive statistics and χ^2 correlation coefficients. The level of significance in the article was assumed to be alpha = 0.05.

Participants of the survey

773 respondents took part in the study, of which 694 questionnaires were included for further analysis, which constitutes 89.8% of all questionnaires

participating in the study (the study omitted the answers of respondents who do not shop online at all). The analyzed demographic variables show that more women (56.9%) than men (43.1%) participated in the study.

In terms of age, respondents were divided into two groups, that is 11–19 (36.3% of respondents) and 20–28 (63.7% of respondents). In terms of education, people with higher education dominated −37.6%, followed by secondary −34% and primary −24.3%, and the least numerous groups of respondents were people in the process of gaining primary education −3.1% and with vocational −0.9%. The largest group, taking into account professional status, were pupils and students −66.1%, half as many people took part in the study and defined their professional status as "professionally active" (32.9%), a few people indicated that they were inactive professionally (0.9%) and temporarily or permanently unable to work (0.1%). The respondents were mainly residents of cities with over 150,000 inhabitants (55.6%), then cities up to 50.000 inhabitants (17.3%), rural areas (16.1%), cities 51.000–100.000 (6.8%), and cities 101.000–150.000 inhabitants (4.2%).

Research results

Delivery methods used when purchasing online

The first problem analyzed was the frequency of online shopping. Thus, people from Generation Z indicated that they most often shop online several times a month (38.5%), then once a month (25.6%), once every three months (14.1%), once a week (10.2%), several times a week (7.2%), once every three times (3.9%), and every day (0.4%). Respondents most often buy clothes and footwear (561 indications), books (387 indications), personal accessories (e.g. watches or jewelry) (353 indications), cosmetics and personal protective equipment (344 indications), electronics and home furnishings (240 indications), multimedia (238 indications), pet food and accessories (154 indications), food (except takeaway food) (98 indications), toys (98 indications), school supplies (except books) (84 indications), and FMCG (fast-moving consumer goods) (e.g. diapers or paper towels) (39 indications).

According to the respondents, the dominant form of collecting purchased parcels is delivery to postal lockers (parcel machines) (82.3%), followed by delivery by courier or postman (12.3%), and delivery to collection points (4.6%). The least frequently used form of delivery was delivery directly from the online store.

The influence of demographic variables on the choice of parcel delivery methods

In order to check whether there is a relationship between the respondents' demographic variables and new forms of delivery of goods purchased on the Internet, a χ^2 test was performed (Table 18.1).

Table 18.1 Selected demographic variables and modern forms of the last-mile delivery

Variable	*Test result*	*p-value**
Drone		
Gender	28.2012	**0.0000**
Age	1.0865	0.2973
Domicile	1.2319	0.8728
Autonomous ground vehicles		
Gender	11.6624	**0.0006**
Age	2.7387	0.0980
Domicile	4.6752	0.3222
Mobile robot		
Gender	2.3181	**0.1279**
Age	3.3375	0.0662
Domicile	2.9789	0.5614
Delivery to the trunk		
Gender	5.6289	**0.0177**
Age	1.0801	0.2987
Domicile	2.1645	0.7055
Crowdsourcing platform		
Gender	13.2802	**0.0003**
Age	0.4570	0.4990
Domicile	2.3221	0.6767

Source: Own study.

* Bold values indicate statistical significance at $p < 0.05$.

The test showed that there is a statistically significant relationship between gender and the possibility of using drone deliveries ($p = 0.0000$). 56.19% of men and only 35.95% of women consider using the drone delivery option. Considering the ground-based autonomous vehicle with lockers, men were also more likely to indicate that they were interested in this type of deliveries: 72.9% of men and 60.5% of women expressed positive opinions on this topic. There is a statistically significant relationship between the variables ($p = 0.0006$). Another solution presented in the study was deliveries to the trunk of a car. Both women (88.61%) and men (82.27%) were not interested in this form of delivery. There is a statistical relationship between the variables ($p = 0.0177$). However, the last modern form of delivery examined was the crowdsourcing platform, where a statistically significant relationship was also found ($p = 0.0003$). 51.84% of men were in favor of using this type of solution and only 37.95% of women. However, no statistically significant relationships were found between the selected forms of delivery and age and place of residence.

Assessment of selected attributes of delivery forms in the-last mile

The aim was to evaluate, at most, pilot solutions implemented in Poland as part of the last-mile logistics (Figure 18.1).

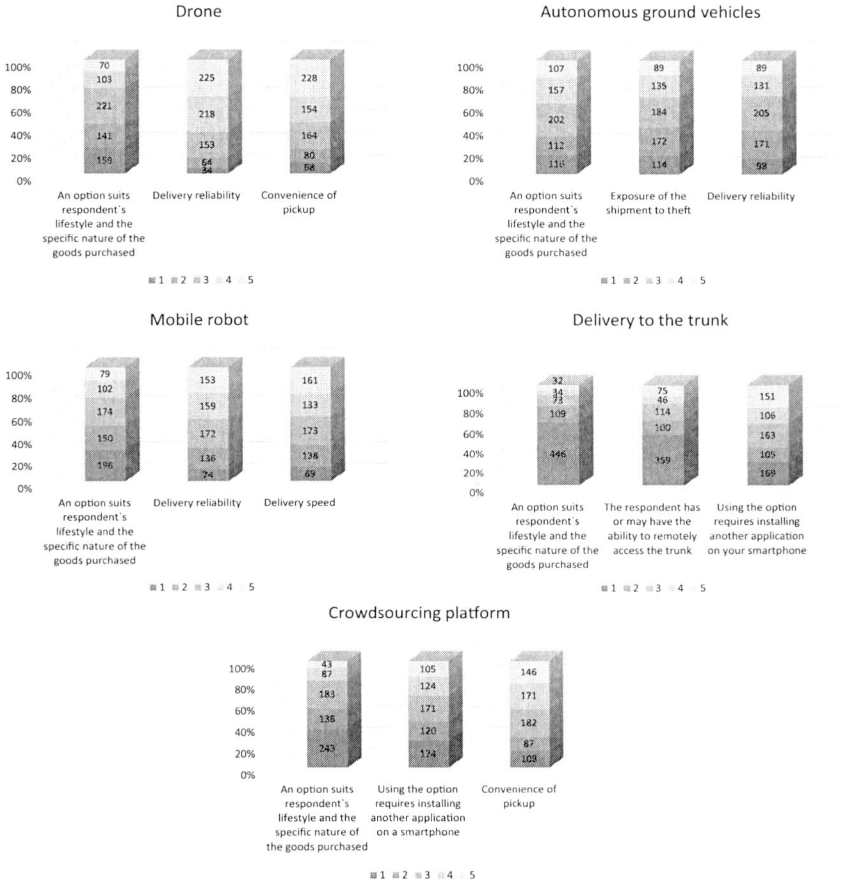

Figure 18.1 Assessment of selected attributes of modern delivery forms.

Source: Own study.

The first form of delivery assessed was a drone. The possibility of convenient pickup was rated the highest (35.85% of the highest responses). Delivery reliability was also rated similarly high (32.42% of the highest ratings). The lowest rating was given to the option relating to lifestyle and the specificity of purchased goods (0.7% of the highest responses). Delivery using autonomous ground vehicles was assessed slightly differently. In that case, there are no such significant differences in assessment. The statements submitted for evaluation were rated most often as "I agree on average" and accounted for 29.1%, 26.51%, and 29.54% of responses, respectively. Mobile robots were rated very similarly. It is worth noting the assessment of their adaptation to the respondents' lifestyle and specificity of shopping –28.53% of respondents disagree with this statement. Delivery to the trunk of a car was rated completely differently. As many as 64.2% of people marked "I do not agree at all" when assessing the adaptation of this option to the lifestyle and specificity of the purchased goods. The statement "the respondent has or will have the ability to remotely access the car" was rated equally low (51.7% of responses "I do not agree at all"). The statement "using this option requires installing another application" was rated more neutrally; each statement received approximately 20% of responses. The last form of delivery assessed was the crowdsourcing platform. The option related to lifestyle and the specificity of purchased goods received the lowest rating (35% of all responses). In turn, the possibility of convenient pickup was rated the highest (21.0% of the highest responses).

Experiences from cooperation with AI in the-last mile

It was examined whether there is a relationship between demographic variables and the fact of conscious communication with AI mechanisms (e.g. chatbot) as an experience from the delivery. For this purpose, the χ^2 test was performed. As in the case of innovative forms of deliveries, a statistically significant relationship occurs only between gender and AI ($\chi^2 = 4.7507$, $p = 0.02929$). 49.83% of men and 41.52% of women communicated with AI.

The last question asked to respondents concerned the feelings aroused by the experience of communicating or corresponding with AI. Multiple answer options could be selected. The following statements received the most responses: "It is difficult to get along with the algorithm because it is programmed for standard answers, not for solving unique problems" (40% of all responses indicated this). About 247 respondents indicated that "I have no opinion on this matter yet" (35% of all responses chose this statement). About 21% of all respondents chose the statement "I have no problems with AI. It is inevitable that machines will replace us). The statements that received the least number of indications were "I feel very good working with modern technologies" (11% of respondents indicated this), "The more modern technologies, the better" (6% of respondents chose this answer), and "AI solves problems better than humans" (4% of respondents agreed with this). There is still a long way to educate and convince consumers to cooperate with AI. This can be concluded from the above results.

Discussion

The presented study is one of the first that directly refers to Generation Z as recipients of deliveries carried out with the increasing use of modern technologies, including AI. Moreover, most scientific articles perceive the youngest (Generation Z) as online consumers (Vieira et al., 2020; Meghisan-Toma et al., 2021; Brand et al., 2022; Puiu et al., 2022).

This article refers rather to delivery experiences, which makes it have a narrower research scope. The results of the diagnostic survey conducted in the chapter can be compared to few findings. A wider selection of delivery options has a positive impact on customer satisfaction, both in cities and in rural areas, although the final satisfaction with the delivery may be different in both groups, as shown by research conducted in Sweden (Vakulenko et al., 2022). The conducted research confirms that parcel lockers are the most popular among the youngest consumers, what was also shown by research conducted in Rome, also among generation Z (Iannaccone et al., 2021).

Conclusions

Based on the collected research material and the analyses made on its basis, it can be noted that:

- Only a small percentage of consumers from the surveyed Generation Z shop online every day, which may have seemed more likely before the research began,
- Courier deliveries or via postman are still popular, perhaps because rural residents do not always have the option of using parcel lockers,
- Regarding the forms of delivery that are very little present (delivery to the trunk of a car) or not at all (delivery by drone), a large distance to these forms of delivery can be assumed; perhaps this is due to lack of knowledge about how this method works (drone) or reluctance to interfere with the privacy of a car or concerns about safety (delivery to the trunk)
- The use of drones for deliveries was rated highly, likely due to their strong presence in the social space. Meanwhile, options such as an autonomous ground vehicle or a mobile robot were rated the lowest, which may be a consequence of the lack of contact with such innovations (respondents may have learned about such possibilities for the first time),
- The option of delivery to the trunk of a car received the lowest rating, even though such an option is already available in Poland (car manufacturers introduce this option by installing appropriate software and providing a mobile application; some retail chains also place purchases directly in the car).

Some limitations were encountered while conducting the research. Due to the relatively new research area of the attitude toward modern technologies used in a specific area of logistics (the last-mile), the questions asked were of a general nature: about shopping habits, delivery preferences, feelings toward selected options, but also about general experience with cooperation with artificial intelligence, for example in communication or filing complaints. These questions should be deepened. Perhaps a complementary method of in-depth interview or even an experiment using a bot communicating with the consumer should be used.

Further limitation is connected with limited access to people under 19, and especially under 15, was limited due to the need to obtain consent from parents to conduct the study. It is worth deepening research on the use of artificial intelligence in the last-mile logistics and extending the research to other age groups, even to seniors (65+). It is one of the main future research target to achieve.

References

Amin, D. (2018). Autonomous Ground Vehicles (AGVS) with Lockers. Retrieved from https://logisticsmgepsupv.wordpress.com/2018/05/14/autonomous-ground-vehicles-agvs-with-lockers/ (Accessed 30 August 2023).

Borghetti, F., Caballini, C., Carboni, A., Grossato, G., Maja, R., & Barabino, B. (2022). The use of drones for last-mile delivery: A numerical case study in Milan, Italy. *Sustainability*, *14*(3), 1766. https://doi.org/10.3390/su14031766.

Brand, B.M., Rausch, T.M., & Brandel, J. (2022). The importance of sustainability aspects when purchasing online: Comparing Generation X and Generation Z. *Sustainability*, *14*(9), 5689. https://doi.org/10.3390/su14095689.

Bray, G., & Cebon, D. (2022). Selection of vehicle size and extent of multi-drop deliveries for autonomous goods vehicles: An assessment of potential for change. *Transportation Research Part E: Logistics and Transportation Review*, *164*, 102806. https://doi.org/10.1016/j.tre.2022.102806.

Chen, Z., & Kong, J. (2023). Research on shared logistics decision based on evolutionary game and income distribution. *Sustainability*, *15*(11), 8621. https://doi.org/10.3390/su15118621.

Collins, A.T. (2015). Behavioural influences on the environmental impact of collection/delivery points. In: B. Fahimnia, M.G.H. Bell, D.A. Hensher, & J. Sarkis (Eds.), *Green Logistics and Transportation. A Sustainable Supply Chain Perspective* (15–34). Cham: Springer. https://doi.org/10.1007/978-3-319-17181-4_2.

Gevaers, R., Van de Voorde, E., & Vanelslander, T. (2011). Characteristics and typology of last-mile logistics from an innovation perspective in an urban context. In: C. Macharis, & S. Melo (Eds.), *City Distribution and Urban Freight Transport* (56–71). Cheltenham: Edward Elgar Publishing, Inc.

Giuffrida, N., Fajardo-Calderin, J., Masegosa, A.D, Werner, F., Steudter, M., & Pilla, F. (2022). Optimization and machine learning applied to last-mile logistics: A review. *Sustainability*, *14*(9), 5329. https://doi.org/10.3390/su14095329.

Gu, Y., & Wallace, S.W. (2021). Operational benefits of autonomous vessels in logistics - A case of autonomous water-taxis in Bergen. *Transportation Research Part E: Logistics and Transportation Review*, *154*, 102456. https://doi.org/10.1016/j.tre.2021.102456.

He, X., He, F., Li, L., Zhang, L., & Xiao, G. (2022). A route network planning method for urban air delivery. *Transportation Research Part E: Logistics and Transportation Review*, *166*, 102872. https://doi.org/10.1016/j.tre.2022.102872.

Iannaccone, G., Marcucci, E., & Gatta, V. (2021). What young E-Consumers want? Forecasting parcel lockers choice in Rome. *Logistics*, *5*(3), 57. https://doi.org/10.3390/logistics5030057.

IBM. (2023). What Is Machine Learning? Retrieved from https://www.ibm.com/topics/machine-learning (Accessed 14 October 2023).

Jaska, E., Werenowska, A., & Balińska, A. (2022). Environmentally and socially sustainable behaviors of Generation Z in Poland stimulated by mobile applications. *Energies*, *15*(21), 7904. https://doi.org/10.3390/en15217904.

Joerss, M., Schroder, J., Neuhaus, F., Klink, C., & Mann, F. (2016). *Parcel Delivery. The Future of Last Mile*. New York: McKinsey & Company.

Jucha, P., & Corejova, T. (2021). Ensuring the logistics of the last mile from the perspective of distribution companies. *Transportation Research Procedia*, *55*, 482–489. https://doi.org/10.1016/j.trpro.2021.07.012.

Jungleworks. (2023). The Rise of Quick Commerce: An Ultimate Guide. Retrieved from https://jungleworks.com/the-rise-of-quick-commerce-an-ultimate-guide/ (Accessed 30 August 2023).

Kahr, M. (2022). Determining locations and layouts for parcel lockers to support supply chain viability at the last mile. *Omega*, *113*, 102721. https://doi.org/10.1016/j.omega.2022.102721.

Kull, T.J., Boyer, K., & Calantone, R. (2007). Last-mile supply chain efficiency: An analysis of learning curves in online ordering. *International Journal of Operations & Production Management*, *27*(4), 409–434. https://doi.org/10.1108/01443570710736985.

Li, F., & Kunze, O.A. (2023). Comparative review of air drones (UAVs) and delivery bots (SUGVs) for automated last mile home delivery. *Logistics*, *7*(2), 21. https://doi.org/10.3390/logistics7020021

Li, Y., Liu, M., & Jiang, D. (2022). Application of unmanned aerial vehicles in logistics: A literature review. *Sustainability*, *14*(21), 14473. https://doi.org/10.3390/su142114473

Liu, Y., Ye, Q., Escribano-Macias, J., Feng, Y., Candela, E., & Angeloudis, P. (2023). Route planning for last-mile deliveries using mobile parcel lockers: A hybrid q-learning network approach. *Transportation Research Part E: Logistics and Transportation Review*, *177*, 103234. https://doi.org/10.1016/j.tre.2023.103234.

Lozzi, G., Iannaccone, G., Maltese, I., Gatta, V., Marcucci, E., & Lozzi, R. (2022). On-Demand Logistics: Solutions, Barriers, and Enablers, *Sustainability*, *14*(15), 9465. https://doi.org/10.3390/su14159465.

Mangiaracina, R., Perego, A., Seghezzi, A., & Tumino, A. (2019). Innovative solutions to increase last-mile delivery efficiency in B2C e-commerce: A literature review. *International Journal of Physical Distribution & Logistics Management*, *49*(9), 901–920. https://doi.org/10.1108/IJPDLM-02-2019-0048.

Meghisan-Toma, G.-M., Puiu, S., Florea, N.M., Meghisan, F., & Doran, D. (2021). Generation Z' young adults and M-Commerce Use in Romania. *Journal of Theoretical and Applied Electronic Commerce Research*. *16*(5), 1458–1471. https://doi.org/10.3390/jtaer16050082.

Özarik, S.S., Veelenturf, L.P., Van Woensel, T., & Laporte, G. (2021). Optimizing E-commerce last-mile vehicle routing and scheduling under uncertain customer presence. *Transportation Research Part E: Logistics and Transportation Review*, *148*, 102263. https://doi.org/10.1016/j.tre.2021.102263.

Patrício, L., Ávila, P., Varela, L., Cruz-Cunha, M.M., Pinto Ferreira, L., Bastos, J., Castro, H., & Silva, H. (2023). Literature review of decision models for the sustainable implementation of Robotic Process Automation. *Procedia Computer Science*, *219*, 870–878. https://doi.org/10.1016/j.procs.2023.01.362.

Puiu, S., Demyen, S., Tănase, A.-C., Vărzaru, A.A., & Bocean, C.G. (2022). Assessing the adoption of mobile technology for commerce by Generation Z. *Electronics*, *11*(6), 866. https://doi.org/10.3390/electronics11060866.

Punakivi, M., & Saranen, J. (2001). Identifying the success factors in E-grocery home delivery. *International Journal of Retail & Distribution Management*, *29*(4), 156–163.

Russo, H. (2019). Amazon comenzó a probar un robot para envíos de productos. Retrieved from https://geeksroom.com/2019/01/amazon-scout/121125/ (Accessed 30 August 2023).

Saigomade, P. (2022). Dostawy paczek za pomocą dronów - czy to jest możliwe? [Parcel deliveries using drones - is it possible?) Retrieved from https://neopak.pl/blog/dostawy-paczek-za-pomoca-dronow-czy-to-jest-mozliwe (Accessed 30 August 2023).

Silva, V., Amaral, A., & Fontes, T. (2023). Sustainable urban last-mile logistics: A systematic literature review. *Sustainability*, *15*(3), 2285. https://doi.org/10.3390/su15032285.

Stokkink, P., & Geroliminis, N. (2023). A continuum approximation approach to the depot location problem in a crowd-shipping system. *Transportation Research Part E: Logistics and Transportation Review*, *176*, 103207. https://doi.org/10.1016/j.tre.2023.103207.

Sułkowski, Ł., Kolasińska-Morawska, K., Brzozowska, M., Morawski, P., & Schroeder, T. (2022). Last mile logistics innovations in the courier-express-parcel sector due to the COVID-19 pandemic. *Sustainability*, *14*(13), 8207. https://doi.org/10.3390/su14138207.

Vakulenko, Y., Arsenovic, J., Hellström, D., & Shams, P. (2022). Does delivery service differentiation matter? Comparing rural to urban e-consumer satisfaction and retention. *Journal of Business Research*, *142*, 476–484. https://doi.org/10.1016/j.jbusres.2021.12.079.

Vieira, J., Frade, R., Ascenso, R., Prates, I., & Martinho, F. (2020). Generation Z and key-factors on E-Commerce: A study on the Portuguese tourism sector. *Administrative Sciences*, *10*(4), 103. https://doi.org/10.3390/admsci10040103.

Wang, X., Zhan, L., Ruan, J., & Zhang, J. (2014). How to choose "last mile" delivery modes for E-fulfillment. *Mathematical Problems in Engineering, 417129*, 1–11. https://doi.org/10.1155/2014%2F417129.

Werenowska, A., & Jaska, E. (2022). Korzystanie z mediów społecznościowych przez pokolenie Z w czasie pandemii COVID-19 [The use of social media by generation during the COVID-19 pandemic]. *Zarządzanie Mediami*, *10*(1), 23–36. https://doi.org/10.4467/23540214zm.22.002.16924.

Wikipedia. (2023). Generation Alpha. Retrieved from https://en.wikipedia.org/wiki/Generation_Alpha (Accessed 14 October 2023).

Zimek, K. (2021). Dostawa prosto do… bagażnika. Skoda testuje nową usługę [Delivery straight to… the trunk. Skoda is testing a new service]. Retrieved from https://moto-high.pl/2021/09/18/skoda-dostawa-do-bagaznika-kurier/ (Accessed 30 August 2023).

19 Trust in artificial intelligence in agriculture

Henryk Runowski and Paulina Kramarz

Introduction

Artificial intelligence (AI) is widely utilized across all sectors of the economy. Technological solutions based on AI have long been employed in manufacturing processes, customer interactions, marketing, and its application is becoming increasingly popular in the agricultural sector. Among such technologies applicable in agriculture, one can distinguish systems for plant and animal disease recognition, animal husbandry automation, animal tracking and welfare maintenance, selective fertilization and chemical spraying, programs and applications aiding decision-making in agriculture, tracking and predicting the impact of irregular weather changes on yields, and soil condition monitoring.

The research objective is to assess the level of trust in AI-based technologies in the realm of agricultural operations in Poland. The limitations on the adoption of these solutions may stem from high costs, the size and structure of agricultural enterprises, infrastructural support, and limited trust. One of the barriers to the proliferation of AI-based technologies in agricultural operations is the reluctance to share information necessary for their functioning, especially concerning decision support systems. Consequently, this research explores how farmers' trust in AI is shaped, their willingness to share information about their operations, and the evaluation of the utility of AI depending on the size of the agricultural enterprise and the production directions pursued.

The possibility of the application of artificial intelligence in agriculture

The history of AI development dates back to the 1940s, but it is only in contemporary times, owing to the proliferation of data volumes, the existence of advanced algorithms, and increased memory and computational power, that it can be fully harnessed (Loudjani et al., 2020). The utilization of AI in agriculture primarily relies on two possibilities: providing recommendations and support to farmers based on data, imagery, and measurements analysis, as well as

DOI: 10.4324/9781032627236-21

employing the labour of AI-driven robots. The former involves the construction of forecasts and decision support. Robots, on the other hand, operate more autonomously, including robots for tasks such as weeding, harvesting, manure removal, pest-spraying drones, and the analysis of crop parameters (Ryan et al., 2021; Ryan, 2022). These functions are made feasible due to the ongoing process of digitalization and the increasing volume of data collected during agricultural production (Cavalcante de Oliveira & Diogne de Souza e Silva, 2023). Algorithms of deep learning and convolutional neural networks (CNN) will play a crucial role in the digitalization of agriculture, enabling image analysis and recognition (Fountas et al., 2020; Figiel, 2022). According to Santos et al. (2020), the most popular outcomes of deep learning technology application in digital modules designed for agriculture encompass disease identification, plant recognition (including weed differentiation), and field surface coverage identification. In the global AI market structure, solutions intended for crop production dominate, with a projected 64% share by 2024. Approximately 19% is expected for technologies supporting animal production, and 15% for indoor agriculture (Figiel, 2022).

AI enables the entry into a new stage of digitizing agriculture, and the possibilities of technologies utilizing its operation are widely recognized and developed. However, in the process of adapting these solutions to practice, numerous problems still arise, related not only to financial aspects but also sociological aspects, including a lack of trust in AI.

In the subject literature, references to potential reasons for perpetuating distrust in AI applied in agriculture can be found. The presented observations are often contradictory, which justifies the need to address this research topic. According to Barret and Rose (2022), the popularization of AI in agriculture will lead to the emergence of social inequalities in rural areas. Some risks may be associated with data security, access to only partial or outdated data, and the implementation of machine learning based on narrow system optimization. There are also concerns about the impact of the intensification of agriculture through AI on biodiversity and the perpetuation of social inequalities, despite the simultaneous improvement in the quality of life for some farmers (Tzachor et al., 2022). According to many authors, AI, however, promotes more efficient resource utilization, optimization of energy use, more precise monitoring of the value chain, which will translate into the care of the natural environment (Bhat & Huang, 2017; Härtel, 2019; European Commission, 2021; Garske et al., 2021).

At the same time, the proliferation of mechanisms based on AI is also associated with positive social and environmental effects. In some regions, one of the motivations is the desire to combat climate change by expanding farmers' knowledge of the use of environmentally friendly digital technologies (Mandapati, 2018). This is also in response to the needs and demands of consumers associating quality with environmental friendliness (Runowski & Kramarz, 2022). AI-based technologies also assist in decision-making regarding the production process, for example, in the context of weather impact (Mandapati, 2018). Bird's-eye view imaging solutions based on drones for monitoring crop conditions enable precise application of chemicals, leading to a reduction in pesticide use by up to 80% (Elbehri & Chestnov, 2021). According to the European Commission's report, AI is a significant factor in achieving the goals of the Common Agricultural Policy 2020+, such as contributing to climate change mitigation, biodiversity protection, enhancement of ecosystem services, habitat and landscape conservation, sustainable development support, and facilitating young farmers' entry into farming (Loudjani et al., 2020). When combined with Big Data and improved connectivity (i.e., with the widespread use of 5G), AI could also enhance tracking of agricultural production to monitor product quality, enforce accountability, and track food distribution and delivery (Elbehri & Chestnov, 2021).

Technological progress can meet the requirements of climate care and resource conservation, but there still remains the problem of support for small-scale farms in accessing such technology. The primary barrier to implementing precision agriculture-related technologies is the reluctance of farmers to adopt solutions they may not be able to efficiently utilize due to their farm's potential. Another issue is the lack of information, which can lead to the selection of unsuitable solutions and, at the same time, perpetuate reluctance to use technology (Bucci et al., 2019). According to Kaushal et al. (2022), providing farmers with complete and up-to-date data reduces risk and saves time. It promotes reduced water consumption, land use optimization, and the development of vertical farming. In many countries, there is a high interest in using simple decision support applications in agriculture, often accessed through mobile phones (Ramcharan et al., 2017; Ramos-Giraldo et al., 2020; Elbehri & Chestnov, 2021).

Research methodology

The instrument /Survey. During the research, an observational and questionnaire method was applied. The aim was to verify whether the topic of AI application in agriculture was of interest to farmers active on social media. Observations were conducted over a period of five months, from April 1 to August 31, 2023, in the largest Polish agricultural groups on Facebook, such as 'Rozmowy o rolnictwie' (131,000 members), 'Forum okiem rolnika' (70,500 members), and 'Rolnictwo cała Polska' (12,200 members). Another method of information gathering was through an account created by the authors on Instagram. By publishing the results of previous research and engaging in discussions under farmers' posts, 193 followers were gathered. In total, 847 accounts were observed, whose owners declared engagement in agricultural activities. Based on the observations, discussions with farmers, and a review of the literature, a survey questionnaire was prepared, and subsequently, an online survey was conducted. The survey was conducted from July 1 to August 31, 2023.

Participants. A purposive sample was used. The criteria included being an owner, co-owner, a person employed in a farm, or someone planning to start an agricultural enterprise. Conducting the research through social media, using an online questionnaire, allowed reaching farmers from various parts of the country. The study included farmers conducting activities in 12 Polish provinces (with no respondents from the Silesian and Opole provinces). A total of 112 properly completed survey questionnaires meeting the sample selection criteria were obtained. Of the respondents, 59% declared having agricultural education (vocational – 2%, technical – 13%, higher – 26%) or being currently enrolled in agricultural studies (18%). Individuals under 35 years of age predominated (75%), which is a result of the research method and should be considered when interpreting the obtained results. The average size of the agricultural farms in the sample was 33 hectares (ranging from 1.2 to 230 hectares for individual farms). Among the forms of activities, grain production was declared by 73% of the respondents, followed by field vegetable production (46%), cattle breeding and husbandry (23%), and oilseed crop production (15%). Due to the significant diversity in the scope of activities, for the purpose of comparisons, the analysis focused on groups centred around the dominant activities among the respondents – cattle breeding, field vegetable cultivation, and grain cultivation. Regarding the analysis related to the sizes of the farms owned by the respondents, a division into two groups was made, taking into account the median value of 16.3 hectares.

Procedure. The online survey questionnaire was sent to 600 randomly selected farmers who had social media accounts and declared engagement in agricultural activities. It was published in agricultural groups on Facebook and on the profile of the authors on Instagram. The responses to the survey, along with a message explaining the purpose and subject of the research, sent directly through private messages, were highly varied. In approximately 5% of cases, responses expressing reluctance to participate in the survey due to its focus on AI were received, while in 10% of cases, refusal was based on other reasons, such as lack of trust, among others. In about 4% of cases, publishing the survey led to direct contact between the participants and the authors, enabling supplementary discussions. Respondents were assured of the full anonymity of the survey, which was guaranteed to them. Before launching the study, the correctness of the survey questionnaire was verified through a pilot study involving 20 farmers from the Mazovian province.

Results

Based on the observations of discussions conducted by farmers on social media platforms, it can be inferred that the topic related to the use of AI in agricultural activities is not very popular. Discussions about AI typically appeared in response to posts promoting technical solutions published by companies specializing in technology and active social media advisors (on Facebook), or in response to posts showcasing its application in farms owned by users actively creating content on Instagram. These discussions usually revolved around solutions with relatively low levels of technological advancement, adapted to the conditions of agricultural operations in Poland. These solutions primarily included GPS systems (discussions about their usefulness and usage-related issues), systems supporting dairy cattle farming, equipment measuring grain parameters during harvesting (as part of a combine harvester system), self-propelled tractors, and fertilization systems with software for dispensing soil fertilizers based on its condition. Reactions to this content varied significantly, and they were not among the most frequently commented on. Responses ranged from showing high interest among farmers in technological innovations to negating the significance of progress in agriculture.

Distrust in AI in agriculture manifested to some extent in the results of a survey. Among the respondents, 31% declared their trust in AI, 23% stated that they do not trust AI, and 46% of the respondents could not assess whether they trust AI-based solutions or not. This can be attributed to the fact that in Polish conditions, the use of AI in agriculture remains a poorly recognized topic, and most agricultural producers have no direct contact with such solutions or the potential for implementing technologically advanced solutions. There are solutions that can be applied without significant investment, but it seems that knowledge of their possibilities remains too low. However, it is worth noting that the percentage expressing trust in AI was higher than the percentage of respondents declaring a lack of trust. Respondents with higher agricultural education and students of agricultural fields showed a slightly greater inclination to trust AI (38% of respondents in this group trusted AI). Among cattle breeders, especially, a higher degree of trust in AI could be distinguished (42%). Among them, only 11% of the respondents declared a lack of trust in AI. In the other groups of respondents separated according to the field of activity (producers of field vegetables and grains), there were no significantly different results compared to the surveyed group as a whole.

A comparison of the opinions of two groups of respondents, divided based on the size of their agricultural holdings (a group of farmers running farms with a size of up to 16.3 hectares and a group of farmers running farms with a size above 16.3 – division established based on the median value), showed that slightly more trust in AI was found among respondents managing larger farms – 33% of them trusted AI, and 19% did not. In the case of smaller agricultural holdings, 29% of the respondents expressed trust, and 27% expressed distrust. However, a high percentage of undecided respondents and small differences in responses within the two groups under investigation do not allow us to unequivocally conclude that trust in AI is determined by the scale of activity. Especially since, in Polish conditions, the size of farms in different regions of the country is conditioned by historical factors, which also entails certain differences in the pace of their socio-economic development.

Subsequently, the respondents were asked to assess their trust in various agricultural technologies utilizing AI on a scale from 1 to 10, where one denoted a lack of trust, and ten represented complete trust (Table 19.1). Respondents had the highest trust in technologies enabling the automation of processes related to animal farming and selective chemical spraying, irrigation, and fertilization (they received average trust ratings of 7.4 out of 10 possible points). The least trust was placed in the capabilities of AI algorithms supporting the selection of optimal plant varieties for planting and technologies enabling the tracking and prediction of the impact of weather changes on yields (trust ratings at 6.2 out of 10 and 6.3 out of 10, respectively).

The analysis of the results revealed a significant variation in opinions regarding the various uses of AI in agriculture, depending on the respondents' previously expressed attitude towards AI. Individuals who trusted AI exhibited a higher level of trust in the various agricultural technologies employing AI. In this group, trust ratings for individual technologies ranged from 7.3 to 8.5 out of 10. The highest trust was for the use of robots for weed control in crops and the automation of processes related to animal farming, while the lowest trust was for tracking and predicting the impact of weather changes on yields. In the case of respondents expressing a lack of trust in AI, trust ratings ranged from 4.4 out of 10 (the use of robots for weed control in crops) to 5.9 out of 10 (selective chemical spraying, irrigation, and fertilization).

Some variation in the assessment of trust in specific types of AI-based technologies was revealed in the division of respondents based on the type of agricultural activity. A group of cattle breeders, vegetable producers, and grain producers were analysed (it is worth noting that 50% of cattle breeders also engaged in grain cultivation, so the results in these two groups showed some similarity). Cattle breeders participating in the survey assessed their trust in AI-based technologies, declaring a high level of trust in technologies related to animal husbandry automation and those that enable tracking animals and caring for their well-being (ratings at 8 points and 7.5 points out of 10, respectively). Vegetable producers showed the highest level of trust in technologies that enable selective spraying of chemicals/fertilization/irrigation (7.9 points out of 10) and the lowest trust in decision-support applications (6.5 points out of 10). In the case of grain producers, the technologies that inspired the most trust were those that enable tracking animals and caring for their well-being, as well as selective spraying of chemicals/fertilization/irrigation (7.7 and 7.4 points out of 10, respectively). The least trust was placed in decision-support algorithms for choosing seed varieties for sowing and tracking the impact of weather changes on crop yields. Decision-support applications were rated best in terms of trust by respondents with the following characteristics: small farm owners, those who trust AI, and grain producers. However, they were not very popular compared to others, even though these technologies, due to their high availability and low cost, could prove to be particularly effective in Polish conditions.

Table 19.1 Trust in technologies using AI in groups of farmers with different characteristics (average values based on trust ratings from 1 to 10, where 1 meant no trust, 10 meant full trust)

Specification	*A*	*B*	*C*	*D*	*E*	*F*	*G*	*H*	*I*	*J*
Total study group	6.8	6.9	6.8	7.4	7.0	6.5	7.4	6.2	6.3	6.9
Farm size										
Up to 16.3 ha	7.0	6.7	6.8	7.2	7.1	6.7	7.4	6.4	6.6	7.2
Over 16.3 ha	6.5	7.2	6.8	7.6	6.9	6.4	7.3	5.9	5.9	6.7
Education										
Agricultural	7.5	7.0	7.0	7.3	6.9	6.5	7.3	6.4	6,1	6.7
Non-agricultural	6.3	6.8	6.6	7.5	7.0	6.6	7.4	6.0	6.4	7.1
Trust in artificial intelligence										
Yes	8.1	7.7	8.5	8.5	8.4	7.3	8.4	6.5	7.3	8.1
No	5.0	5.4	4.3	5.5	4.7	4.7	5.9	4.8	4.5	5.2
Hard to say	6.8	7.1	6.9	7.7	7.2	6.9	7.4	6.7	6.5	7.0
Scope of agricultural activity										
Cattle breeding	7.2	7.1	7.4	8.0	7.5	6.0	7.3	6.0	5.7	6.8
Production of field vegetables	7.2	7.2	7.1	7.5	6.9	6.5	7.9	6.6	6.6	7.0
Cereal production	6.6	6.9	6.8	7.7	7.1	6.7	7.4	6,1	6.2	6.9

Source: own study. Legend: A – systems for recognizing plant diseases and damage caused by pests; B – use of robots in fruit and vegetable harvesting; C – use of robots to remove weeds from crops; D – automation of animal breeding; E – tracking animals and taking care of their welfare; F – programs and applications for decision support in agriculture; G – selective spraying of chemicals/fertilization/irrigation; H – artificial intelligence algorithms enabling the selection of the optimal seed variety for sowing; I – tracking and predicting the impact of irregular weather changes on crops; J – processing data collected by drones to monitor the condition of crops and soils.

During in-depth interviews with respondents, observations were made that could indicate the reason for the limited popularity of these solutions. According to one of the respondents: '*Decision support applications are not created by farmers, so they are not intuitive for them.*' Discussions with respondents also pointed to the limited capabilities of some applications due to the insufficient scope of information processed within them: '*I used the Plantix application, but it does not include apple varieties. Even the difference in the colour of their leaves makes it difficult to define diseases. Probably after implementing the appropriate data, it could work correctly.*'

Respondents were asked to indicate which types of AI-based technologies they found most effective from their perspective. The majority pointed to the use of precision agriculture technologies (74%), automation of animal production, and automation of plant production (66% and 65%, respectively). Below 50% were noted for technologies related to monitoring the state of animals. The utility of predictive technologies was noted at the level of 40%, and decision-support applications at the level of 37%. At the same time, it should be noted that these technologies are not very popular in Polish conditions, so this assessment may be more strongly influenced by the beliefs of respondents rather than practical experiences. These experiences were extremely varied. In in-depth discussions with respondents after completing the survey, the following comments were made:

> Two weeks ago, a specialist in soil moisture measurement devices contacted me. From his information, only a small part of farmers show interest in such technology tests. I suggested some solutions; I see the application of AI in regulating the irrigation process, which can save water.

'*Everything depends on what this AI is specifically supposed to serve and whether it will be used for even greater control by the state and for imposing fines,*'

> It will be difficult for me to trust a tool whose mechanisms of operation I cannot adjust or control myself. If, for example, AI is supposed to decide what protective measures I can use, I am sure it will aim to encourage purchases from a specific manufacturer.

About 5% of the respondents participating in the study stated that none of the AI-based technologies are useful in agriculture. The largest group of respondents holding this view were farmers who declared that they do not trust AI (23%). The usefulness of decision-support applications was noted by vegetable producers (80% of responses), and the monitoring of the state of animals by cattle breeders (73% of responses). At the same time, 85% of cattle breeders indicated the usefulness of automating animal production.

Since the operation of applications that enable prediction or decision support depends on the ability to obtain information provided by farmers, respondents were asked about their willingness to provide such information. Among all the respondents, 45% declared a high or very high willingness to provide information about their activities that could be useful in the operation of decision support applications (in the case of 6%, the willingness was assessed as very high). The difficulty in declaring a willingness to provide information was experienced by 41% of respondents, and a low or very low willingness was expressed by a total of 14% of those surveyed.

Discussion

Doidge et al. (2023) point to the multidimensionality of perceiving ethics in the implementation of AI-based agricultural solutions. For example, animal behaviour research favours their health and well-being, but for some, increased frequency of subjecting them to medical procedures (which results from constant monitoring) may be associated negatively. Similarly, various social, ethical, and ideological factors influencing farmers' decisions regarding the adoption of digital solutions are difficult to classify into just two approaches – those inclined to adopt them and those rejecting them. Not all technologies are perceived in the same way – for example, milking automation does not raise mistrust, while other elements of animal care automation are considered to restrict the freedom of animals and affect the human-animal relationship (Bos et al., 2018; Doidge et al., 2023). Observations conducted by Doidge et al. (2023) indicated that there is a group of farmers who reject the possibility of using ready-made digital solutions but at the same time use digital technology to complement their knowledge and achieve a similar effect. Survey results indicated that the entrenched view of AI (in terms of trust) among the respondents influenced the perception of all types of this technology. Almost 50% of respondents who participated in the survey could not assess whether they trust AI-based solutions. Technologies better recognized – related to precision agriculture or animal production automation – enjoyed the highest level of trust. Respondents' assessments may have been influenced by their experiences with individual AI systems or their lack thereof. According to Herhem et al. (2017), mistrust can also result from concerns about measurement errors of AI sensors, and developing proven safeguards would improve trust levels.

AI enables the achievement of goals in agriculture that would be beyond human capabilities. This is possible through processing countless amounts of data and transforming them into solutions. Although the transfer of agricultural data raises some objections, it is worth noting that their acquisition took place before the digitization stage, through the collection of measurements directly by agricultural advisors. The effects of using sensors that collect data and allow machines to adapt to the soil condition, plants, or the presence of weeds reflect activities that could be carried out by humans but with much less precision and accuracy. The heterogeneity of data sources and the multitrack way of their transformation create the impression of a lack of control over this process, which is a source of concerns (EPRS, 2023). Farmers who participated in the survey showed a fairly high willingness to provide information that could then be processed by AI algorithms in decision support systems and forecasting. However, it should be noted that the study was conducted among respondents present on social media, with access to information, most of them interested in AI. The usefulness of decision support applications was more often indicated by owners of smaller agricultural farms, which may be related to their search for easily accessible means to support their work without generating high expenses. Research conducted in Sweden showed that factors encouraging farmers to be interested in AI-based technologies include a desire to increase the profitability of production, improve comfort, and reduce working time. High market variability (making investments in agricultural technology risky), high cost, lack of technical knowledge to operate devices, and difficulties in acquiring current knowledge on the subject were mentioned as barriers. There were opinions that technologies at the time of market introduction are not fully developed. Negative experiences in dealing with device vendors were also cited as reasons for lack of trust (Carlberg & Jerhamre, 2021). About 5% of respondents in the survey stated that none of the AI-based technologies are useful in agricultural activities. The assessment of usefulness depended on the production profile. The lack of trust in algorithms allowing the selection of the most optimal seed variety among grain producers turned out to be surprising (however, this was a large group – 80% of all respondents, so general trust in AI technology could influence the opinion). According to many authors, trust plays an important role in both the positive attitude towards phenomena and in sharing experiences (Grudzewski et al., 2007; Paliszkiewicz, 2013). The limitation in the use of AI solutions and their full recognition may not only be due to reluctance to use them or lack of funds, but also to the incompatibility of systems from different manufacturers. Applications and other systems are developing so quickly that continuous technical knowledge supplementation is still needed (Catraio, 2021).

Conclusion

AI promotes making accurate decisions in agriculture, increasing efficiency, improving precision agriculture systems, and supports sustainable development processes. It is increasingly seen as a way to improve farmers' income and well-being, and the trends in technology development and research directions indicate that it will be an important factor in the competitiveness of agriculture in the future.

Implications of the Study: The conducted research shows that Polish farmers are very diverse in their receptivity to modern technological solutions, including those related to the use of AI. A significant percentage of respondents exhibit uncertainty in their assessment. Among the surveyed farmers, a tendency to trust AI systems was revealed, which are closer to the direction of their activities, especially among cattle breeders who value the usefulness of solutions enabling animal monitoring and production process automation. Overlooked forms of AI utilization in agriculture include decision support applications. This solution, which has a positive impact on production efficiency worldwide, may raise mistrust due to uncertainty about data processing mechanisms. It can be assumed that the primary barrier to the adaptation of simple AI-based solutions (requiring no significant financial investment) in rural areas in Poland will be a lack of trust and limited access to information about how they work.

Limitations of the Study and Future Directions: The size of the research sample is a limitation of the study. Numerous refusals to participate in the study were motivated by reluctance to the topic of using AI in agriculture. Therefore, it should be kept in mind that the presented results may not fully reflect the extent of mistrust in AI and technological progress. The significant diversity of respondents' opinions and a significant number of refusals to participate in the study indicate that attitudes towards technological progress may result from socio-economic factors that show significant regional differences in the country – which should be the subject of further research. The presented study can be treated as a pilot for research taking into account local conditions in assessing the propensity to adapt AI solutions in agriculture.

References

Barret, H., & Rose, C. (2022). Perceptions of the Fourth Agricultural Revolution: What's In, What's Out, and What Consequences Are Anticipated? *Sociologia Ruralis*, 62(2), 162–189.

Bhat, S. A., & Huang, N. (2017). Big Data and AI Revolution in Precision Agriculture: Survey and Challenges. *IEEE Access*, 9(99), 110209–110222. DOI: 10.1109/ACCESS.2021.3102227.

Bos, J. M., Bovenkerk, B., Feindt, P. H., & Van Dam, Y. K. (2018). The Quantified Animal: Precision Livestock Farming and the Ethical Implications of Objectification. *Food Ethics*, 2, 77–92. DOI: 10.1007/s41055-018-00029-x.

Bucci, G., Bentivoglio, D., & Finco, A. (2019). Factors Affecting ICT Adoption in Agriculture: A Case Study in Italy. *Quality - Access to Success*, 20(2), 122–129.

Carlberg C. J. C., & Jerhamre, E. (2021). *Artificial Intelligence in Agriculture Opportunities and Challenges.* Civilingenjörsprogrammet inom System i Teknik och Samhälle, Upsala.

Catraio, R. A. (2021). Wie revolutioniert KI in der Landwirtschaft die Agrarindustrie? Retrieved from: https://cynoteck.com/de/blog-post/how-is-ai-in-agriculture-revolutionizing-the-agriculture-industry/.

Cavalcante de Oliveira, R., & Diogne de Souza e Silva, R. (2023). Artificial Intelligence in Agriculture: Benefits, Challenges, and Trends. *Aplied Sciences*, 13(13), 1–17. DOI: 10.3390/app13137405.Doidge, C., Frössling, J., Dórea, F. C., Ordell, A., Vidal, G., & Kaler, J. (2023). Social and Ethical Implications of Data and Technology Use on

Farms: A Qualitative Study of Swedish Dairy and Pig Farmers. *Frontiers in Veterinary Science*, 10, 1–16. DOI: 10.3389/fvets.2023.1171107 1–16.

Elbehri, A., & Chestnov, R. (eds.) (2021). *Digital Agriculture in Action – Artificial Intelligence for Agriculture*. FAO and ITU, Bangkok.

EPRS. (2023). *Artificial Intelligence in the Agri-Food Sector Applications, Risks and Impacts*. European Parliamentary Research Service Scientific Foresight Unit (STOA), Brussels.

European Commission. (2021). *Communication from the Commission to the European Parliament, the Council, the European Economic and Social Committee and the Committee of the Regions: 2030 Digital Compass: The European Way for the Digital Decade*. European Commission, Brussels.Figiel, S. (2022). Development of Artificial Intelligence and Potential Impact of Its Applications in Agriculture on Labor Use and Productivity. *Problems of Agricultural Economics*, 373(4), 5–21. DOI:10.30858/zer/153583.

Fountas, S., Espejo Garcia, B., Kasimati, A., Mylonas, M., & Darra, N. (2020). The Future of Digital Agriculture: Technologies and Opportunities. *IT Professional*, 22(1), 24–28.

Garske, B., Bau, A., & Ekardt, F. (2021). Digitalization and AI in European Agriculture: A Strategy for Achieving Climate and Biodiversity Targets? *Sustainability*, 13(9), 1–21. DOI: 10.3390/su13094652 1–21.

Grudzewski, W. M., Hejduk, I. K., Sankowska, A., & Wańtuchowicz, M. (2007). *Zarządzanie zaufaniem w organizacjach wirtualnych*, [*Trust Management in Virtual Work Environments*]. Difin, Warsaw.

Härtel, I. (2019). Agrar-Digitalrecht für eine nachhaltige Landwirtschaft 4.0. *Natur und Recht*, 41, 577–586.

Herhem, T., Rooijakkers, L., Berckmans, D., Pena Fernández, A., Norton, T., Berckmans, D., & Vranken, E. (2017). Appropriate Data Visualisation Is Key to Precision Livestock Farming Acceptance. *Computers and Electronics in Agriculture*, 138, 1–10.

Kaushal, S., Kumar, S., & Tabrez, S. (2022). International Journal of Science and Research (IJSR). *Artificial Intelligence in Agriculture*, 11(5), 1682–1688.

Loudjani, P., Devos, W., Baruth, B., & Lemoine, G. (2020). *AIA: Artificial Intelligence and EU Agriculture*. Joint Research Centre, Brussels.

Mandapati, J. M. (2018). Communication Strategies for Building Climate-Smart Farming Communities. In: Leal Filho, W., Manolas, E., Azul, A., Azeiteiro, U. & McGhie, H. (eds.) *Handbook of Climate Change Communication. Vol. 2. Climate Change Management*. Springer, Cham. DOI:10.1007/978-3-319-70066-3_24.

Paliszkiewicz, J. O. (2013). *Zaufanie w zarządzaniu* [*Trust in management*]. Wydawnictwo Naukowe PWN. Warsaw.

Ramcharan, A., Baranowski, K., McCloskey, P., Ahmed, B., Legg, J., & Hughes, D. P. (2017). Deep Learning for Image-Based Cassava Disease Detection. *Frontier in Plant Sci*ence, 8, 1852–1852.

Ramos-Giraldo, P., Reberg-Horton, C., Locke, A., Mirsky, S., & Lobaton, E. (2020). Drought Stress Detection using Low-Cost Computer Vision Systems and Machine Learning Techniques. *IT Professional*, 22(3), 27–29.

Runowski, H., & Kramarz, P. (2022). Factors Determinating the Perception of Agricultural Products. *Annals of the Polish Association of Agricultural and Agribusiness Economists,* XXIV(4), 155–168.

Ryan, M. (2022). The Social and Ethical Impacts of Artificial Intelligence in Agriculture: Mapping the Agricultural AI Literature. *AI and Society*, 38(6), 2473–2485.

Ryan, M., Van der Burg, S., & Bogaardt, M. (2021). Identifying Key Ethical Debates for Autonomous Robots in Agri-Food: A Research Agenda. *AI and Ethics*, 2(3), 493–507. DOI:10.1007/s43681-021-00104-w.

Santos, L., Oliveira, P. M., Dos Santos, F. N., & Shinde, P. (2020). Deep Learning Applications in Agriculture: A Schort Review. *Robot 2019: Fourth Iberian Robotics Conference*, 139–151. DOI:10.1007/978-3-030-35990-4_12.

Tzachor, A., Devare, M., King, B., Avin, S., & Ó hÉigeartaigh, S. (2022). Responsible Artificial Intelligence in Agriculture Requires Systemic Understanding of Risks and Externalities. *Nature Machine Intelligence*, 4, 104–109.

20 Trust in public sector AI

The case of Finland's Aurora national program

Sorin Dan, Johanna Kalliokoski and Khuram Shahzad

Introduction

Governments at national, regional, and local levels have taken steps to experiment with and implement artificial intelligence (hereafter AI) technologies to improve the organization and provision of government services across different sectors and policy areas (Misuraca & van Noordt, 2020). AI is part of a growing trend in public management scholarship and practice, which see AI as one technology that is part of a set of 'highly integrative and immersive technology' (Kim et al., 2022, p. 362) that are contributing to a deepening of digital government. Smart technology, i.e. cloud computing, big data analytics, social media, the Internet of Things (IoT), and AI, may operate according to government as a platform (GaaP) principles (O'Reilly, 2011; Brown et al., 2017) that value leaner interfaces, user-centricity, and a different role for the government which becomes 'a convener and an enabler rather than the first mover of civic action' (O'Reilly, 2011, p. 15). Such an approach to public management claims to enhance public trust through the use of inter-connected digital infrastructure that creates new and rich information flows, leading to increased accessibility of government information and services (Cordella & Paletti, 2019).

AI is assumed to lead to a redesign of service delivery processes toward a renewed and more trustworthy model that places the service user at the center (O'Reilly, 2011). Increased citizens' demand for efficient and personalized public services has surged public sector investments in new AI-based digital technologies. Therefore, how AI is designed and implemented in practice in the complex world of public sector organizations, which oftentimes lack citizens' trust, becomes essential to achieving this aspiration. Majorly, citizens are unaware of how the AI system makes decisions that affect them, leading to mistrust (Wang et al., 2021). To foster trust in government organizations, it necessitates ensuring that AI-driven public services can satisfy the citizen's expectations and ensure transparency and accountability.

Yet, despite a significant interest in AI from academics and policymakers alike, there is insufficient research on the implications of using AI for trust in

DOI: 10.4324/9781032627236-22

governmental organizations that operate in a sensitive political environment and are under public scrutiny (Vogl et al., 2020; Zuiderwijk et al., 2021). Although some studies (e.g. Wirtz & Müller, 2019) have attempted to point out the influence of AI on public value creation theoretically, there remains a significant lack of an empirical framework examining AI implementation in the public sector to foster citizen trust.

This chapter addresses the implications for the trust of experimenting with AI in public sector organizations by exploring the National Artificial Intelligence Programme AuroraAI (hereafter AuroraAI) developed by the Government of Finland. This program ran from January 2020 until December 2022 (Ministry of Finance, 2023).

Trust and public sector AI

The research on trust and public sector AI builds on the extensive literature on trust that is found in the field of public administration and management (Nyhan, 2000; Van de Walle et al., 2008). This research is also influenced by the digital government research on citizen and user centricity in digital and smart services and government as a platform (Rana et al., 2013; Lee & Lee, 2014; Pope, 2019). How to design and provide public services in a way that responds to citizens' needs and adds value to them is an important theme within this literature. Digital government scholars have argued that Web 2.0 has created the conditions for adopting a platform thinking in public governance that better integrates service provision across different providers, yet to this day its application has largely remained government-centric instead of citizen-centric (Anttiroiko et al., 2014; Sigwejo & Pather, 2016).

The use of ubiquity principles and intelligent solutions may address this challenge and better personalize the design and delivery of public services (de Sousa et al., 2019). Big data and data analytics reinforced by AI techniques may facilitate the creation of new public service models that place citizens at the center and that incorporate a greater emphasis on prevention and responsiveness (Wirtz & Müller, 2019). There is some empirical evidence that points in this direction, yet the existing evidence is limited and contradictory. Zheng et al. (2018) find that the use of AI can lead to service integration, improved civil service satisfaction, and, importantly, reduced waiting times for citizens and users. Automation of service delivery processes can free up case workers' time, which can be used to provide more personalized service in high-risk cases where it is most needed. Coupled with the adoption of self-service delivery through chatbots, this service model can enhance the availability and centricity of public services (Eggers et al., 2017). AI-enhanced data analytics can provide predictive information across policy areas and types of services that can be used to develop better preventive healthcare, educational, and employment services (Fernandes et al., 2018; Sun & Medaglia, 2019). Yet, the extent to which AI-supported techniques can result in greater citizen centricity and trust depends on the nature and complexity of service considering that AI is best suited for managing caseloads and administrative procedures that are labor intensive, time-consuming, and which require the processing of a large volume of data (Kleinberg et al., 2018).

Furthermore, AI holds great promise for effective government services, improved decision-making processes, and optimizing resource allocation to interact with citizens and deliver services (Wang et al., 2021). Existing studies (e.g. Wirtz & Müller, 2019; Sun & Medaglia, 2019) have highlighted the potential positive influence of AI on public value creation, as the goal of governmental organizations is to improve citizen-government interaction by developing citizens' trust in their delivered services. Trust in public sector AI, often referred to as public trust, is considered a key indicator of public value and explains the level of confidence citizens have in the government's procedures for AI-based services (Wang et al., 2021). Citizens tend to have faith in AI-based services' ability to address their concerns while acting in their best interests. This leads to citizen acceptance of such digital services by the government, which enhances the public tendency to interact with government organizations (Hu et al., 2021), as well as trust. Several scholars (e.g. Bertot et al., 2010) have already highlighted the importance of information and communication technologies (ICTs) in fostering public trust, as such technologies offer citizens timely information which is a key factor to develop citizen's perception of transparency and trust in government (Wang et al., 2021). Prior research has also highlighted some AI implementation challenges in the public sector, such as AI algorithms have the potential to unintentionally create bias, perpetuate historical inequalities, exhibit a preference for specific political perspectives, or strengthen unwanted behaviors (Janssen & Kuk, 2016), leading to creating uncertainty among citizens. While Wang et al. (2021) found no significant relationship between the effective use of AI and trust in government, suggesting the inconsistencies in value orientation between governments (intended purpose to promote IT) and citizens' behavior (i.e. perception of meeting the needs).

Thus, despite significant interest in AI from academics and policymakers alike, there is insufficient research on the implications of using AI for trust in governmental organizations that operate in a sensitive political environment and are under public scrutiny (Vogl et al., 2020; Zuiderwijk et al., 2021). Existing research majorly remained scattered and somewhat limited with more theoretical analysis of AI and public trust while also highlighting the related challenges of AI implementation, leading to uncertainty. Therefore, it is important to address such limitations in existing research as the implications of AI to develop trust in governmental organizations is a critical endeavor that corroborates the responsible use of AI technologies in safeguarding citizens' rights and interests.

Methodology

This chapter employs a qualitative single case study research design. As the chapter addresses a real-life phenomenon, a case study design is beneficial for exploring its implications as applicable to the novel setting of public service delivery (Eisenhardt, 1989). Data collection involved semi-structured interviews with key individuals involved in program design, ideation, and implementation, and the analysis of relevant policy documents and existing evaluations. The interviewees represent different professional backgrounds, such as ethics experts, digital analysts, and other technology professionals in municipal governments and ministries (see Appendix).

All interviewees were chosen on purpose and invited to participate in the study due to their involvement and/or knowledge in the case study. The interviews were held online during March-May 2020. Selected follow-up interviewing was conducted in February 2021 to assess further developments in implementation in municipal government that took place between May 2020 and February 2021. All interviews were recorded and transcribed verbatim, consisting of 42 pages of transcribed material. In addition, we performed an analysis of relevant documents produced by different actors, such as government, consultancies, and academic institutions.

Finland's AuroraAI and the quest for trust in public sector AI

The Finnish context is conducive to AI experimentation. As a Nordic welfare state, local governments and well-being services counties are legally mandated to provide a considerable array of services in a relatively sparsely populated geographic area. Finland boasts a high degree of digital literacy and innovation and has been ranked first for two years in a row in the European Commission's Digital Economy and Society Index (European Commission, 2020). However, the aging of the population coupled with fiscal pressures on state finances, have determined policymakers to utilize the potential of AI for public service improvement. This interest is also motivated by the realization that the current public service model is not only fiscally unsustainable but also ineffective at using large amounts of existing data for analytical and predictive purposes to improve policymaking and citizen outcomes (AuroraAI Development and Implementation Plan, 2019).

To address these challenges and opportunities, over the past few years, the Finnish authorities turned to AI to examine its possible societal role. This led to the publication of several policy documents such as the 2017 Research and Innovation Council's report which proposed an ambitious vision and a roadmap according to which Finland aspires to become 'the most attractive and competent environment for experimentation and innovation' by 2030 (Research and Innovation Council Finland, 2017, p. 1). This vision emphasizes the role of data and AI as key drivers for building a platform economy and boosting digital innovation in public services. As a result, Finland became one of the first countries to launch a national AI program, entitled Finland's Artificial Intelligence Programme. During 2017–2019, the program published a series of reports including *Finland's Age of Artificial Intelligence*, which assessed the importance and likely impact of AI and proposed recommendations to help the nation succeed in the age of AI (Ministry of Economic Affairs and Employment, 2017).

The AI developments surveyed above culminated with the launch of a dedicated public sector AI program, AuroraAI, which was a national, multi-annual program that is officially defined as 'a concept for human-centric and ethical society in the age of artificial intelligence' (AuroraAI Development and Implementation Plan, 2019, p. 9). The ambition of AuroraAI is to transform Finland into a 'human-centric society'. It sought to achieve this by means of a new service model that is based on a 'human-centric service market' that gravitates around particular life events and business activity (ibid., p. 12). It uses an artificial neural network (ANN), 'the AuroraAI network', which is defined as a 'decentralized open network of smart services and applications' that connects different service providers (public, private and non-governmental) (ibid., p. 9). Its citizen-centric focus and service orientation is clear: 'AuroraAI seeks to provide a holistic set of personalized AI-driven services for citizens and businesses in a way that is human-centric and works towards their wellbeing as its ultimate goal' (OECD AI Policy Observatory, 2021).

Data is constructed around people's service needs, user segments, and service history logs, which enables the AI system to suggest personalized and targeted services seamlessly. It aims to facilitate both an improved user experience and more effective services that respond to changing and complex service requirements and preferences. This seamless service experience means that it is possible to use AuroraAI in multiple ways such as in the form of a website, email, chatbot conversation, or other messaging tool, much like in the case of the internet (AuroraAI Development and Implementation Plan, 2019). Instead of having one single AI system, it is suggested that AuroraAI would form a network of smart services that can better adapt to technological developments (Cybercom Finland Oy, 2019). Through the management of their personal data, service users are supposed to gain more holistic access to available services and better 'situational awareness' of their multi-faceted well-being. The program defined situational awareness as 'a data-based model of describing statuses, attributes, trends, and system dynamics relating to customers or the analyzed phenomenon in sufficiently realistic detail to enable its use in customer/phenomenon-focused management' (AuroraAI Development and Implementation Plan, 2019, p. 9). Through 'situational management' users can access tailored public services based on their service chain history instead of them 'figuring out' what services they need or without a public authority suggesting such service.

Public benefits, public risks, and ethical dilemmas

We present our exploratory results of the implications of AuroraAI for trust by grouping them into three categories: public benefits, public risks, and ethical dilemmas.

Public benefits

The first expected public benefit of AuroraAI is cost reduction in public services. By better customizing service provision, it is possible to reduce costs over the medium and long term through the optimization of service offerings. This involves delivering the right service at the right time and to the right person in a citizen-centric way. Within AuroraAI, cost reduction, however, is not a stand-alone objective; rather it is closely connected to the expectation that effectiveness will improve if services proactively address citizens' needs. Using data analytics supported by AI, service providers can harness data about citizens' holistic well-being and provide service recommendations that can improve well-being, as this interviewee explains:

> *The effect of current public service structures on society is not strong enough and that needs to be boosted and the only way to boost it is to go human-centric, which in this particular context means creating all these snapshots of states of well-being in certain population segments and quite possibly with individuals even, and then building these service ecosystems around those segments to provide services specifically targeted, and preferably proactive services, to address people's needs maybe even before they realize they have those needs.*

Based on personalized data, it is possible to identify risk factors in advance design tailored interventions, and address wellbeing. In this service model, cost reductions may result from the fact that it costs less to prevent an unfavorable life situation using AI than to respond to it once it has occurred (Sun & Medaglia, 2019). Happier and healthier people will also use the service less, which may bring about savings over the long run. In addition, efficient optimization of service provision and available resources could reduce spoilage and oversupply (AuroraAI Development and Implementation Plan, 2019). AuroraAI's service chatbots and automated user-experience interface may reduce the working hours of public service employees, who could use the saved time to manage higher-risk cases. These changes have the potential to improve citizen-centricity and service effectiveness to the extent that case workers' time is dedicated to people who need it the most and benefit most from more personalized care.

Another public benefit concerns improved user experience. If service provision happens mostly digitally, then it means that it would be more flexibly available than today: AuroraAI could be accessed anytime and from anywhere. This can reduce the need for other means of communication, which are time-consuming for both service providers and users. In addition, it can also make the process faster for the user: instead of waiting in line on the phone or spending time searching on the internet, AuroraAI can provide guidance almost immediately. While the current suomi.fi service brings together information from several public services to a single website. The AuroraAI experience would be different for the user due to an automated customization process. Moreover, user experience is more seamless as AuroraAI integrates service offerings from different providers to create a holistic service plan and address, in a coordinated way, a particular life event. In this renewed service model, the user doesn't need to be aware of all the services or procedures that are available to choose the most suitable one, which simplifies user experience:

> *If the system succeeds... then of course it will make public services qualitatively better. It will be a fluent and flexible way to combine all kinds of sources of information and if the system is developed in the right way, it can indeed help you to fix your problems in a better way than it has been originally possible.*

As an AI public service program, AuroraAI can add significant public value by using large amounts of user data that currently many different Finnish public service organizations are legally mandated to collect. However, this data is not integrated across different service areas and public organizations and is insufficiently used in policymaking. This deprives policymakers and service providers of useful knowledge that may significantly improve policy effectiveness, as this interviewee notes:

> *As you have noticed, I'm... a bit cynical about how public administration has used data so far. There's much room for improvement there... For example, social [and] healthcare sector organizations have enormous amounts of data, but we don't really from, my point of view, at the moment have smart ways to understand the data and to use the data. (...) If we have, for example, data about some issue over ten or fifteen years... we can use that to also predict what's going to happen in the future.*

However, existing GDPR and national legislation currently do not allow user-level data sharing across different organizations. To address this legal challenge, the Finnish law of automated decision-making in public services is currently under examination (Vainio et al., 2020). If these are resolved, AuroraAI could constitute a more effective way to utilize the data and use it to predict changes in population segments. In addition, the AuroraAI system gathers data itself in the form of history logs to customize service provision according to service needs identified in population segments:

> *Perhaps it also gives feedback to the public sector authorities in a way that has not been possible before because all the records about uses can be analyzed later and used for correcting the system. It can use the feedback information to develop the system better and better in the long run.*

Public risks

AuroraAI is a system that lives on data, which includes both personal and individual data and data derived from population clusters. Without this data, the system cannot function and its success depends on the accuracy and quality of data that it receives and generates. This gives rise to several public risks and ethical dilemmas. First, there is a risk that the estimated cost reduction is not as significant as predicted. If AuroraAI does not acquire quality data to target high-risk cases, then it would serve those individuals who are already doing well and who may be more willing to share their personal data given that data sharing is voluntary. In Finland, most social and healthcare sector costs are used for serving a small part of the population (Valtiontalouden tarkastusvirasto, 2017). This implies that without effective targeting of people in need, cost reduction is unlikely to be significant. Yet, this proactive targeting may be difficult to realize, and it depends on the extent to which people in need make use of AuroraAI's service model. This public risk challenges the very notions of human centricity and societal inclusion and raises the question of how to design AI systems in a way that they are appropriate and effective for the most vulnerable members of society.

Another significant public risk lies in the clustering of the data and the creation of user segments, a process that can be disempowering in itself (Dan & Kalliokoski, 2023). Algorithms in AI applications may contain bias due to errors and may result in unfavorable outcomes for some users (Panch et al., 2019). In addition, despite the existence of big data (which may not be accurate enough), it is difficult to confirm that individuals accurately match the user segments with which they are compared. This is particularly problematic in the case of AuroraAI which aims for a holistic concept that incorporates various facets of well-being (economic, social, emotional, spiritual, etc.). Moreover, it raises the question of how being a 'person at risk' or otherwise shapes the user experience of the system:

> *The relationship between segment and individual has never been clearly defined..., and can I deny that I'm part of some segment? Can I say no, I'm not part of that segment? You're misreading me, or even if I am statistically part of a segment and the data categorizes me into a segment it's some segment that I don't want to be a part of... at the moment.*

A significant risk is posed by legal bottlenecks concerning the full implementation of AuroraAI. In the context of Finland, this risk is particularly relevant considering that similar hurdles led in 2019 to the resignation of the government following the crash of a major social and healthcare reform (Hiilamo, 2020). Relatedly, policymakers' quest for creating citizen-centric services by means of AI capabilities is equally challenging considering that AuroraAI's service network consists of providers and organizations (public, private, and non-governmental) that have different organizational values, norms, and operating procedures. In this case, it can be challenging to provide holistic support to the individual while sharing responsibilities and assessing performance in a multi-actor network:

> *The Ministry of Finance wants to build a human-centric society in which AI is used for the good of the individuals. But all the organizations that are in the network have their own responsibilities. There are laws that say that okay Kela [Finland's Social Insurance Institution], that's your role. Vero [Finnish Tax Agency], that's your role. City of Tampere. You have your responsibilities. How do you turn this into a human-centric society?*

The multi-actor system also raises public risks related to privacy and data management. AuroraAI is assessed as mostly secure for the user by the interviewees, yet data security depends on how each participating organization handles user data. For the system to assess an individual's complex situation accurately and to intervene proactively, it needs relevant, holistic data from different sources. This poses questions concerning user consent and control of the data, especially given that the data may 'speak against the user', which can undermine the incentive to provide reliable data. AuroraAI architects claim that all personal data that citizens provide to the system is anonymized and forms a digital twin that is not identical to the data provider (AuroraAI Development and Implementation Plan, 2019). However, previous research has shown that anonymity may be undermined when combining data from multiple sources, which is particularly relevant for a sparsely populated country such as Finland (Green et al., 2017; Honkanen, 2020). In addition, interviewees noted that it might be tempting for the state to use the system for surveillance purposes, given its proactive stance that targets behavioral change (Dan & Kalliokoski, 2023):

> *One big concern that is all the time existing is how to prevent the formation of a big brother society.*

Ethical dilemmas

Ethical dilemmas involve decision-making between at least two ethical imperatives, none of which is preferable or acceptable (Yung, 2016). Our empirical results point to several ethical dilemmas.

A significant ethical dilemma concerns the personal data conundrum and its implications for effective citizen-centric services. While it is possible to use

AuroraAI without sharing personal data, users cannot use its full citizen-centric functionalities unless they share at least some personal data that is relevant and accurate. There are trade-offs between the type and amount of personal data that is shared and the accuracy of service recommendations that AuroraAI delivers in response. 'Anonymous browsing' is possible within AuroraAI, according to the interviewees, but results may lack precision and personalization, which undermines the very concept of the program. Moreover, to ensure a degree of anonymity (which it can be argued it cannot be absolute), AuroraAI needs to securely have in place technical requirements such as blocking IP addresses and phone IDs and refraining from using identifying cookies. Individuals need to consent to the processing of their data to use AuroraAI and they are responsible for what they share with the system, yet they need to understand its functionalities to decide what data they share about themselves and for what purposes. The dilemma concerns the extent to which the service providers who make up the AuroraAI network should be legally mandated to use population data for issuing service recommendations given possible public benefits and risks that may be different from one individual to another.

> *The thing is that the [artificial] neural network alone will not solve anything. It will never make things better than they are now. It will need good quality data and it will also need... a certain amount of that data. Not necessarily open data [which means] data from all the organizations that is shared with everyone, but the [artificial] neural network cannot function without quality data.*

This implies that for AuroraAI to be effective, particularly in the case of vulnerable citizens, Finnish people have to trust and accept the service model and its data management principles and procedures. This brings us to the next dilemma, which is that some users might have difficulties understanding the AuroraAI concept and to whom they give their personal information and consent given its multi-actor logic. AuroraAI can become a black box, unintelligible not only to its designers and architects but particularly to the most vulnerable members of society who may need it the most, including children, immigrants, and persons with disabilities (see also Haataja & Latvanen, 2019). It should be noted that even though Finland is a digitally advanced nation and in general possesses above-average digital skills, not all citizens have the same level of digital literacy to make use of AuroraAI capabilities.

Another salient ethical dilemma that results from the interviews lies with the predictive features of AuroraAI. By using data analytics to analyze social and health datasets, it is possible to predict risk factors for specific population segments based on historical data. Although the Finnish legislation does not allow this use of data, the ethical dilemma concerns whether this should be allowed, given its possible public benefits and risks. Such predictions can have an impact on how people live their lives. Even if the aim is to improve well-being for the common good, an ethical dilemma remains concerning whether AI should be affecting people's behavior in the first place and whether people have the right not to know the predicted trend if they so wish.

Although the usage of AuroraAI is not planned to be mandatory, nor does any individual have in most cases any responsibility to follow the suggestions created by the system, it must be acknowledged that AuroraAI is most likely going to be trusted by people who use it, and it may well have an impact on people and their decision making. If created and used properly, an AI system may be less biased and contain more information than, for example, in the case of a human counselor (Kleinberg et al., 2018), but an ethical dilemma is apparent concerning the system's power to 'voice' recommendations and the implications that this new reliance on an AI system has over the medium and long term. Will it only assist with personalized service recommendations or will it go further and further, like a train going on a one-way track, to decisively influence and determine lives and destinies?

> *(...) In Aurora there's this application that they're creating now in which young people [students] around the age of 14–15 are advised that you should apply to this school or you should go to study this and then the question that raises is... how much the state actually controls the future of those kids (....). It is a system that categorizes those kids into A type and B type and so on. That's the danger (...).*

There is an inherent danger that the use of AI and data analytics becomes so essential to policymaking, and perhaps the only approach to policymaking, that governments may wish to overstep their boundaries. This concern was well reflected in the interviews:

> *It will be so tempting for the public sector to cut costs and do all kinds of things to make the sector more efficient by using the data in a way that makes it as a surveillance society and whether or not I give my permission for them to do so doesn't make the surveillance any less of a surveillance.*

If citizens consent to the use of data, which in a high-trust Nordic welfare state such as Finland, is likely to be the case, then where does the limit go? A regulatory framework and democratic processes that follow transparency and accountability principles can provide checks and balances (Floridi et al., 2018; Wirtz & Müller, 2019), yet public governance arrangements may become so entrenched in AI that they become oblivious to correcting their own faults. In so doing they may form what elsewhere we call *'surveillance public governance'* (Dan & Kalliokoski, 2023), a system in which power relations between the state, networked service providers, and service users are significantly shaped by the use of AI, to their own benefit and risk.

Conclusions

This chapter has explored the relationship between public sector AI and trust by analyzing the AuroraAI National Programme implemented by the Government of Finland. Our main findings show that the creation of AI-supported services is likely to lead to public benefits in terms of systemic cost reduction and improved user experience and service effectiveness. This results from the delivery of tailored service recommendations and offerings through the AuroraAI network according to service needs for particular life events and population segments. Cost reductions and improved effectiveness may result from the use of predictive capabilities within AuroraAI based on which policymakers and service providers can issue service recommendations ahead of time that prevent the occurrence of unfavorable and costly future life events.

Nevertheless, our results show that the attainment of these benefits is fraught with public risks and ethical dilemmas that are inherent in AI technologies. Public risks that are apparent in the context of AuroraAI include the unequal use of the system across population groups with those who need it the most being at risk of using it the least, and thus benefitting from it, the least. This is particularly relevant given the data intensity of the program, which requires accurate, personalized high-quality data to be effective. Without the provision, sharing, and availability of such data, the very concept of citizen-centric public services is at risk. An additional significant public risk relates to the clustering of the data and creation of population segments, which can be both disempowering and difficult to apply considering that inclusion in population clusters can be an imprecise process that can result in error. AuroraAI has raised ethical dilemmas related to the government's use of AI, which despite its aim of improving the common good, can lead to public surveillance at the expense of individual liberty (Dan & Kalliokoski, 2023).

References

Anttiroiko, A. V., Valkama, P., & Bailey, S. J. (2014). Smart cities in the new service economy: Building platforms for smart services. *AI & Society*, *29*, 323–334.

AuroraAI Development and Implementation Plan. (2019). *AuroraAI – Towards a human-centric society. Development and implementation plan 2019–2023 based on the preliminary study on the Aurora national artificial intelligence programme*. Ministry of Finance, Helsinki, Finland.

Bertot, J. C., Jaeger, P. T., & Grimes, J. M. (2010). Using ICTs to create a culture of transparency: E-government and social media as openness and anti-corruption tools for societies. *Government Information Quarterly*, *27*(3), 264–271.

Brown, A., Fishenden, J., Thompson, M., & Venters, W. (2017). Appraising the impact and role of platform models and Government as a Platform (GaaP) in UK Government public service reform: Towards a Platform Assessment Framework (PAF). *Government Information Quarterly*, *34*(2), 167–182.

Cordella, A., & Paletti, A. (2019). Government as a platform, orchestration, and public value creation: The Italian case. *Government Information Quarterly*, *36*(4), 1–15.

Cybercom Finland Oy. (2019). *Arkkitehtuurikuvaus AuroraAI: Työpaketti 4* [Architectural description of AuroraAI: Work package 4]. Report.

Dan, S., & Kalliokoski, J. (2023). "Surveillance Public Governance": Towards a new public sector reform trajectory? Working paper, School of Management, University of Vaasa, Vaasa, Finland.

de Sousa, W. G., de Melo, E. R. P., Bermejo, P. H. D. S., Farias, R. A. S., & Gomes, A. O. (2019). How and Where is artificial intelligence in the public sector going? A literature review and research agenda. *Government Information Quarterly, 36*(4), 1–14.

Eggers, W. D., Fishman, T., & Kishani, P. (2017). AI-augmented human services: Using cognitive technologies to transform program delivery. In R. Moses, Nikitia Garia, and Abrar Khan (eds.), *Deloitte Development LCC*. Accessed 18 July 2023. https://www2.deloitte.com/ insights/us/en/industry/public-v.sector/artificial-intelligence-technologies-human-services-pro grams.html.

Eisenhardt, K. M. (1989). Building theories from case study research. *Academy of Management Review*, *14*(4), 532–550.

European Commission. (2020). *Digital economy and society index 2020*. European Commission, Brussels.

Fernandes, E., Holanda, M., Victorino, M., Borges, V., Carvalho, R., & Van Erven, G. (2018). Educational data mining: Predictive analysis of academic performance of public school students in the capital of Brazil. *Journal of Business Research, 94*, 335–343.

Floridi, L., Cowls, J., Beltrametti, M., Cowls, J., Beltrametti, M., Chatila, R., Chazerand, P., Dignum, V., Luetge, C., Madelin, R., Pagallo, U., Rossi, F., Schafer, B., Valcke, P., & Vayena, E. (2018). AI4People—An ethical framework for a good AI society: Opportunities, risks, principles, and recommendations. *Minds & Machines, 28*, 689–707. https://doi.org/10.1007/s11023-018-9482-5

Green, B., Cunningham, G., Ekblaw, A., Kominers, P., Linzer, A., & Crawford, S. (2017). *Open data privacy*. Berkman Klein Center for Internet & Society Research Publication. http://nrs.harvard.edu/urn-3:HUL.InstRepos:30340010.

Haataja, M., & Latvanen, M. (2019). *AuroraAI-esiselvityshanke - Etiikkatyökokonaisuuden suositukset* [AuroraAI preliminary study – Recommendations of the Ethics-work package]. Report.

Hiilamo, H. (2020). Why did social and healthcare services reform fail in Finland? *Socialmedicinsk Tidskrift*, *97*(3), 433–441.

Honkanen, M. (2020). *Omadata osana AuroraAI-hanketta* [MyData as a part of AuroraAI-project]. Report.

Hu, Q., Lu, Y., Pan, Z., Gong, Y., & Yang, Z. (2021). Can AI artifacts influence human cognition? The effects of artificial autonomy in intelligent personal assistants. *International Journal of Information Management*, *56*, 102250.

Janssen, M., & Kuk, G. (2016). The challenges and limits of big data algorithms in technocratic governance. *Government Information Quarterly*, *33*(3), 371–377.

Kleinberg, J., Lakkaraju, H., Leskovec, J., Ludwig, J., & Mullainathan, S. (2018). Human decisions and machine predictions. *The Quarterly Journal of Economics*, *133*(1), 237–293.

Kim, S., Andersen, K. N., & Lee, J. (2022). Platform government in the era of smart technology. *Public Administration Review*, *82*, 362–368.

Lee, J., & Lee, H. (2014). Developing and validating a citizen-centric typology for smart city services. *Government Information Quarterly, 31*(1), 93–105.
Ministry of Economic Affairs and Employment. (2017). *Finland's age of artificial intelligence - Turning Finland into a leading country in the application of artificial intelligence. Objective and recommendations for measures.* Publications of Ministry of Economic Affairs and Employment, MEAE reports 41/2017. Helsinki, Finland.
Ministry of Finance. (2023). *National artificial intelligence programme AuroraAI.* Retrieved 13.9.2023. https://vm.fi/en/national-artificial-intelligence-programme-auroraai.Misuraca, G., & van Noordt, C. (2020). *Overview of the use and impact of AI in public services in the EU.* Publications Office of the European Union, Luxembourg.
Nyhan, R. C. (2000). Changing the paradigm: Trust and its role in public sector organizations. *The American Review of Public Administration, 30*(1), 87–109.
OECD AI Policy Observatory (2021). *AuroraAI.* Retrieved 15.2.2021. https://oecd.ai/dashboards/policy-initiatives/2019-data-policyInitiatives-24206.
O'Reilly, T. (2011). Government as a platform. *Innovations: Technology, Governance, Globalization, 6*(1), 13–40.
Panch, T., Mattie, H., & Atun, R. (2019). Artificial intelligence and algorithmic bias: Implications for health systems. *Journal of Global Health, 9*(2), 020318.
Pope, R. (2019). *Playbook: Government as a platform.* Ash Center for Democratic Governance and Innovation, Harvard Kennedy School.
Rana, N. P., Dwivedi, Y. K., & Williams, M. D. (2013). Evaluating alternative theoretical models for examining citizen centric adoption of e-government. *Transforming Government: People, Process and Policy, 7*(1), 27–49.
Research and Innovation Council Finland. (2017). *Vision and road map of the research and innovation council Finland.*
Sigwejo, A., & Pather, S. (2016). A citizen-centric framework for assessing E-government effectiveness. *The Electronic Journal of Information Systems in Developing Countries, 74*(1), 1–27.
Sun, T. Q., & Medaglia, R. (2019). Mapping the challenges of Artificial Intelligence in the public sector: Evidence from public healthcare. *Government Information Quarterly, 36*(2), 368–383.
Vainio, N., Tarkka, V., & Jaatinen, T. (2020). *Assessment memorandum on the need to regulate automated decision-making within public administration in general legislation.* Publications of the Ministry of Justice, Reports and guidelines 2020:14. Helsinki, Finland.
Valtiontalouden tarkastusvirasto (2017). *Tuloksellisuustarkastuskertomus - Paljon palveluja tarvitsevat ja käyttävät asiakkaat perusterveydenhuollossa* [Performance audit report - Frequent users of basic health care services]. Valtiontalouden tarkastusviraston tarkastuskertomukset 11/2017. Helsinki, Finland.
Van de Walle, S., Van Roosbroek, S., & Bouckaert, G. (2008). Trust in the public sector: Is there any evidence for a long-term decline? *International Review of Administrative Sciences, 74*(1), 47–64.
Vogl, T. M., Seidelin, C., Ganesh, B., & Bright, J. (2020). Smart technology and the emergence of algorithmic bureaucracy: Artificial intelligence in UK local authorities. *Public Administration Review, 80*(6), 946–961.
Wang, C., Teo, T. S., & Janssen, M. (2021). Public and private value creation using artificial intelligence: An empirical study of AI voice robot users in Chinese public sector. *International Journal of Information Management, 61*, 102401.

Wirtz, B. W., & Müller, W. M. (2019). An integrated artificial intelligence framework for public management. *Public Management Review, 21*(7), 1076–1100. https://doi.org/10.1080/14719037.2018.1549268.

Yung, B. (2016). Introduction: Ethical dilemmas, social values, and public policy: The context of governance and citizenship. In: Yung, B. & Kam, Por Y. (eds.) *Ethical dilemmas in public policy. Governance and citizenship in Asia*, pp. 1–19, Springer, Singapore.

Zheng, Y., Yu, H., Cui, L., Miao, C., Leung, C., & Yang, Q. (2018). SmartHS: An AI platform for improving government service provision. Paper presented at the 32nd AAAI Conference on Artificial Intelligence, 7704–7711.

Zuiderwijk, A., Chen, Y.-C., & Salem, F. (2021). Implications of the use of artificial intelligence in public governance: A systematic literature review and a research agenda. *Government Information Quarterly, 38*(3), 101577.

Appendix:

Table 20.1 Key interviews

Interview	*Timing*	*Male (M)/female (F)*	*Area of expertise*
1	March 2020	M	Local governance (LG)
2	May 2020	M	Digitalization (D)
3	May 2020	M	Technology, ethics (TE)
4	May 2020	F	Science, ethics (SE)
5	February 2021	M	Local governance (LG)

21 Trust in artificial intelligence in regional development

Edyta Skarzyńska

Introduction

In today's dynamic world, artificial intelligence (AI) is becoming a key element in the transformation of societies and economies around the world. In the context of regional development, trust in AI is becoming a central issue, as technology can play an important role in shaping the future of regions through innovation, operational efficiency, and sustainability (Lazzeretti et al., 2023). One of the main challenges we face is building public trust in AI. The public is often concerned about the loss of jobs, invasion of privacy, and ethical aspects of using AI. To integrate AI into regional development processes, it is necessary to focus on transparency, ethics, and public engagement. People should understand how algorithms make decisions and what data are used to train them. Organizations developing AI projects should operate in a spirit of openness, sharing information about their activities and the technologies they use. Regional development should consider social and cultural values, and AI projects should not violate people's fundamental rights and dignity. Introducing ethical principles as an integral part of AI projects will help build a positive image of the technology and encourage its acceptance. Public consultation, stakeholder dialogues, and broad community involvement will help avoid misunderstandings and identify potential concerns about the technology. Educating the public about AI's benefits and potential challenges is also a key element. Evolutionary perspectives, too, have long assumed that a region's innovative outcomes depend on the exchange and recombination of local knowledge bases, defining trajectories of technological change that are both path-dependent and place-dependent (Bellandi et al., 2018). In the context of technological transformation, a region's ability to integrate new technologies can be seen as the result of the region's cumulative technical capabilities (Cohen & Levinthal, 1990; Giuliani, 2005). Since each new technology is associated with a specific set of know-how and information, regions are assumed to absorb and exploit new potentials depending on their previous technological capabilities accumulated in their knowledge bases (Boschma, 2017; Corradini, 2019). The assimilation of new technical knowledge is strongly based on optimal

DOI: 10.4324/9781032627236-23

levels of cognitive proximity, enabling both exploratory and exploitative learning processes (Nooteboom et al., 2007). High mental proximity between regional knowledge sources and new technological knowledge can determine more effective absorptive capacity in regions (Isaksen & Trippl, 2016). It is, therefore, important to study confidence in AI affecting regional development. It plays a key role in decision-making processes at the economic and social levels and AI's impact on the lives of individuals and communities, which is increasingly permeating from the institutional level to the economic, political, cultural, and social groups. The themes that emerge from a recent analysis of the literature are Industry 4.0, smart cities, big data, AI, and related technologies (robotization, IoT, augmented, and virtual reality). Among the few studies on regional development related to AI, the authors note that successful AI regions are those where AI technologies have a deep-rooted place in their knowledge space. In contrast, a study by Balland and Boschma (2021) focuses on regional knowledge production related to Industry 4.0 technology, including AI. The study confirms that the topic of AI is still new in the field of regional science, contributing to the identification of the most intriguing issues and future research trends aimed at developing this new research field.

Literature review

Trust in artificial intelligence

The uniqueness of AI lies in the integration of its functions with human-like capabilities (Krafft et al., 2020). This study adopts the definition of AI proposed by the OECD (2019, pp. 23–24): "An artificial intelligence system is a machine-based system capable of making predictions, recommendations or decisions affecting real or virtual environments, according to preset human goals. AI systems are designed to operate at different levels of autonomy." Unlike traditional technologies, where users have full control over operations, AI's ability to operate autonomously carries risk and uncertainty for users. Trust in AI is a key aspect affecting its widespread adoption and effectiveness in various areas of life (Choung et al., 2023). Today's AI systems, based on deep machine learning, require trust from users, policymakers, and society as a whole. Users expect AI systems to be adequately protected against attacks and dangers. With growing awareness of AI's potential power and impact as a social engineering system, there is great emphasis on developing trustworthy solutions in this field. In recent years, various institutions such as corporations, governments (e.g., OECD, EU), technology companies (e.g., Google and Microsoft), and professional associations (e.g., IEEE) have developed frameworks for the development and implementation of trustworthy AI. It is widely recommended that aspects of trustworthiness be built into the system from the initial development stage (Jobin et al., 2019; Hagendorff, 2020). Recent research on trust in AI has focused on identifying factors that influence the development of trustworthy

AI systems (Gillath et al., 2021; Lockey et al., 2021; Rheu et al., 2021). Shin's most recent research has focused on the relationship between FATEx, trust, user perceptions (Shin, 2021b, 2022), and acceptance (Shin, 2021a). Overall, Shin's findings show that FATEx is a fundamental requirement for trustworthy AI, and trust plays a key role in shaping users' positive perceptions and experiences with AI technologies (Choung et al., 2023). Building trust in AI requires a holistic approach, considering technical, ethical, social, and legal aspects. Open communication, transparency, and awareness are key to maintaining and developing a positive public attitude toward AI.

Key areas affecting regional development

Artificial intelligence

In recent years, there has been rapid development of AI, which includes both an increase in the allocation of resources for it and significant advances in its performance. Increased investment is stimulating the rapid development of AI technical and capabilities (Furman & Seamans, 2019). Today's AI is playing an important role in shaping the future of work (Brynjolfsson & Mitchell, 2017), stock market activities (Kirilenko et al., 2017), medical advances (Hosny et al., 2018), transportation systems (Bonnefon et al., 2016) the future of warfare (Russell et al., 2015) and the way society is governed (Taddeo & Floridi, 2018). On the one hand, implementing AI has the potential to reduce human error and eliminate human bias (Frank et al., 2019). On the other hand, however, recent examples suggest that AI technologies can be introduced without considering existing social biases or social issues. These examples also highlight the extensive applications of the technology, leading some to describe it as a general-purpose technology (Cicerone et al., 2023). AI is increasingly important in regional development, affecting various socioeconomic aspects. Its applications span a wide range of fields, such as education, health, transportation, and public administration. It catalyzes innovation, introduces new technologies, streamlines processes, and creates new opportunities for local communities. One of the key areas is the development of AI skills, which can help make regions more competitive in the global labor market. As a result, integrating AI in a regional context is becoming an important driver of innovation and socioeconomic development. Regions can use AI to counteract the limited innovation capabilities resulting from a lack of specialization and combinatorialism (Cicerone et al., 2023).

Industry 4.0

The term Industry 4.0 describes the comprehensive metamorphosis of industrial production resulting from implementing new technologies and the Internet. This is generating advances in AI, an unparalleled ability to process massive amounts of data (Gottge et al., 2020), and an accelerated pace of innovation (Siltori et al., 2021). Regional studies are increasingly looking at the spatial implications of digital technological change in the manufacturing sector, putting it on the research agenda (De Propris & Bailey, 2020). One of the main streams of research focuses on the role of cities, first, as centers of innovation and early adoption (Clark et al., 2018) and second, because Industry 4.0 is assumed to foster the creation of new forms of manufacturing value, especially in urban spaces. The adoption of Industry 4.0 is a complex and multifaceted phenomenon characterized by various stages, such as intensive exploration, research, and moments of adjustment and consolidation (Culot et al., 2020). In the initial phase of intensive exploration, the study found that human trust was a key factor influencing the decision to adopt Industry 4.0 (Alarcon et al., 2021). Deciding on trust as part of a group reasoning process is susceptible to various constraints, such as efficiency (Yost & Cheng, 2021), negotiation, and satisfaction (Yao & Storme, 2021), as well as education, knowledge, familiarity, benefits, preference ranking, and verbal elements (Ushada et al., 2021).

Smart cities

Despite years of discussion about "smart cities," the concept remains ambiguous. In earlier studies, many terms seem to be used interchangeably with "smart," such as "smart" or "digital" (Alizadeh, 2017). The field of smart cities, as a transdisciplinary field, is often combined with various topics such as urban science, ICT (Angelidou, 2014), IoT (Scuotto et al., 2016), and healthcare (Aborokbah et al., 2018). In addition, smart cities include a variety of technologies, such as AI (Srivastava et al., 2017), transportation (Masek et al., 2016), and even government regulation (Hartswood et al., 2014). From a data perspective, most studies have identified experimental and pilot smart use cases, such as the implementation of Internet of Things applications (Bresciani et al., 2018), data analysis in the context of social media (Batty et al., 2012), open innovation practices and crowdsourcing (Cohen et al., 2016), and the use of open government data (Pereira et al., 2017). With the growing importance of big data analytics, Kitchin (2014) concludes that an information-rich city can effectively use advanced data analytics to understand, monitor, regulate, and plan. Nevertheless, there are controversial ethical issues and criticisms regarding the rapid development of urban big data. Taylor (2016) argues that the precise information on human mobility generated by smart devices and big data sets can lead to overreaction to conflicts and migrations, and there is a need to strike the right balance between privacy and security. In China, Wu et al. (2018) identify several risks associated with developing smart cities with Chinese characteristics, including information security, emergency response, and technological limitations. The public's unease may relate to various attitudes, such as "fear," "anxiety," "confusion," "skepticism," "powerlessness" and "ignorance," as noted in studies (e.g., Butot et al., 2020; Cabalquinto & Hutchins, 2020). Additionally, citizen dissatisfaction may manifest itself through acts of "resignation," "resistance," "sabotage," or "marginalization" toward the smart city concept and its implementation, as indicated by other studies (e.g., Amsellem, 2021; van Twist et al., 2023). It is, therefore, crucial to examine the confidence level in the changes being made. The process of introducing technologies such as ride-sharing, autonomous vehicles, and smart infrastructure is transforming the face of cities. Many researchers are now focusing on developing the concept of smart cities and related urban innovations. However, the economic geography of the technology sector in an urban context still needs to be sufficiently understood (Adler & Florida, 2021).

Related technologies (robotization, IoT, augmented, and virtual reality)

There is a growing interest in rethinking and transforming urban infrastructure and urban social life, which is being made possible by advances in robotics and autonomous systems (Macrorie et al., 2021). This trend is particularly noticeable in the growth of the literature on drones, other uncrewed aerial vehicles (UAVs), and autonomous vehicles (AVs) (Bissell, 2018; Garrett & Anderson, 2018). However, the potential for social robotics in the urban context is much broader, with robots replacing or supporting current human tasks such as maintaining order and security, delivering goods and food, maintenance, construction, personal assistance, and healthcare. Opportunities for extensive urban restructuring using robotics reflect a new generation of this technology, based on advanced AI and machine learning. These aspects are linked to information gathering and socio-technical platforms that use robotics to expand and reconnect service infrastructure (e.g., Frank et al., 2019). Robotics applications in an urban context have the potential to bring long-term financial savings, but at the same time, they are costly and carry risks in configuration. Visions of developed robotic urbanism are becoming more numerous, and research is needed to thoroughly understand the possibilities, realities, and implications of this new phase of urban restructuring. The idea of an interconnecting network, in which physical objects are connected online via the Internet, is referred to by the term "Internet of Things" (IoT). IoT introduces a new model of digital communication, where everyday objects are equipped to communicate with each other and users (Zanella et al., 2014). Reflections on the Internet of Things and smart cities describe a vision of a modern world where millions of connected devices can communicate effectively, helping to increase productivity, promote sustainability, and improve the quality of human life. It has been suggested that widespread adoption of this technology depends largely on the availability of power and infrastructure to enable communication between devices (Chambers & Evans, 2020).

Methodology

The survey was conducted in Poland. Data collection took place from August 2023 to September 2023. All data was collected via a survey questionnaire using Forms. Participants were recruited using the snowball method – that is, through social media channels (e.g., distributing the survey on popular Facebook pages and profiles on Instagram, which proved extremely valuable for recruitment). The questionnaire was conducted anonymously and comprised two sections. Initially, participants were asked to complete a custom survey to gather socio-demographic information such as age, gender, education level, and marital status. Subsequently, their confidence in applying AI in regional development was assessed. A total of 362 responses were received, and the socio-demographic characteristics of the overall sample were as follows: 233 respondents were female (64.4%) and 129 were male (35.6%). The age range was 19–65 years (M = 29.47, SD = 11.56). The educational level was as follows: 81 respondents (22.4%) had completed high school, 25 (7%) had vocational education, 115 (31.8%) had a bachelor's/engineering degree, 139 (38.4%) had a master's degree, and 2 (0.5%) had completed a doctorate. Analyzing the marital status of respondents, 112 (30.9%) were single, 188 (51.9%) were married or in a stable relationship, 15 (4.1%) were separated or divorced, 10 (2.7%) were widowed, and 37 (10.2%) preferred not to specify.

Results

AI system designers and engineers have identified several technical characteristics essential for system reliability. At the time of this writing, nine identified characteristics define the trustworthiness of an AI system: Accuracy, Reliability, Resilience, Objectivity, Security, Explainability, Safety, Accountability, and Privacy (NIST, 2019). From an engineering perspective, an AI system needs these characteristics to be trustworthy. Respondents evaluated the identified trustworthiness characteristics of AI systems and urban technology categories developed in the Adler and Florida (2021) study (Table 21.1). The highest scores were for explainability (4.02) and objectivity (3.49). Privacy (2.11) and accountability (2.56) received the lowest average. In the analysis of urban technology categories, on-demand delivery services received the highest score, and urban engagement technologies received the lowest score. Mobile apps that allow people to get around cities faster and directly address the trade-off between agglomeration and congestion received an average score of 3.76. Shared spaces, or apps that rationalize excess urban space, including services like Airbnb and services like Open Table, were rated 3.96 by respondents. On-demand delivery services: On-demand delivery services such as Instacart, Uber Eats, and Door Dash allow consumers and businesses to access goods without having to move were

Table 21.1 Evaluation of perceptions of AI and urban technology verticals using a 5-point Likert scale

Specification	*M*	*Me*	*Mo*
Accuracy	3.15	3	3
Reliability	3.28	3	3
Resiliency	3.49	4	4
Objectivity	3.07	3	3'
Security	2.77	3	3
Explainability	4.02	4	4
Safety	2.62	3	3
Accountability	2.56	3	3
Privacy	2.11	2	1
Urban Technology Verticals Affecting Development			
Mobile applications	3.76	4	4
Shared spaces	3.96	4	4
On-demand delivery services	4.23	4	4
Smart infrastructure	3.41	3	3
Real estate technology	3.57	4	3
Automated service work	2.87	3	3
Vertical farming	2.32	2	2
Technologies of urban engagement	2.12	2	1

Source: Autor elaboration.

Notes: M – mean, Me – median, Mo – mode.

rated at 4.23, while smart infrastructure or information updates to the physical infrastructure of cities should be considered urban technologies received a score of 3.41. These technologies allow urban infrastructure to collect and respond to information about its use. Real estate technology or prop tech, that is, companies that develop new organizational forms for the real estate industry, should be considered urban technologies received a score of 3.57, and automated service work, particularly automated retail, for example, is a way out of this quagmire. Amazon's checkout-free stores, which use geo-tracking and mobile payment technologies to fully automate frontline work, are rated at 2.87. Vertical farming/urban farming is responsible for developing vertical farming technologies and is also an example of agglomeration-friendly innovation, rated at 2.32. Urban engagement technologies, which is how buyers and sellers can connect on urban platforms for verified location-based transactional activities. Urban platforms are channels for non-market interaction, is rated at 2.12.

Since trust in AI used in regional development is a complex, multidimensional construct that is not directly identifiable, its level in the completed survey was measured by relying on lower-level research constructs composed of identifiable variables (general technological trust and trust in advanced technology used in regional development). During the measurement, respondents were asked to respond to the questions assigned to each category (data security, transparency, public participation, technology ethics, institutional support, accessibility and equity, education and training, impact monitoring, and long-term commitment) by selecting a specific response category on a 0–10 scale, where 0 meant "Completely disagree," and 10 meant "Completely agree." Based on the survey (Figure 21.1), the confidence level in AI used in regional development was 5. 74 (M = 5.52, Me = 5.83, SD = 1.93).

Such a result should be interpreted cautiously; it indicates respondents' moderately high confidence in AI used in regional development. Referring to the sub-assessments for the nine areas studied, it is worth noting that confidence was supported mainly by accessibility equality and long-term commitment. Advanced technologies are accessible to all segments of society, and their implementation does not exacerbate existing inequalities. A commitment to long-term cooperation and commitment from technology companies, the government, and other partners in regional development significantly influences the level of trust. The lowest ratings were given to data security (4.48), transparency (4.75), and institutional support (4.8). Problems were perceived in the provision of effective measures to protect personal data and sensitive information, the lack of sharing of information on technology functions, algorithms, and processes that increase transparency and allow local communities to understand the benefits and risks of technology better, and insufficient support from government institutions and community organizations that oversee, regulate and control technology implementation.

Figure 21.1 Trust in artificial intelligence.

Source: Autor elaboration.

Summary

Trust in AI in regional development is key to successfully integrating technology into society. Increasing transparency, basing activities on sound ethical principles, and actively engaging the public are key elements in building a positive image of AI. As technology advances, it is necessary to simultaneously shape public acceptance and understanding of AI's role in regional development. Intelligent systems can contribute to sustainable regional development by optimizing processes, improving energy efficiency, developing infrastructure, and supporting sectors such as health and education. Through data analysis, AI can provide vital information on the needs of local communities, allowing resources to be directed to the right places. Involving local communities in the decision-making process for AI implementation is a key part of building trust. Public consultation, stakeholder dialogues, and broad community involvement will help avoid misunderstandings and identify potential concerns about the technology. Educating the public about AI's benefits and potential challenges is also a key element. Due to advances in AI, a more sophisticated understanding of trust in these systems is needed. The growing impact of AI on daily life is forecast to continue in the coming days. In the context of the unclear mechanisms underlying AI, the relevance of trust in AI is also likely to increase. AI manifested in many areas used in regional development should be easy to use, useful, and trustworthy for potential users. This study has some limitations that can be used as research directions for the future. The study focused on the impact of the main areas affecting trust in AI in regional development. Future research could focus on specific aspects to explore the role of trust and additional critical factors in adopting and using technologies in regional development. For any online panel, the participants in our study may only partially reflect the general population. The fact that all respondents had access to computers and the Internet introduces some sample bias toward greater technology acceptance. Respondents in online surveys are likely to have expanded access to various forms of information technology and related experiences. Therefore, future research should focus on analyzing more representative samples, considering varying experience levels with technology.

References

Aborokbah, M. M., Al-Mutairi, S., Sangaiah, A. K., & Samuel, O. W. (2018). Adaptive context aware decision computing paradigm for intensive health care delivery in smart cities—A case analysis. *Sustainable Cities and Society*, *41*, 919–924.

Adler, P., & Florida, R. (2021). The rise of urban tech: How innovations for cities come from cities. *Regional Studies*, *55*(10–11), 1787–1800.

Alarcon, G. M., Gibson, A. M., Jessup, S. A., & Capiola, A. (2021). Exploring the differential effects of trust violations in human-human and human-robot interactions. *Applied Ergonomics*, *93*, 103350.

Alizadeh, T. (2017). An investigation of IBM's smarter cites challenge: What do participating cities want? *Cities*, *63*, 70–80.

Amsellem, A. (2021). The noise of silent machines: A case study of LinkNYC. *Surveillance & Society*, *19*(2), 168–186.

Angelidou, M. (2014). Smart city policies: A spatial approach. *Cities*, *41*, S3–S11.

Balland, P. A., & Boschma, R. (2021). Complementary interregional linkages and smart specialisation: An empirical study on European regions. *Regional Studies*, *55*(6), 1059–1070.

Batty, M., Axhausen, K. W., Giannotti, F., Pozdnoukhov, A., Bazzani, A., Wachowicz, M., Ouzounis, G., & Portugali, Y. (2012). Smart cities of the future. *The European Physical Journal Special Topics*, *214*, 481–518.

Bellandi, M., Santini, E., & Vecciolini, C. (2018). Learning, unlearning and forgetting processes in industrial districts. *Cambridge Journal of Economics*, *42*(6), 1671–1685.

Bissell, D. (2018). Automation interrupted: How autonomous vehicle accidents transform the material politics of automation. *Political Geography*, *65*, 57–66.

Bonnefon, J. F., Shariff, A., & Rahwan, I. (2016). The social dilemma of autonomous vehicles. *Science*, *352*(6293), 1573–1576.

Boschma, R. (2017). Relatedness as driver of regional diversification: A research agenda. *Regional Studies*, *51*(3), 351–364.

Bresciani, S., Ferraris, A., & Del Giudice, M. (2018). The management of organizational ambidexterity through alliances in a new context of analysis: Internet of Things (IoT) smart city projects. *Technological Forecasting and Social Change*, *136*, 331–338.

Brynjolfsson, E., & Mitchell, T. (2017). What can machine learning do? Workforce implications. *Science*, *358*(6370), 1530–1534.

Butot, V., Bayerl, P. S., Jacobs, G., & de Haan, F. (2020). Citizen repertoires of smart urban safety: Perspectives from Rotterdam, the Netherlands. *Technological Forecasting and Social Change*, *158*, 120164.

Cabalquinto, E., & Hutchins, B. (2020). '"It should allow me to opt in or opt out": Investigating smartphone use and the contending attitudes of commuters towards geolocation data collection. *Telematics and Informatics*, *51*, 101403.

Chambers, J., & Evans, J. (2020). Informal urbanism and the Internet of Things: Reliability, trust and the reconfiguration of infrastructure. *Urban Studies*, *57*(14), 2918–2935.

Choung, H., David, P., & Ross, A. (2023). Trust in AI and its role in the acceptance of AI technologies. *International Journal of Human–Computer Interaction*, *39*(9), 1727–1739.

Cicerone, G., Faggian, A., Montresor, S., & Rentocchini, F. (2023). Regional artificial intelligence and the geography of environmental technologies: Does local AI knowledge help regional green-tech specialization? *Regional Studies*, *57*(2), 330–343.

Clark, J., Harrison, J., & Miguelez, E. (2018). Connecting cities, revitalizing regions: The centrality of cities to regional development. *Regional Studies*, *52*(8), 1025–1028.

Cohen, B., Almirall, E., & Chesbrough, H. (2016). The city as a lab: Open innovation meets the collaborative economy. *California Management Review*, *59*(1), 5–13.

Cohen, W. M., & Levinthal, D. A. (1990). Absorptive capacity: A new perspective on learning and innovation. *Administrative Science Quarterly*, *35*(1), 128–152.

Corradini, C. (2019). Location determinants of green technological entry: Evidence from European regions. *Small Business Economics*, *52*(4), 845–858.

Culot, G., Nassimbeni, G., Orzes, G., & Sartor, M. (2020). Behind the definition of Industry 4.0: Analysis and open questions. *International Journal of Production Economics, 226*, 107617.

De Propris, L., & Bailey, D. (2020). *Industry 4.0 and regional transformations*. Taylor & Francis.

Frank, M. R., Wang, D., Cebrian, M., & Rahwan, I. (2019). The evolution of citation graphs in artificial intelligence research. *Nature Machine Intelligence, 1*(2), 79–85.

Furman, J., & Seamans, R. (2019). AI and the economy. *Innovation Policy and the Economy, 19*(1), 161–191.

Garrett, B., & Anderson, K. (2018). Drone methodologies: Taking flight in human and physical geography. *Transactions of the Institute of British Geographers, 43*(3), 341–359.

Gillath, O., Ai, T., Branicky, M. S., Keshmiri, S., Davison, R. B., & Spaulding, R. (2021). Attachment and trust in artificial intelligence. *Computers in Human Behavior, 115*, 106607.

Giuliani, E. (2005). Cluster absorptive capacity: why do some clusters forge ahead and others lag behind? *European Urban and Regional Studies, 12*(3), 269–288.

Gottge, S., Menzel, T., & Forslund, H. (2020). Industry 4.0 technologies in the purchasing process. *Industrial Management & Data Systems, 120*(4), 730–748.

Hagendorff, T. (2020). The ethics of AI ethics: An evaluation of guidelines. *Minds and Machines, 30*(1), 99–120.

Hartswood, M., Grimpe, B., Jirotka, M., & Anderson, S. (2014). Towards the ethical governance of smart society. *Social Collective Intelligence: Combining the Powers of Humans and Machines to Build a Smarter Society*, 3–30.

Hosny, A., Parmar, C., Quackenbush, J., & Schwartz, L. H. (2018). HJ and Aerts. *Artificial Intelligence in Radiology, Nature Reviews Cancer, 18*(8), 500–510.

Isaksen, A., & Trippl, M. (2016). Path development in different regional innovation systems. A conceptual analysis. In D. Parrilli, R. Dahl Fitjar, & A. Rodriguez-Pose (Eds.), *Innovation drivers and regional innovation strategies* (pp. 66–84). New York and London: Routledge.

Jobin, A., Ienca, M., & Vayena, E. (2019). The global landscape of AI ethics guidelines. *Nature Machine Intelligence, 1*(9), 389–399.

Kirilenko, A., Kyle, A. S., Samadi, M., & Tuzun, T. (2017). The flash crash: High-frequency trading in an electronic market. *The Journal of Finance, 72*(3), 967–998.

Kitchin, R. (2014). The real-time city? Big data and smart urbanism. *GeoJournal, 79*(1), 1–14.

Krafft, P. M., Young, M., Katell, M., Huang, K., & Bugingo, G. (2020). Defining AI in policy versus practice. In *Proceedings of the AAAI/ACM Conference on AI, Ethics, and Society*, 72–78.

Lazzeretti, L., Innocenti, N., Nannelli, M., & Oliva, S. (2023). The emergence of artificial intelligence in the regional sciences: a literature review. *European Planning Studies, 31*(7), 1304–1324.

Lockey, S., Gillespie, N., Holm, D., & Someh, I. A. (2021). A review of trust in artificial intelligence: Challenges, vulnerabilities and future directions. *Hawaii International Conference on System Sciences.*

Macrorie, R., Marvin, S., & While, A. (2021). Robotics and automation in the city: A research agenda. *Urban Geography*, *42*(2), 197–217.

Masek, P., Masek, J., Frantik, P., Fujdiak, R., Ometov, A., Hosek, J., Abdreev, S., Mlynek, P., & Misurec, J. (2016). A harmonized perspective on transportation management in smart cities: The novel IoT-driven environment for road traffic modeling. *Sensors*, *16*(11), 1872.

National Institute of Standards and Technology (2019) "US LEADERSHIP IN AI: A Plan for Federal Engagement in Developing Technical Standards and Related Tools, Prepared in response to Executive Order 13859."

Nooteboom, B., Van Haverbeke, W., Duysters, G., Gilsing, V., & Van den Oord, A. (2007). Optimal cognitive distance and absorptive capacity. *Research Policy*, *36*(7), 1016–1034.

OECD (2019). *Artificial intelligence in society*. OECD Publishing.

Pereira, G. V., Macadar, M. A., Luciano, E. M., & Testa, M. G. (2017). Delivering public value through open government data initiatives in a Smart City context. *Information Systems Frontiers*, *19*, 213–229.

Rheu, M., Shin, J. Y., Peng, W., & Huh-Yoo, J. (2021). Systematic review: Trust-building factors and implications for conversational agent design. *International Journal of Human–Computer Interaction*, *37*(1), 81–96.

Russell, S., Hauert, S., Altman, R., & Veloso, M. (2015). Ethics of artificial intelligence. *Nature*, *521*(7553), 415–416.

Scuotto, V., Ferraris, A., & Bresciani, S. (2016). Internet of things: Applications and challenges in smart cities: a case study of IBM smart city projects. *Business Process Management Journal*, *22*(2), 357–367.

Shin, D. (2021a). The effects of explainability and causability on perception, trust, and acceptance: Implications for explainable AI. *International Journal of Human-Computer Studies*, *146*, 102551.

Shin, D. (2021b). Why does explainability matter in news analytic systems? Proposing explainable analytic journalism. *Journalism Studies*, *22*(8), 1047–1065.

Shin, D. (2022). How do people judge the credibility of algorithmic sources? *Ai & Society*, *37*, 81–96.

Siltori, P. F., Anholon, R., Rampasso, I. S., Quelhas, O. L., Santa-Eulalia, L. A., & Leal Filho, W. (2021). Industry 4.0 and corporate sustainability: An exploratory analysis of possible impacts in the Brazilian context. *Technological Forecasting and Social Change*, *167*, 120741.

Srivastava, S., Bisht, A., & Narayan, N. (2017, January). Safety and security in smart cities using artificial intelligence—A review. In *Cloud Computing, Data Science & Engineering-Confluence, 2017 7th International Conference on* (pp. 130–133). IEEE.

Taddeo, M., & Floridi, L. (2018). How AI can be a force for good. *Science*, *361*(6404), 751–752.

Taylor, M. (2016). Lifshitz holography. *Classical and Quantum Gravity*, *33*(3), 033001.

Ushada, M., Wijayanto, T., Trapsilawati, F., & Okayama, T. (2021). Modeling SMEs' Trust in the Implementation of Industry 4.0 using Kansei engineering and artificial neural network: Food and beverage SMEs context. *Journal of Engineering & Technological Sciences*, *53*(2), 227–246.

van Twist, A., Ruijer, E., & Meijer, A. (2023). Smart cities & citizen discontent: A systematic review of the literature. *Government Information Quarterly*, *40*(2), 101799.

Wu, Y., Zhang, W., Shen, J., Mo, Z., & Peng, Y. (2018). Smart city with Chinese characteristics against the background of big data: Idea, action and risk. *Journal of Cleaner Production*, *173*, 60–66.

Yao, J., & Storme, M. (2021). Trust building via negotiation: Immediate versus lingering effects of general trust and negotiator satisfaction. *Group Decision and Negotiation*, *30*(3), 507–528.

Yost, E., & Cheng, Y. (2021). Customers' risk perception and dine-out motivation during a pandemic: Insight for the restaurant industry. *International Journal of Hospitality Management*, *95*, 102889.

Zanella, A., Bui, N., Castellani, A., Vangelista, L., & Zorzi, M. (2014). Internet of things for smart cities. *IEEE Internet of Things Journal*, *1*(1), 22–32.

22 The impact of an artificial intelligence on the information asymmetry and the trust of the actors of the insurance market

Irena Jędrzejczyk and Wojciech Pizło

Introduction

Insurers belong to the institutions of the public trust – this is an assumption not only of the scientific insurance theory but also of the international and national law regulations. The main task of insurers is to provide insurance coverage, which requires their constant readiness to pay insurance benefits (including compensation) and maintain an adequate level of financial liquidity. Insurance buyers purchase a sense of security, trusting that the insurer will protect their lives and health, their property and their businesses against the financial consequences of risk. For the economy, the insurers are the biggest suppliers of capital, investing their own profits and accumulated premium resources on other financial markets, and by purchasing treasury bonds they become creditors with a significant influence on state policy. The Act on Insurance and Reinsurance Activities (2015, Article 4(1)) states "insurance activities are activities of a public trust character, requiring appropriate qualifications, experience and credibility." Compliance with legal provisions is supervised and monitored, and it is based on numerous procedures, including those related to the protection of personal data, antimoney laundering, the quality and adequacy of the offered insurance products, as well as to complaint procedures.

The most important factor in building trust in relations between insurance market participants is their equal access to information. We refer in this chapter to the theory of information asymmetry, formulated by the Nobel Prize winner George Akerlof (1970), which proves that insurance market participants do not have equal access to information. The problem of information asymmetry in the relations between the insurer and the insurance services consumer originates from the consumer concealing information about the size and structure of risk from the insurer in order to reduce the insurance premium. The result of this type of unethical practices, used by consumers as insurance services buyers, is underinsurance, that is a service purchase and sale transaction do not cover the actual size of potential damage. Another manifestation of information asymmetry in these relations is the practice of extorting undue compensation by the insured

DOI: 10.4324/9781032627236-24

from the insurer. On the other hand, there are reprehensible practices of excluding the insurer's liability for a part of the risk accepted for insurance, by means of the abusive clauses, manipulation of the provisions of the General Terms and Conditions of Insurance, techniques unknown to the consumer, to avoid or delay the payment of compensation due.

The process of digitization of insurance is a completely new source of information asymmetry in the relationship between the insurer and the consumer of insurance services, and it includes the use of artificial intelligence (AI). The strategic documents (Resolution no. 196, 2020) indicate the importance of AI applications at the state level, in particular in the public or market sector; especially those recognized as publicly trusted. It is therefore justified to choose the insurance sector as the subject of the confidence analysis for AI applications. The basic problem to be solved in this chapter is the violation of basic insurance functions due to the deepening information asymmetry in favor of insurers thanks to the use of AI. The aim of the chapter is to assess the direction of the impact of AI on the differences in access to the benefits of digitalization between the Insurer and the insured, in further considerations called the client. The term "client" is understood as a person seeking insurance protection, the policyholder or the insured. A person seeking insurance protection means a person who has expressed to the insurance distributor the will to conclude an insurance contract (Act of December 15, 2017 on insurance distribution, Journal of Laws 2017 item 2486, Article 3(1)(10) and (13)).

The following hypotheses were adopted for verification in this study:

H1: The equal access to information by insurance market participants is a factor in building trust in relations between them;

H2: the insurer's information advantage over other insurance market actors, and especially over the client, reinforces as a result of AI applications;

H3: the growing information asymmetry leads to impairment of some insurance functions as a mechanism protecting the insured person against risk.

In the current scientific debates on the importance of AI applications in insurance, included in the related literature, the transfer of benefits from these applications to individual links in the value chain was a characteristic feature. In this article, the subject of the study will be both the benefits and costs of AI applications in insurance, related to the functions of the insurance and not to the value chain.

Review of the literature and of the key research centers

The projects, which focused on the applications of AI for risk assessment and insurance policy valuation were selected from many research projects currently implemented or completed at the turn of the second and third decades of the

21st century. They illustrate the benefits of using AI to alleviate information asymmetry between the parties to an insurance contract in the best way.

A review of the scientific literature on the impact of AI on information asymmetry in insurance leads to the conclusion that these are relatively new publications, coming from the beginning of the second decade of the 21st century. Researchers focus mainly on information asymmetry, but their work does not focus on the application of AI in the insurance sector. Since 2020, during the COVID-19 pandemic, many articles have been published in which interest in the impact of AI applications in insurance on the asymmetry of information has been revealed by many authors who previously ignored this topic. Hiroshi Nakaso (Daiwa Institute of Research, https://www.dir.co.jp) created a bridge between the earliest period and the post-pandemic years. Christian Wiens, (https://www.insurtechinsights.com/europe/speakers/christian-wiens/, 2023) uses artificial intelligence in his research to assess risk, but also to personalize insurance offers for customers. The works of the following authors or their teams illustrate the problems of the impact of artificial intelligence on the asymmetry of information in the insurance market:

- Maas (2019) seeks answers to a research question on how the use of machine learning can affect the risk selection process in the insurance industry and in the asymmetry of information between seller and customer;
- Salchegger, Wiedenmann and Widemann (2019), argue that the main problem and challenge in AI innovation for the next decade is databases;
- Salant and Sweeting (2020), describe a market experiment to investigate the impact of artificial intelligence on the insurance market and information asymmetry;
- Faggini, Corazza and Zaccaroni (2020), study the use of artificial intelligence in detecting insurance fraud, based on game theory in their research, which can reduce information asymmetry and increase customer confidence in the insurance market;
- Li, Birge and Haugh (2020) explore how machine learning can help reduce information asymmetry in the insurance market and improve the quality of risk assessment;
- Liu, Liu and Wang (2020) reviewing the literature on the use of AI in the insurance sector, including its impact on information asymmetry.

Methodology

Two types of measuring instruments were used in the research. Participants were guaranteed anonymity during the study. The basic research instrument was a standardized interview addressed to persons with expert status resulting from their position in the business environment or due to professional or scientific achievements) who participated in any edition of Fintech & InsurTech Digital from 2019 to 2022 and sat on the jury of the "Friendly Insurance Company" - a competition organized by Gazeta Bankowa from 2020 to 2022. In the tabular presentation of the results of own research, the framework borrowed from Go Digital and Be Digital (Digitization of the insurance sector in Poland, 2018, p. 4) will be used, which contains expert opinions relating to the level of digitization of customer experience and within the organization's insurance (digitization of operations and business models). The most serious limitation of the research tools used is their focus on the insurer-client relationship, ignoring the insurer's relations with employees, agents, brokers or reinsurers, as well as relations with digital technology providers, that is FinTech entities.

Results of the research procedure

The results of the research procedure were compared according to the insurance function criterion and the direction of AI impact was assessed as strengthening a given function (+) or violating this function (−) (Table 22.1).

The results of the analysis of standardized interview questionnaires confirm the first hypothesis, because, according to the experts interviewed, building a client's confidence in new technologies on the insurance market is a preliminary condition for their dissemination. Clients are encouraged to perform some insurance (business) operations themselves by an insurer that has invested in digital solutions. Such operations include, among others: independent selection of the offered insurance product and conclusion of an insurance contract online, independent reporting of damage, handling of policy renewal and ongoing management of the policy. Clients have access to this type of business operations thanks to tools such as mobile applications, online forms and dedicated portals, chatbots and virtual assistants. According to the experts interviewed, clients are highly differentiated due to their digital competences. In addition, their concerns about the security of data, especially financial data, are growing, and there is a fear of a potential identity theft. The following statements demonstrate the asymmetry of digital competences between insurers and their clients, time pressure, and sources of lack of trust:

- *In our motor insurance and real estate insurance (including apartments), we use artificial intelligence provided by Tractable. After attaching photos of the damaged vehicle (or apartment), clients will respond within 30 seconds on average. They receive a ready cost estimation and a proposal to complete the liquidation process. This solution allows clients to view and accept or reject the instant repair cost calculation directly in the application, with a fully automated process;*
- *Cybercriminals can use AI technology to create sophisticated frauds in the insurance market. They can create fake profiles, generate authentic-looking emails or calls in order to mislead clients and extort data or money;*

Table 22.1 The impact of AI applications on information asymmetry in insurance in Poland and on insurance functions

Function of the insurance	*Assessment of the direction of the impact of AI on the correctness of the client–insurer relationship*		*Opinions communicated by participants in a study on the impact of AI applications on insurance functions*
	Go digital client	*Be digital insurer*	
Preventive	+	+	Applications of the Internet of Things with AI in preventing diseases, accidents, thefts, breakdowns, etc.
	+	+	Financial support for the state or client in disease prevention, construction of protective infrastructure (e.g. against floods, fire protection, and others)
	+	+	Supporting activities aimed at cybersecurity
	+	+	Cyber assistance is a subscription product
	+	+	Incorporating AI into the early warning system
	+	+	AI as a crisis management tool
Protective			INSURANCE DISTRIBUTION AND COOPERATION WITH AN INTERMEDIARY PHASE
	+	+	The use of AI by brokers and agents to examine client needs and personalize the offer for him
			PHASE OF PREPARATION AND IMPLEMENTATION OF THE INSURANCE CONTRACT
	+	+	– possibility of remote consultation and concluding (or renewing) a simple contract online
	–	+	– negative risk selection by Insurers to reduce loss ratio and improve technical result
	–	+	– using prohibitive prices to avoid accepting the largest risks for insurance, which may exclude some clients from insurance (e.g. elderly and/or sick people)
	–	+	– possibility of overinsurance, i.e. collecting an insurance premium that is inappropriately higher in relation to the size and types of risk accepted from the client
			– possibility of obtaining external data about the client

(*Continued*)

Table 22.1 (Continued)

Function of the insurance	*Assessment of the direction of the impact of AI on the correctness of the client–insurer relationship*		*Opinions communicated by participants in a study on the impact of AI applications on insurance functions*
	Go digital client	*Be digital insurer*	
Compensative			DAMAGE SETTING AND PAYMENT OF COMPENSATION PHASE
	+	+	– possibility of remote reporting and verification of damage based on electronic documentation (photos, maps, etc.)
	+	+	– possibility of cooperation with public services such as police, fire brigade, hospitals, etc.
	+	+	– transparency in payments, preventing compensation fraud,
	–	+	– time discipline in the payment of the so-called the "undisputed amount" of compensation
Stabilizing (financial)	–	+	– The AI revolution is changing the insurance business in Poland. AI offers enormous opportunities to change the business model and improve revenues or margins,
	+	+	– We can boast of an extremely solid capital position, with a solvency ratio of 228% (221% FY2022), also thanks to the strong share of normalized capital generation.
	–	+	– Subject to significant fluctuations in interest rates and market developments, and taking into account the current extreme weather conditions, we expect to achieve annual gross profit in the range of EUR 700–750 million in accordance with IFRS 17/9

Source: The Authors' own study based on own research results.

- *Clients willingly share their sensitive data with AI in exchange for the promise of a more affordable policy;*
- *The most widespread and accepted data-driven insurance option is telematics insurance, in which driver behavior is recorded. The application helps to improve the quality of your driving style by analyzing five key parameters: speed, acceleration, braking, cornering, and phone use while driving.*

The answers obtained in the interview confirm the second hypothesis, because the current direction of impact of AI applications in insurance increases the bargaining power of insurers in the insurance services market against all its other actors. It also increases the advantage of insurers over the client, thanks to the deepening asymmetry of information in favor of insurers. To illustrate the theorem about the deepening inequality of the parties to the insurance contract, examples of some statements given in the interview will be provided below, namely:

- *So far, insurers were obliged to explain the logic of decisions in the insurer-client relationship. As users of AI, insurers may have difficulty with determining the appropriate level of trust in it because they do not know what it can do and how well it can perform. As a result, they begin to over-rely on it and accept its incorrect recommendations;*
- *AI used to collect, store and process information, and this way – personal data, on an unprecedented scale. The risk of leakage is the main source of consumers' anxiety and distrust. To guarantee the security of processed data, companies collecting it usually decide to implement AI systems on their own servers;*
- *The decision-making process of AI algorithms is difficult to understand, contrary to the expectation of possibility to audit algorithms in the insurance sector as a strictly regulated and supervised activity, subject to principles such as transparency and explainability;*
- *The operation of the AI system depends largely on the data on which it learns. If it is trained on inaccurate, biased, or plagiarized data, it will produce undesirable results, even if it is designed well technically. Therefore, effective voicebots and chatbots for the insurance industry should be trained on domain data and records of conversations with clients.*

The third hypothesis is also confirmed, because the deepening asymmetry of information leads to impairment of certain insurance functions as a mechanism intended to protect the insured against risk. The subjective and objective scope of protection is decreasing due to the negative selection of risk that is less profitable for insurers, made with the help of AI, which, operating on large databases, has a better understanding of the size and structure of the client's risk and a better ability to estimate its profitability. In this way, the basic function of insurance, namely the protective function, is violated. Therefore, the gap between the size of the risk that the client needs to transfer to the insurer and the part of the risk whose acceptance for insurance is recognized as profitable for the insurer using AI is growing. Taking into consideration the risk profile and the client profile, it occurs that some client groups are exposed to risks that are unprofitable for insurers and are therefore excluded from insurance coverage. To illustrate the claim that the deepening asymmetry of information leads to impairment of some insurance functions, below are examples of some statements provided in the interview, and namely:

- *Both insurers and brokers more and more often offer the additional support to corporate clients as a part of AI-based monitoring and prediction systems. The company's consent to use the suggested solutions sometimes allows to reduce the premium, especially when telemetry solutions are used. Sometimes, if we deal with a risk that is initially unacceptable to the company, it is even crucial to concluding an insurance contract.*
- *The AI revolution is changing the insurance business in Poland. AI offers enormous opportunities to change the business model and improve revenues or margins,*
- *We have defined several business needs around our products and now use AI in our internal CRM solution. Our team created an AI algorithm to predict sales results based on current data recorded in the CRM system. We gave it enough data to run specific forecasts and now we are happy to use the AI elements presented in the view of potential clients in our CRM system.*

The compensation function may also be violated as a result of insurers' desire to pay the so-called undisputed amount, that is lower than the amount of compensation due.

Similarly, there are cases of the applications of AI supporting the investment policy of insurers and bringing a higher rate of return on investments of the insurance premium by the insurer - but they do not translate directly into the benefit of the client. The arguments supporting this interpretation are stated below.

Discussion

The first attempts to define and apply AI were made in the Definition of Artificial Intelligence document and the recommendations of the AIGO group (OECD). An attempt to define AI can be found in the soon-to-be-published opinion of the European Economic and Social Committee entitled "Digitization and innovative business models in the European financial sector – implications for employment and clients" (COM, 2017). A year later, the document was published: "FinTech Action Plan: For more competitive and innovative European financial sector, (European Commission Brussels, 8.3.2018, COM(2018) 109 final, Communication from the Commission to the European Parliament, The Council, The European Central Bank, The European Economic and Social Committee and the Committee of The regions, Brussels, 8/03/2018, COM (2018) 109 final). In 2021, the proposal for a regulation of the European Parliament and of the Council of 21 April 2021 establishing harmonized rules on AI (Artificial Intelligence Act) and amending certain legislative acts of the Union was published [COM(2021) 206 final AI Act; hereinafter: AIA].

An outline of the use of AI in practice was presented by the European Commission for the Effectiveness of Justice in the European Ethics Charter (Felten, Raj and Seamans, 2023). In Poland, recommendations for the practical application are created within the Working Group on Artificial Intelligence (GRAI) at the Chancellery of the Prime Minister of the Republic of Poland.

Since none of the documents mentioned above has developed a uniform and generally accepted definition of artificial intelligence, for the purposes of this study the least controversial concept of an "artificial intelligence system" will be used, understood here as a system based on the concept of a machine that can influence the environment by formulating recommendations and predictions or decisions about a given set of goals. As evidenced by the results derived from our own research presented in these considerations, in their statements, the interview participants emphasized that the primary goal of insurance companies is profit and, in the long run, an increase in the company's value. Therefore, it is reasonable to conclude that the set of goals for AI will include maximizing the insurer's profit and, while waiting for the company's value to increase, collecting and investing the largest possible premium resources.

The collected insurance premium invested by insurers, for example in safe financial instruments such as State Treasury bonds, may encourage the State, as an institution, to generate more internal public debt, which, after exceeding a certain ceiling, is not a positive phenomenon for the balance of public finances. Moreover, the insurance sector, as one of the largest creditors, has a huge influence on the legislation in a country with such high debt in order to further strengthen its position, including its advantage over the client. As an evidence of the influence of the insurance business on legislation, the so-called compulsory insurance can serve, provided for by insurance law (Act of May 22, 2003 on compulsory insurance, the Insurance Guarantee Fund and the Polish Motor Insurers' Bureau), which the client is statutorily forced to purchase regardless of the quality and price of this type of products offered on the insurance market.

To sum up, the financial stabilization function from the point of view of the client's interests as a citizen and as a payer of insurance premiums cannot be clearly interpreted positively.

Referring to the difference in the level of digital competences between the insurer and the client on the insurance market, it is also necessary to indicate the level of the insurer's ability to manage his own sensitive data and manage access to data that concerns him and which is used by the insurer with the help of AI on a much larger scale than before the digital transformation.

In building the insurer's advantage over the client, even the legal obligations of insurers in the field of liability for data, traceability, and historiography arising in Europe from the legal acts of the GDPR and the IDD directive are not a sufficient brake.

The Regulation on the protection of natural persons with regard to the processing of personal data and on the free movement of such data (2016) and the

Regulation on the respect for private life and the protection of personal data in electronic communications (2017) are documents of key importance for the rights of consumers of insurance services to protection one's data.

At the same time, it can be reasonably assumed that the fact that an insurer obtained the client's consent to access his data freely in exchange for a lower policy price (lower insurance premium) and/or a gift in the form of a telematics gadget, as disclosed in the interview, are ways to circumvent Art. 6(1) and Art. 22, section 2 of the Regulation of 2016.

As Jerzy Łańcucki writes in his article (2018, p. 13), the Polish Chamber of Insurance requests, on behalf of insurers, the changes in national legal provisions so that insurers gain the access to such large databases of their clients as the PESEL database, the CEPIK database (Central Vehicle Register). and Communication), database of the National Health Fund (NFZ), database of the Social Insurance Institution (ZUS), database of the Agricultural Social Insurance Fund (KRUS), and database of the National Labor Inspectorate (PIP). The same author points out (Łańcucki, 2018, p. 9) that companies such as Google, Apple, Facebook, Amazon ("GAFA") have a lot of information about individual clients, which can potentially be used to estimate risk or to design insurance products.

It is surprising that the supervisory authority, that is the Polish Financial Supervision Authority (KNF), does not defend the weaker side of the insurance market, that is the client, considering that some conditions set by the GDPR constitute a barrier to the development of insurers by limiting their free access to client data.

Once again, the State does not protect the weaker side of the market, i.e. the client, when it extends the implementation time of the requirements of the EU Digital Services Regulation (DSA), which entered into force in Europe on November 16, 2022. For insurers and their indirect service providers, implementation has been postponed until February 2024.

Conclusion

So far, in the scientific discussions and studies on the importance of AI applications in insurance included in the literature on the subject, identified benefits from these applications were related to individual links in the value chain. In this chapter, the subject of the study was both the benefits and costs of AI applications in insurance, related not to the value chain but to the insurance function, which is a certain element of novelty in the approach to the topic.

This new approach made it possible to diagnose, based on the opinions revealed by interview participants, that AI applications may violate the basic functions of insurance, in particular the protective function through an improved process of negative risk selection due to the criterion of its profitability for the insurer. The exclusion from the protection of entire groups of people at high risk (e.g. old people, sick people, people working in dangerous professions, etc.), for example by insurers applying high prohibitive prices for a personalized insurance product may be a subsequent effect of this process. Violations may also concern the compensation function of insurance, especially when the set of goals within the AI algorithm includes the so-called undisputed amount of compensation, which is less than the full compensation due to the client. The undisputed amount of compensation is the amount acceptable to the insurer, he has no doubts, and he is eager to pay it to the insured (or beneficiary). In order to obtain the full compensation due from insurance, the client must know first that he has so far received an amount lower (undisputed) than the amount due, second, he must know the complaints procedure against the insurer's decision, and thirdly, in response to this decision, he must take complaint actions. Violations of the client's rights may also relate to the stabilization function (financial and investment), if the insured person does not participate in the profits from investing the collected premium as part of the insurers' investment policy, or if insurers use the State Treasury bonds held in their portfolio to enact national law to their advantage, for example to introduce compulsory insurance (compulsory insurance) or subsidies (subsidies) to the price of insurance policies in the agriculture.

A function clearly enhanced by the use of AI in insurance, to the benefit of both, the client and the insurer, is the preventive function.

According to the considerations contained in the study, the most serious problem to be solved that requires further research is still the deepening information asymmetry in favor of insurers thanks to the use of AI in the business models. Insurers gain an advantage over clients thanks to the difference in the level of digital competences, the competences regarding the client data management and obtaining this data (sometimes in a controversial way), the difference in the level of knowledge and application of legal provisions, and also thanks to the weakness of state authorities in the field of consumer protection of insurance services.

Literature

Act of 22 May 2003 on compulsory insurance, the Insurance Guarantee Fund and the Polish Motor Insurers' Office [Ustawa o ubezpieczeniach obowiązkowych z dnia 22 maja 2003 r., Ubezpieczeniowym Funduszu Gwarancyjnym i Polskim Biurze Ubezpieczycieli Komunikacyjnych Dz.U. z 2003 r. Nr 124, poz.1152].

Act of December 15, 2017 on insurance distribution, Journal of Laws 2017 item 2486 [Ustawa z dnia 15 grudnia 2017 r. o dystrybucji ubezpieczeń, Dz.U. 2017 poz. 2486]

Act of September 11, 2015 on insurance and reinsurance activities [Ustawa o działalności ubezpieczeniowej i reasekuracyjnej z dnia 11 września 2015 r. Dz.U. z 2015 r. poz. 1804 z późniejszymi zmianami]

Act on insurance and reinsurance activities Journal of Laws 2023.0.656, i.e. - Act of September 11, 2015 on insurance and reinsurance activities [Ustawa o działalności ubezpieczeniowej i reasekuracyjnej. Dz.U.2023.0.656 t.j. - Ustawa z dnia 11 września 2015 r.].

Akerlof, A. G. (1970). The Market for 'Lemons': Quality Uncertainty and the Market Mechanism. *The Quarterly Journal of Economics*, 84(3), 488–500. https://doi.org/10.2307/1879431

Communication from the Commission to the European Parliament, the Council, the European Central Bank, the European Economic and Social Committee and the Committee of the Regions. *FinTech Action Plan: For more competitive and innovative European financial sector* (Brussels, 8.03.2018, COM (2018) 109 final)

Digitization of the insurance sector in Poland [Cyfryzacja sektora ubezpieczeń w Polsce]. Raport przygotowany przez Accenture we współpracy z Polską Izbą Ubezpieczeń pod Honorowym Patronatem Ministerstwa Cyfryzacji RP, Warszawa listopad 2018.

European Commission, Brussels, 10.1.2017, COM(2017) 10 final 2017/0003 (COD). Proposal for a regulation of the European Parliament and of the Council concerning the respect for private life and the protection of personal data in electronic communications and repealing Directive 2002/58/EC (Regulation on Privacy and Electronic Communications) [Rozporządzenie Parlamentu Europejskiego i Rady w sprawie poszanowania życia prywatnego oraz ochrony danych osobowych w łączności elektronicznej, uchylające Dyrektywę 2002/58/WE - rozporządzenie w sprawie prywatności i łączności elektronicznej (COM, 2017 (10))].

European Commission Brussels, 8.3.2018, COM(2018) 109 final, Communication from the Commission to the European Parliament, The Council, The European Central Bank, The European Economic and Social Committee and the Committee of The regions, Brussels, 8/03/2018, COM (2018) 109 final;

Faggini, F., Corazza, M., & Zaccaroni, G. (2020). Artificial Intelligence and Insurance: A Game-Theoretic Approach to Fraud Detection. *European Journal of Operational Research*, 283(3), 1083–1095.

Felten, E. W., Raj, M., & Seamans, R. (2023). How will Language Modelers like ChatGPT Affect Occupations and Industries? *https://arxiv.org.* https://doi.org/10.48550/arXiv.2303.01157.

Łańcucki, J. (2018). Customer in the Digital Insurance Market [Klient na cyfrowym rynku ubezpieczeniowym]. *Prawo Reasekuracyjne,* 2(95), 3–14.

Li, J., Birge, J. R., & Haugh, M. (2020). The Impact of Machine Learning on Asymmetric Information in Insurance Markets. *Operations Research,* 68(6), 1648–1666.

Liu, X., Liu, Y., & Wang, W. (2020). Artificial Intelligence and Insurance: A Systematic Review and Future Research Agenda. *Journal of Risk and Insurance,* 87(4), 1005–1048.

Maas, P. (2019). Artificial Intelligence and Insurance: The Impact of Machine Learning on Risk Selection. *Journal of Risk and Insurance,* 86(1), 5–37.

Opinion of the European Economic and Social Committee on the 'Digitalisation and innovative business models in the European financial sector, impact on employment and customers' (own-initiative opinion). (2017/C 246/02). Cyfryzacja i innowacyjne modele biznesowe w europejskim sektorze finansowym – skutki dla zatrudnienia i klientów, Opinia Europejskiego Komitetu Ekonomiczno-Społecznego, Dz. Urz.UE C 246 z 28.07.2017

Proposal for a REGULATION OF THE EUROPEAN PARLIAMENT AND OF THE COUNCIL concerning the respect for private life and the protection of personal data in electronic communications and repealing Directive 2002/58/EC (Regulation on Privacy and Electronic Communications). COM/2017/010 final -2017/03 (COD).

Proposal for a Regulation of the European Parliament and of the Council of 21 April 2021 laying down harmonised rules on artificial intelligence (Artificial Intelligence Act) and amending certain legislative acts of the Union (Proposal for a Regulation laying down harmonised rules on artificial intelligence -Artificial Intelligence Act - and amending certain legislative acts of the Union). COM (2021) 206 final -2021/04 (COD).

Regulations (EU) 2016/679 of The European Parliament and of The Council of 27 April 2016 on the protection of natural persons with regard to the processing of personal data and on the free movement of such data, and repealing Directive 95/46/EC (General Data Protection Regulation) [Rozporządzenie Parlamentu Europejskiego i Rady (UE) 2016/679 z dnia 27 kwietnia 2016 r. w sprawie ochrony osób fizycznych w związku z przetwarzaniem danych osobowych i w sprawie swobodnego przepływu takich danych oraz uchylenia dyrektywy 95/46/WE (ogólne rozporządzenie o ochronie danych), Dz.U. UE L 119 z 4.05.2016].

RESOLUTION NO. 196 OF THE COUNCIL OF MINISTERS of December 28, Policy for the Development of Artificial Intelligence in Poland from 2020, 66–74.

Salant, Y., & Sweeting, A. (2020). Artificial Intelligence and Insurance: A Market Experiment. *Journal of Political Economy*, 128(8), 3049–3084.

Salchegger, M., Wiedenmann, Th., & Widemann, Ph. (2019). *Risk and compliance implications of AI in the insurance indus*try. Deloitte Touche Tohmatsu Limited. https://www2.deloitte.com

23 Collaborative implementation framework for building trust in software robots

Case of Posti Group

Damian Kedziora, Artur Modliński, Kari Smolander, Wojciech Piotrowicz and Rehan Syed

Introduction

Robotic Process Automation (RPA) is a technological innovation of colossal growth and business hype that has recently impacted digital transformations at hundreds of organisations. Understood as a low-code piece of software developed to automatically perform tasks previously done by human workers (Kortesalmi et al., 2023), it can be used for improving operational efficiency, risk management, compliance, and quality of work, as it may be implemented and integrated fast with already existing systems (Syed et al., 2020). Software robots can also get integrated with elements of other technologies (such as task or process mining, optical character recognition, machine learning, and natural language processing), evolving into intelligent automation (IA) or cognitive automation (CA) solutions (Fantina, 2022). By transforming corporate culture with RPA, only an effective change management strategy can shift the mindset of employees, helping to bridge the gap between software robots and business (Siemon and Kedziora, 2023). As genuine worries of job losses might likely arise among the staff. To mitigate the risk of negative RPA perception among employees, efficient collaboration between management, employees, and the implementing unit is critical. It is therefore crucial for companies piloting RPA to concentrate on sustainable transition management, allowing them to focus on customer value co-creation, at the same time transforming job profiles from monotonous into more creative, value-adding, and challenging (Kedziora and Kiviranta, 2018).

The motivation for our study comes from the fact that regardless of the multiple benefits RPA brings to organisations, there are also trust-associated risks (Enríquez et al., 2020). Modliński et al. (Modlinski et al., 2022) discovered that companies that robotised their operational tasks may experience a situation when they must discontinue in-operation robots and return to manual processing. Lamberton et al. (Lamberton et al., 2017) proposed ten reasons for failed RPA adoption, yet with no reference to employee sentiment, focusing on such

DOI: 10.4324/9781032627236-25

aspects as process choice, implementation methodology, infrastructural faults, financials, and ownership of robots. The systematic literature review provided by Syed et al. indicated the significant gap in a methodological approach that would increase the chance of its successful implementation (Syed et al., 2020), claiming that 'guideline or best practices for benefits realisation of RPA deployment (from adoption to delivery) rarely exists' (p. 4). Smids et al. (Smids et al., 2020) reflected on how employees treat robots they work with, and recommended further exploring the question if robots can be valuable colleagues at work. Seiffer et al. recommended a deeper exploration of the understanding of employees' behaviours from the perspective of consequences for RPA implementations (Seiffer et al., 2021). Considering the indicated gaps and recommendations, the starting point for our exploratory research was the following question:

RQ: How to implement RPA technology from the perspective of employee adoption and trust?

Research design

Our study adopted an interpretative paradigm by collecting individual experiences and trying to find meanings that represent their social reality (Miller, 2005). We used non-normative/descriptive method to avoid unnecessary valuation by applying retrospective-praxeography with interviews and secondary data analysis (Czarniawska-Joerges, 2014). We applied the approach developed by Mol and Hatty (2005), starting with analysing the company's training outcome and meeting protocol documents from RPA's introduction, followed by unstructured interviews that got transcribed and independently analysed by researchers to increase its objectivity. The authors manually coded the studied narrations based on the research scope and aggregated them into 'cause and effect' sequences.

As for the case company selection, we used purposive sampling (Polkinghorne, 2005) with three main principles. First, we sought an information-rich case (Patton, 1990) in an industry where worker jobs are vulnerable to automation. Second, we investigated a knowledge-work organisation considering automation solutions. Thirdly, we sought a company that had not terminated any robots, indicating that the case company had successfully automated some manual tasks. As reflexive narratives can enhance employee experiences in organisations (Boncori and Smith, 2019), our research focused on implementing the first software robot at the case organisation.

When it comes to our case organisation, Posti Group is a leading postal and logistics company in Finland, established in the 16th century and currently operating in ten countries of Nordic, Central, and Eastern Europe. The RPA initiative at Posti was started in the summer of 2017 by evaluation and selection of vendors and by building an Intelligent Automation Centre of Excellence (IA CoE). Operational efficiency, quality, lead time, and cost reduction were its key drivers. Our data collection included online surveys, interviews, internal documents, and non-participant observation at Posti's premises. It took place during the first half of 2020, with the survey participants (P1–P7), and interviewees (P8–12) listed below:

1 P1, Specialist, 16 years of total experience, one year at Posti, Payroll Unit
2 P2, Process Owner, three years of total experience, three years at Posti, Finance Unit
3 P3, Specialist, 27 years of total experience, eight years at Posti, HR Unit
4 P4, Process Owner, seven years of total experience, three years at Posti, Sales/Channel Unit
5 P5, Manager, 17 years of total experience, 15 years at Posti, HR Unit
6 P6, Specialist, 20 years of total experience, eight years at Posti, Admin Unit
7 P7, Development Manager, 12 years of total experience, three years at Posti, Finance Unit
8 P8, Head of Intelligent Automation, located in Helsinki, Finland
9 P9, RTR Finance Process Owner, Finance Development, located in Tallin, Estonia
10 P10, RTR Specialist, Finance and Sourcing Service Area, located in Tallin, Estonia
11 P11, RTR Team Lead, Finance and Sourcing Service Area, located in Tallin, Estonia
12 P12, RTR Specialist, Finance and Sourcing Service Area, located in Tallin, Estonia

Introducing robots as a cross-team, collaborative process

The Accounting Department (AD), a multinational team of people with expertise in finance and accounting, was the first unit at Posti, where RPA was introduced. AD was selected to implement the first robot voluntarily, as the hallmark of work at this department was mundane, time-consuming, routine tasks. Before the final decision about RPA's adoption was made, due diligence with an external consulting company was conducted. As a result, the robotics team (RT) was founded as part of the IA CoE, which employed specialists in business process improvement, IT, and RPA. In contrast to the cases of RPA's adoption discussed in previous sections, Posti decided to adopt RPA with their implementation design, based on the collaboration and knowledge exchange between all agents involved at each step of the process.

Actors involved

There were three actors involved in automating the first task at Posti: (1) Delivery Team (DT), (2) Robotic Team (RT), and (3) Software Robots (SR). DT possessed subject knowledge on accounting processes. The primary responsibilities of DT were to propose and select tasks that may get automated, provide descriptions of the task, support RT in prototyping, and suggest changes that may later improve the robot's performance. RT represented experts in RPA who were supposed to teach DT about the possibilities and scope of RPA, develop a prototype of a robot, run implementations, and fix errors. Surprisingly, we found that our interviewees perceived robots as actors in this process, attributing their roles typical for teammates (humans): P4: *We thought of robots together, and there are no errors anymore. We are really happy to have these employees (robots) in our team.* P3: *The robot was like my student. I still was checking it. I am glad to see that everything is done. I was proud of the robot because I am always eager to learn new things*. These findings follow both the Computers Are Social Actors (CASA) and Ikea theories. CASA claims that people use the same heuristics with computers while talking to humans, empowering machines that may lead to higher trust and performance (Nass and Moon, 2000). The Ikea effect claims that when people (co)create something, they are more attached to its outcomes (Norton et al., 2012).

Implementation framework

Introducing RPA at Posti consisted of six stages. In the first five stages, two actors were involved: DT and RT. At the last stage, as a result of cross-team collaboration, a new actor appeared: a software robot. Interestingly, all methodologies of RPA adoption discussed in the previous section followed the top-down approach, managers decided which tasks should be robotised. At Posti, the managers allowed employees to choose the task to be robotised (bottom-up) and run the implementation collaboratively.

Stage I: Brainstorming

Starting with an orientation meeting organised by the board of managers, employees of each department got acquainted with RPA strategic vision and goals of Posti. Employees learned how robots work, which types of tasks have automation potential, and what benefits such change may bring. Next, employees were asked to discuss how their tasks may have been automated in their teams. Based on the automation criteria provided by RT, DT was asked to write down tasks they perceive as (1) simple and (2) annoying. Then, DT brainstormed and decided which task should get robotised in the first place. Interestingly, the criteria were not fixed regarding the cost or economic benefits of automation. The scepticism and fear of robots stealing jobs had appeared in the team at this stage. Posti employees called it 'Pre-robot fear.' At this stage, the central role of management was to cope with this fear by (1) organising clarification meetings, (2) sending brochures about the benefits of RPA for individual workers, and (3) inviting people from other companies or departments where people may freely exchange experience and clear doubts. In contrast to earlier methodologies, the management board does not decide on which tasks should be robotised but invites mid-level managers to board meetings where the intentions are presented. Mid-level managers are responsible for organising the meeting within their teams, presenting the concept of RPA, and discussing whether their teammates perceive any of their tasks as suitable for automation. At Posti, the accounting department started with an open discussion between workers and the manager: P9: *People were proposing something automated or done with a robot. It was a joint discussion regarding this (the first task to be robotised).* When DT proposes tasks to be robotised, RT was asked for task assessment on both (1) benefits that robotised tasks may bring and (2) functional check whether tasks can get robotised with RPA technology. The DT brainstorming and RT evaluation led to the selection of a task that is acceptable for automation by both teams. Its important part was to obtain the consent of all team members, which was crucial to overcoming the 'Pre-robot fear'. Nevertheless, such task selection may last quite long, especially when team members have doubts and dilemmas. At Posti, the most common criteria for task selection were: (1) perceived simplicity, (2) perceived annoyance the task brings, and (3) time consumption. The simpler, more annoying, and more time-consuming the task, the stronger the intention to robotise it.

Stage II: Task description

In the second stage, DT collaborates with RT to collect as many details about the chosen task as possible. To facilitate the knowledge exchange, stage two was divided into two substages: (1) Providing a description, and (2) Clarifying the scope. The second stage starts with a meeting between DT and RT. DT is asked to provide details about the task selected, incl. who was involved in performing the task manually so far, how people did it, possible difficulties, performance, and place of the task at the Posti value chain. When RT believes they understand both tasks and their context, they prepare their flow map. When it is ready, RT presents it to DT to check if their understanding is correct and explains how the task will change when the robot is introduced. RT is expecting feedback to improve their map. When both teams have no more comments, robot implementation enters the third stage. At this stage, the organisation already has a description of the task to be automated, its map, and its delivery context.

Stage III: Prototyping

RT prepares the robot's prototype based on the formerly collected documents. The third stage consists of two substages: (1) programming the robot and (2) preparing the robot for testing. The case of Posti shows that at this stage, the dilemma and doubts regarding the functionality of the robot may still appear.

The programming is based on the formerly collected documentation. We found that the map acts as a reference when RT configures the bot and then gets stored as a key document for future changes, as the processes at Posti may change over time. Programming is the most technically advanced stage, based mainly on the competence of RT. It finishes when RT believes the robot is ready to work. When the robot is ready, RT prepares the first instructions for DT, which may be used during both the testing and implementation stages. Instructions are stored for all in one easily accessible document. It is crucial that instructions are transparent for non-IT workers to avoid situations when DT contacts RT regularly, not understanding the errors and discrepancies. The case of Posti suggests that RT should avoid sharing the instructions before the Testing stage, as documents where the context of automation is not explained during face-to-face meetings may raise confusion and stress among DT.

Stage IV: Testing

When both the robot and instructions for DT are ready, the fourth stage begins with an alignment meeting. This stage is focused on error detection, robot improvement, and mutual learning. It consists of two iterative substages: (1) performing testing and (2) improving the robot. Performing testing is initiated with the cross-team meeting where both the robot and instruction draft is introduced to DT. RT tests the robot together with DT, acquainting them with the possibilities the robot offers. We found that instructions at Posti may have various forms: written text, pictures, and/or slides. At this stage, the second wave of fear among DT appears. Posti calls it 'Pre-test fear' as people are afraid that they will not manage to learn how to work with the robot. This is also the moment when employees realise that technology may replace them. After the DT tests the robot's performance, RT collects the errors, complaints, and inconsistencies identified by DT. In the case of Posti, the iteration of this stage consisted of four elements: (1) testing the robot by DT, (2) collecting feedback by RT, (3) improving the robot by RT, (4) explaining the changes to DT. The process ends when both DT and RT recognise the robot as usable in the field. During the 'Improving the robot' substage, RT continues its technical work to eliminate the shortcomings identified earlier.

Stage V: Collaborative deployment

The critical assumption of Posti's methodology is the collaborative deployment of robots. At the fifth stage, both DT and RT are asked to fix a date when a robot will be implemented. Interestingly, all team members (DT and RT) should be in the office on that day for improved communication. The case of Posti suggests that on the first day, DT had plenty of questions, and they work together with RT to dispel doubts and problems as fast as possible. As the first errors appear at this stage, they may negatively impact company performance if not corrected immediately. During the first days, DT meets with RT at the end of the working day to sum up the challenges they face and provide stories linked to robot performance. It is worth mentioning that a moment when RT leaves DT is critical for the robot's implementation, as stress and fear in DT arise for the third time. Posti's employees call it 'Use fear' and link it to self-empowerment psychology; DT is already integrated with RT and may be afraid that they will not control the situation fast enough to be alone. According to our research, one of the most common solutions used to decrease this wave of fear is the 'red button'. "Red button" is a possibility the company offers to DT when an emergency appears to call RPA experts from Robotic Teams when they are already delegated to other tasks. Each team member of DT should have easy access to RT members (for instance, by Skype or some internal communication tool).

Stage VI: Performance improvement

At the final stage, RT helps DT improve the robot's performance. It is divided into two substages: (1) adjustment and (2) maintenance. When critical errors are eliminated, non-routine/unusual errors may still occur, and both teams must collaborate to further adjust robots. The new robot may produce errors even months after its implementation, and both DT and RT should be prepared for such scenarios, both for non-routine and random events and changes in the internal processes of Posti. Adopting robots into organisational space demands rethinking all processes and critical events that people face on their own in the past. In contrast to common hopes, RPA does not solve all problems without human assistance. Contrary to common stereotypes, robots need constant attention, improvements, and changes. They evolve together with the organisation and team itself and must be adapted to the changing environment. P11: *Now (after two years from implementation), we still do some statistics and follow what robots are doing, checking for possible improvements (their performance).* Robots may become the tool that encourages collaboration between employees in DT and across teams. People 'teaching' robots in the beginning suggested that they feel more attached to the robot, team, and organisation. Despite initial fears and a lack of trust towards the robots, the collaborative approach to robot implementation allowed the team to overcome some negative attitudes and sow optimism concerning the prospects of new robots being implemented in their organisation.

Stages and actions of RPA implementation framework

Introducing the robot to Posti consisted of six main stages taken by DT, RT, and SR. Interestingly, the implementation of software robots at Posti were designed with a bottom-up approach, since managers allowed the employees to choose the task to be robotised and design the robot collaboratively (Table 23.1).

Table 23.1 CRI (Collaborative Robot Implementation) framework

Stage	*Substage*	*Action*	*Challenges linked to fear*
Brainstorming	Discussion among team members	– presenting RPA assumptions – identifying development team (DT) – writing down the tasks that DT wish to robotised	**First wave of fear (Pre-robot fear)**
	Tasks' selection	– selecting a pool of tasks that may be robotised – evaluating proposed tasks by RT – choosing a task that has the highest potential to be robotised	
Task description	Providing task's description	– describing the task – mapping the task	
	Clarify the scope of the process	fixing the interdependencies between the task and internal processes	
Prototype	Programming robot	constructing robot	
	Preparing for testing	preparing instruction for DT	
Testing	Performing testing	– presenting robot to DT – clearing up the doubts of DT – collecting shortcomings	**Second wave of fear (Pre-test fear)**
	Improving	– reconstructing robot – preparing new instructions	
Collaborative implementation	Collaborating implementation	– teaching DT to work with robots – collecting shortcomings	**Third wave of fear (fear of use)**
Performance improvement	Adjustment	adjusting robot to changes	
	Maintenance	integrating the team around the robot	

Conclusions

Our research discovered that involving all employees was crucial to facilitating collaboration and understanding of RPA among all team members. This was particularly important, as trust and attitude towards the bots, as well as relationships between engaged stakeholders have evolved over time. As Posti's case uncovered, RPA implementation may start with a worry that robots will entirely replace people in the workforce so trust in technology before the implementation may be low. Discussions with managers and collaborative meetings convinced employees that RPA may be a company-wide opportunity. Our research contributes by focusing on employee perspectives rather than company benefits when implementing such technology. Our findings show that Posti's strategy was bottom-up and involved employees from the start. In earlier studies, it was explained that automation may support organisation (Cooper et al., 2022), yet our case suggests that this relationship is two-sided: organisation can also support automation (Kedziora et al., 2021). Our next theoretical contribution is the concept of '3 waves of fear' (3WoF) to explain three different types of bias and fear at various stages of RPA implementation. The fear of mass automation has been discussed by Spencer (2018), yet our findings reflect on the evolution of fears along the implementation process. The RPA case of Posti suggests to practitioners that employees may display concerns towards software robots, yet collaborative efforts can resolve them and build trust around technology. Moreover, the concept of 'red button', being the possibility to call RPA experts from Robotic Teams in critical situations, creates a sense of comfort among workers observing the daily performance of RPA. Another novelty extending the available body of knowledge is the approach of Posti to RPA drivers and automation case selection. Posti employees noticed the opportunity brought by RPA: to remove time-consuming, repetitive, and manual tasks. Kedziora et al. (2021) indicated that companies select tasks for automation based on pure commercial aspects, such as cost reductions and quality improvements. Managers may forecast the Return on Investment (ROI) of each implementation, considering financial benefits, yet our study builds upon the available theory with new perspectives: focusing on how employees perceive particular tasks, how much they are involved in it, what is the context of those tasks and their place in the entire value chain, and effectively whether employees approve to pass this task along to the bot.

In summary, our paper offers three key contributions to the theory on RPA implementation, where its successful adoption is proven by no later need to terminate or discontinue robots (Modlinski et al., 2022). Firstly, it modified earlier work that looked at implementation stages, with actions, actors, and behaviours identified for each stage. Secondly, our research uncovered three waves of fear among employees at different stages of implementation. It is a novel contribution to the formerly studied RPA adoption failures, such as the ten reasons listed by Lamberton et al. (2017). In our case, the fears were reduced as implementation progressed and trust was built. Finally, our study presents a novel approach to implementation drivers, countering the view of top-down implementation, stressing the need for collaboration and involvement of employees from the early stages, and allowing a bottom-up approach.

Our findings also offered practical implications: despite possible differences in organisational cultures, the need for collaborative development of bots may reduce the risk of RPA failure, while implementation stages could be used as starting point for implementation design. As for the limitations, our research was focused on the early stages of RPA's adoption and limited to one case only. In the future, our objective is to scale this research and compare the results with other experiences and cases. Hence, it is recommended to further explore the applicability of our findings with a larger sample, within more companies and sectors. Moreover, the authors encourage applying the discovered guidelines against implementations of other low-code technologies at multinational companies, as well as their positioning in organisational strategy, as some goals and behaviours might differ due to contextual, cultural, and organisational conditions.

References

Boncori, I., & Smith, C. (2019). I lost my baby today: Embodied writing and learning in organizations. *Management Learning, 50*(1), 74–86.

Cooper, L. A., Holderness, D. K., Sorensen, T. L., & Wood, D. A. (2022). Perceptions of robotic process automation in Big 4 public accounting firms: Do firm leaders and lower-level employees agree? *Journal of Emerging Technologies in Accounting, 19*(1), 33–51.

Czarniawska-Joerges, B. (2014). *A Theory of Organizing* (2nd ed.). Edward Elgar, Cheltenham.

Enríquez, J. G., Ramirez, A. J., Mayo, F. J. D., & García-García, J. A. (2020). Robotic process automation: A scientific and industrial systematic mapping study. *IEEE Access, 8*, 39113–39129.

Fantina, R. (2022). *Introducing Robotic Process Automation to Your Organization: A Guide for Business Leaders*. Apress, Berkeley, CA.

Kedziora, D., & Kiviranta, H.-M. (2018). Digital business value creation with robotic process automation (RPA) in northern and central Europe. *Management 13*(2), 161–174.

Kedziora, D., Leivonen, A., Piotrowicz, W., & Öörni, A. (2021). Robotic process automation (RPA) implementation drivers: Evidence of selected Nordic companies. *Issues in Information Systems*, 22(2), 21–40.

Kortesalmi, H., Aunimo, L., & Kedziora, D. (2023). RPA experiments in SMEs through a collaborative network. In L. M. Camarinha-Matos, X. Boucher, & A. Ortiz (Eds.), *Collaborative Networks in Digitalization and Society 5.0. PRO-VE 2023. IFIP Advances in Information and Communication Technology,* 688. Springer, Cham.

Lamberton, C., Brigo, D., & Hoy, D. (2017). Impact of robotics, RPA, and AI on the insurance industry: Challenges and opportunities. *Journal of Financial Perspectives, 4*, 8–20.

Miller, K. (2005). Interpretive perspectives on theory development. *Communication Theories: Perspectives, Processes, and Contexts*, McGraw-Hill, London.

Modlinski, A., Kedziora, D., Jiménez-Ramírez, A., & del-Río-Ortega, A. (2022). Rolling back to manual work: An exploratory research on robotic process re-manualization. In A. Marrella, R. Matulevicius, R. Gabryelczyk, B. Axmann, V. Bosilj-Vuksic, W. Gaaloul, M. I. Stemberger, A. Ko, & Q. Lu (Eds.), *Lecture Notes in Business Information Processing, Volume 459*, 154–169. Springer, Münster.

Mol, A., & Hatty, S. E. (2005). The body multiple: Ontology in medical practice. *Journal of Contemporary Ethnography, 34*(1), 101–112.

Nass, C., & Moon, Y. (2000). Machines and mindlessness: Social responses to computers. *Journal of Social Issues, 56*(1), 81–103.

Norton, M. I., Mochon, D., & Ariely, D. (2012). The IKEA effect: When labor leads to love. *Journal of Consumer Psychology, 22*(3), 453–460.

Patton, M. Q. (1990). *Qualitative Evaluation and Research Methods*. SAGE Publications, Inc., Newbury Park, CA.

Polkinghorne, D. E. (2005). Language and meaning: Data collection in qualitative research. *Journal of Counseling Psychology, 52*(2), 137.

Seiffer, A., Gnewuch, U., Maedche, A. (2021). Understanding employee responses to software robots: a systematic literature review. In: *ICIS 2021 Proceedings*, Austin, TX, 20–24 September 2021. Available at: https://aisel.aisnet.org/icis2021/is_future_work/is_future_work/5

Siemon, D., & Kedziora, D. (2023). From accountant to software developer – Transforming employees with robotic process automation (RPA). *Proceedings of the 56th Hawaii International Conference on System Sciences (HICSS),* pp. 4244–4253.

Smids, J., Nyholm, S., & Berkers, H. (2020). Robots in the workplace: A threat to – or opportunity for – meaningful work? *Philosophy & Technology, 33*(3), 503–522.

Spencer, D. A. (2018). Fear and hope in an age of mass automation: Debating the future of work. *New Technology, Work, and Employment, 33*(1), 1–12.

Syed, R., Suriadi, S., Adams, M., Bandara, W., Leemans, S. J. J., Ouyang, C., ter Hofstede, A. H. M., van de Weerd, I., Wynn, M. T., & Reijers, H. A. (2020). Robotic process automation: Contemporary themes and challenges. *Computers in Industry, 115*, 103162.

24 Unpacking the black box

Exploring the intersection of trust and machine learning

Caner Asbaş, H. Cenk Sözen and Şule Erdem Tuzlukaya

Introduction

The effort to bring the abilities of analyzing, synthesizing, interpreting, inferring, thinking, and evaluating human beings to computers and information and communication technology products in an artificial and virtual way is known as artificial intelligence (Araujo, Helberger, Kruikemeier & de Vreese, 2020; Gillath et al., 2021). Studied applications include computer vision, language translations, machine vision, chat robots, voice assistants, autonomous driving, smart homes and smart production systems, and recommendation robots. Moreover, methodologies heavily influenced by fundamental mathematical and geometric approaches, computational algorithms, and computer technologies often employ qualitative methods in areas such as natural language processing. Unlike traditional paradigms of the past, artificial intelligence applications operate adaptively, incorporating various feedback loops during their performance to achieve more accurate and successful computation results. Such an approach gives artificial intelligence the ability to learn from the data fed into its operations processes, which emphasizes the transfer of human learning ability and behavior to virtual systems (Gillath et al., 2021). At this point, artificial intelligence refers to the virtual systems' mimicry of human intelligence, whereas machine learning refers to the technology and algorithms used during this mimicry and transfer, though they are sometimes used interchangeably or separately to express the same meaning. As human roles are increasingly taken over by cyber-physical systems that develop unique behaviors through machine learning, a variety of novel trust issues have come to the forefront for discussion (Araujo et al., 2020; Asan, Bayrak & Choudhury, 2020; Winfield & Jirotka, 2018).

The source of these issues can be linked to a variety of factors, including informatics dominated by machine learning. These factors encompass systems devoid of emotion, solely reliant on information and data, as well as systems requiring trust because they develop knowledge and attitudes based on predefined machine learning algorithms (Ryan, 2020). Furthermore, unlike humans, who can make instant and detailed decisions based on real-time observations and

DOI: 10.4324/9781032627236-26

their own set of attitudes, robotic and automated systems cannot replicate the emotional and social cooperative behaviors.

As machine learning becomes increasingly common in our daily lives, it's important to consider the implications of this technology on trust. Machine learning algorithms are described as "black boxes" because to the fact that they can be challenging to understand, giving users little insight into how decisions are being made. This lack of transparency can reduce trust in the technology, especially when it comes to sensitive areas such as healthcare, finance, and law enforcement (Asan et al., 2020; Felzmann, Villaronga, Lutz & Tamo-Larrieux, 2019).

This chapter will investigate the relationship between trust and machine learning, focusing on the elements that contribute to trust in this technology and the potential results of a lack of trust. Additionally, we will consider the role that human-centered design and ethical considerations might help create machine learning systems that are transparent and trustworthy. By exploring the intersection of trust and machine learning, we pave the way for future, where these technologies are not only trusted but also reliable and ultimately beneficial for society at large.

Artificial intelligence paradigm

Three distinct approaches can be employed to delve deeper into the conceptual domain of artificial intelligence and engage in a more comprehensive discussion of the topic. As one of the initial approaches, analytical methods can typically be grouped into two distinct categories (Asjad, Alam & Hasan, 2018). These methods rely on fundamental mathematical concepts, follow linear paths to produce results, and lack feedback mechanisms to adjust outcomes. Examples of traditional methods include statistics for interpreting large datasets, regression for observing variable relationships, linear algebra for solving systems, and calculus for studying continuity and integral equations.

According to the methodological approach, artificial intelligence methods constitute the second category. These methods benefit from numerical algorithms, advanced mathematical models, and computational techniques beside basic mathematical approaches. Artificial intelligence methods also apply an adaptive approach that includes some feedback loops or correction methods to adjust their outcomes with respect to the realized or expected results in order to achieve higher success rates. This adaptive approach provides artificial intelligence methods with the ability to learn through their operations on the given data. These processes of artificial intelligence methods, and in some cases directly themselves, are referred to as machine leaning, to emphasize their ability to be trained. On the other hand, while artificial intelligence refers to the transfer of human thinking, learning and problem-solving competencies to information and communication systems and simulations there, machine learning actually points out the algorithms and approaches implemented in artificial intelligence technologies.

In addition to aforementioned numerical analysis methods, some qualitative analysis methods have been also developed in the literature as such discourse analysis, semantic analysis, narrative analysis, grammatical analysis, thematic analysis, and content analysis, These analyses, which aims to interpret verbal and nonnumerical data generally, can also be effective in study sphere of quantitative analysis in some cases by enabling the formation of a multidisciplinary approach in artificial intelligence applications such as natural language processing.

Various machine learning algorithms have been defined and developed in the literature with different advantages and superiorities for artificial intelligence applications, and these developments still continue to constitute an important research topic. Support Vector Machine (SVM), Decision Trees, k-Nearest Neighbors, k-Means, Artificial Neural Network (ANN), Principal Component Analysis (PCA), and Convolutional Neural Network (CNN) can be listed examples of these different algorithms that are studied and applied in the literature most frequently. In the methodological approach, artificial intelligence phenomenon is considered as a mathematics-based method with highly advanced techniques, ignoring its philosophical and logical foundations.

The second approach that can be applied to analyze the conceptual position of artificial intelligence is the epistemological perspective (Wheeler & Pereira, 2004). In this approach, the transfer of human thinking abilities and learning abilities to computers and all other information and communication technology-related devices can be given as the definition of artificial intelligence. From this perspective, the first black box in artificial intelligence stems from the opacity offered by most machine learning models (Carabantes, 2019). The opacity of artificial intelligence stems from its predominantly closed-system operations. An additional matter of concern delves into the nature and source of knowledge, a fundamental subject within the domain of epistemology. The concept of artificial intelligence, where a "machine" can learn and acquire knowledge, raises questions about the nature, source and epistemological underpinnings of this knowledge. However, what sets this concept apart is that entity acquiring knowledge is not human being or a human consciousness This circumstance calls for specific adjustments to traditional epistemological frameworks, as McCarthy (1981) has pointed out.

The third approach is to examine conceptual position of artificial intelligence with the philosophy of logic. In this context, the concept of logical truth points out to propositions that are true, independently from the truth or falsehood of the propositions that constitute them. (Quine, 1986). The accuracy of artificial intelligence is thus measured by its logicality, in other words, its success rate. Thus, in terms of trust, the fact that artificial intelligence is logical, in other words, its mathematical foundations are established in a correct and functioning way, leads us to trust artificial intelligence. It is easily possible to argue that this approach has a direct and close connection with the methodological approach described above.

Artificial intelligence finds extensive applications in both everyday life and organizational-industrial settings. Among its fundamental and widely adopted uses, one of the most common is classification tasks. In these applications, artificial intelligence is used to recognize the differences between specified features and classify provided samples into different categories accordingly. Classification applications through artificial intelligence are implemented for a variety of purposes, from segmenting company customers to identifying faulty mass production parts in manufacturing. Another artificial intelligence application is pattern recognition. Regularities and similar structures in data sets are called patterns. (Aggarwal & Subbian, 2014; Albaity & Mahmood, 2022; Ali & Mustafa, 2023; Barron-Estrada, Zatarain-Cabada, Romero-Polo & Monroy, 2021). Pattern recognition can be defined as the automatic capture of these regularities and repetitive or similar structures in datasets and networks. The integration of artificial intelligence methods into pattern recognition applications, and their use often in combination with signal processing applications, has enabled important practice, including many security applications such as facial recognition and fingerprint matching. One of the application areas of artificial intelligence is also link prediction. With link prediction, especially in networks, links that are unknown or are likely to be formed but have not yet been formed are tried to be estimated. (Clauset, Moore & Newman, 2008; Dunlavy, Kolda & Acar, 2011; Haghani & Keyvanpour, 2017; Liben-Nowell & Kleinberg, 2007; Lü & Zhou, 2011). In connection with link prediction and classification, artificial intelligence has also enabled the development of recommendation systems. Recommendation systems can benefit various areas, from recommending products on shopping websites to connecting potential friends on social media or offering personalized content like movies, music and TV series on streaming platforms. As mentioned above, another important artificial intelligence application is natural language processing. Translators, chatbots and search engines can be designed by natural language processing methods and principles. The fact that natural language processing applications generally include human interaction at higher levels makes this area an important subject of investigation for trust. One of the relatively new areas where artificial intelligence has the opportunity to be applied is cybersecurity. By using applications such as pattern recognition, classification and link prediction in cyberthreat identification, proactive methods have been developed in addition to traditional cyber-defense tools, and in some cases, even overall cybersecurity management can be left to artificial intelligence. An other application of artificial intelligence that is gaining importance is time series analysis. The use of artificial intelligence and artificial intelligence-supported decision systems is increasing, especially in the field of finance, in transactions such as stock market, bonds and exchange rate forecasts and in making investment decisions, and the decisions taken in this field are increasingly transferred to artificial intelligence. Another artificial intelligence application that may be related to the issue of trust is autonomous vehicles. For many vehicles including trains, cars and airplanes, artificial intelligence is now a support system for drivers in some cases, and sometimes it directly replaces the driver and performs the driving itself, excluding human factor totally.

Trust

Throughout human history, trust has been a basic idea, and several examples from early cultures attest to its significance (Mayer, Davis & Schoorman, 1995; Paliszkiewicz, Goluchowski & Chen, 2022). For instance, trust has been pivotal in shaping human societies. In ancient Mesopotamia, the Code of Ur-Nammu established trust through written laws. Ancient Egypt relied on trust among laborers and rulers for grand construction projects like the pyramids. Ancient Greece embraced trust in democratic governance, and Confucianism in ancient China emphasized virtuous conduct and trustworthiness. In the Roman Republic, trust was vital for governance, while the Silk Road's success depended on trust among diverse traders. These historical examples illustrate the enduring significance of trust in human interactions, societies, and economies (Rousseau, Sitkin, Burt & Camerer, 1998).

The historical development of trust in social sciences literature has also evolved through various phases. In its early roots, trust was explored philosophically, from ancient Greece to Enlightenment thinkers like Locke and Hobbes. The 19th century marked the emergence of sociology, delving into trust's role in societal order, notably through Emile Durkheim's work. The 20th century saw trust analyzed in economics and game theory by scholars like Arrow and Nash, highlighting its significance in economic transactions and cooperation. Trust later extended to institutional contexts, influencing political behavior and organizational dynamics.

As the 21st century dawned, trust took on new dimensions. The digital age brought concerns about online trust and data privacy. Technology-driven research explored trust in artificial intelligence and specifically autonomous systems. Globalization posed questions about trust in international relations, and crises such as the financial meltdown and COVID-19 pandemic raised issues of trust during uncertainty. In parallel, trust dynamics in social media and digital communities became a critical focus. The evolution of trust in social sciences literature reflects its adaptability to shifting societal, economic, and technological landscapes, from its philosophical origins to its multifaceted 21st century manifestations.

According to Guerrero-Cusumano (2023) trust can be used to minimize the entropy of a society. Entropy is a measure of disorder and trust could make business transaction, political activities, news, and media less complex. Entropy can be used as a direct measure of complexity, where the smaller value of entropy indicates lower system complexity, whereas its larger value indicates higher system complexity.

New management paradigms have been developed by the digital economy to take advantage of opportunities and address problems, although much of their effectiveness depends on mutual trust between the parties or on the underlying technology. As a result, contextual needs, enabling factors, and "actors" in commercial or social interactions all contribute to the very complex and multidimensional phenomenon known as digital trust (Paliszkiewicz & Chen, 2023).

The foundation of trust is not just technology, but a variety of factors. The phenomena of trust are transdisciplinary and is influenced by both quantifiable and intangible factors. Trust involves the belief that someone will carefully consider your unique perspective and evaluate any issue related to you before arriving at a judgment or conclusion. It is closely connected to how the individual positions the person they trust, encompassing intangible elements and evolving over time. The establishment of trust relies on the initial impression it creates, especially when it relates to matters of significant importance to one or more parties. In such instances, trust can endure and be attributed to a wide range of behaviors exhibited by the trusted party. According to Kozuch (2022) due to the complexity of this fundamental human phenomena, there are many different ways to understand and define trust. According to one perspective, having faith in someone or something is a behavioral choice made by the individual. Trust is viewed in this context as an intentional decision involving the evaluation of dependability and reliability. From another perspective, choosing, evaluating, or preferring to rely on someone or something is seen as a cognitive activity occurring within. When it comes to giving trust, it highlights the person's reasoning and decision-making. Additionally, the terms "trust" and "trustworthiness" are interchangeable, suggesting that trust is acquired over time through a pattern of reliable behavior. This point of view emphasizes how vital behavior and moral character are to developing and preserving confidence. Integral to all human relationships, trust is based on traits like honesty, dependability, openness, and vulnerability. It is essential to cooperation, collaboration, and general well-being in interpersonal interactions, business collaborations, and societal organizations.

A multifaceted idea, trust in machine learning includes the technology itself, the data it uses, and the judgments it makes. Reliability, openness, fairness, ethics, and other relevant challenges must be addressed in order to establish and retain trust in artificial intelligence systems. In order for machine learning to be responsibly and ethically used across a variety of sectors as technology develops, trust-building research and practices will be essential.

Recently, the definition of trust had to change once more. The growing usage of social media, artificial intelligence, machine learning, and recommendation systems was one of the change-causing elements. Even routine daily tasks were increasingly controlled by machines and computers, necessitating interaction not only with other system and network users but also with automata, software, intelligent agents, and complex IT systems (Dziczkowski, Głowania & Zacny, 2022).

Trust is very hard to measure, that makes it vital also. Trust should be considered a key feature of anything involving software. Research indicates that loss of trust can have extremely significant consequences. Security and confidentiality are important factors, and process responsibilities should be determined internally and externally.

Few technological concepts have so thoroughly captured our curiosity and imagination as machine learning. Its capacity to sort through enormous datasets, reveal hidden patterns, and produce forecasts has not only transformed entire industries, but also fundamentally changed how we make decisions and consider the potential of automation. This extraordinary capacity for transformation has been accompanied by a recurrent conundrum, though: trust. How can we put our faith in algorithms' judgments, especially when they act as mysterious "black boxes"?

Trusting the machine

As we negotiate the complex realm of machine learning, the idea of trust assumes a central role in the ever-changing technological landscape. This mysterious field has transformed numerous businesses, altered how we make decisions, and ushered in a new era of automation. However, it raises serious concerns regarding the dependability of algorithms, particularly when they function as mysterious "black boxes."

Navigating the data symphony

Data has taken on the role of the heartbeat of our interconnected society in this age of digital revolution. It moves across our systems, devices, and applications, arranging a complicated symphony of data. The subtitle alludes to the dynamic nature of data, which isn't just static but continually changing and influencing choices, discoveries, and experiences. Understanding and using the power of data is crucial for organizations, researchers, and individuals alike, just as a good director may lead an orchestra to make melodic tunes. It serves as a reminder of the opportunities and difficulties associated with organizing, analyzing, and interpreting data in a society where knowledge is the basis of advancement.

Several crucial factors are at the core of machine learning trust. Data quality is, first and foremost, the foundation upon which trust is formed. It is crucial that training data are accurate; models that use biased, imperfect, or noisy data may produce results that are skewed in addition to being incorrect. A crucial component in building confidence is ensuring data quality through careful data quality assurance procedures.

Additionally, trust covers data privacy, a crucial issue. It is not just a legal responsibility, but also an ethical requirement, to protect sensitive data. In order to preserve trust in how data is handled and used, privacy-preserving mechanisms like differential privacy must be used.

The accuracy of the judgments and suggestions that machine learning generates is at the heart of how much faith to place in it. Three important factors are included in this. First and foremost, it is critical to combat bias and advance fairness in algorithms. Due to the propensity of machine learning algorithms to inherit biases from their training data, results may occasionally be both unfair and discriminating. Second, trust is fundamentally based on the question of accountability. It is crucial to recognize who is accountable for the choices made by machine learning algorithms, including responsibility for any biases or inaccuracies included in forecasts. Finally, ethical issues play a significant role in the area of trust.

Fostering human engagement

To increase confidence in machine learning, the human factor is crucial. It is not only a technological problem; it is fundamentally a human problem since it involves how people perceive and interact with these systems. User trust depends on making sure that users have a thorough understanding of the capabilities and limitations of machine learning systems, which encourages confidence in their use. Additionally, a key component in building trust is the inclusion of user feedback in iterative model improvement procedures. This includes setting up efficient channels for users to communicate any problems or worries they have, resulting in a dynamic feedback loop that encourages ongoing improvement and dependability in machine learning systems.

Conclusion

The relationship between machine learning and trust is a very complex and highly dynamic field that requires a deep investigation in a wide range of areas such as data quality for artificial intelligence applications such as natural language processing, data privacy for artificial intelligence applications such as pattern recognition and cybersecurity and trust in algorithms, accuracy for artificial intelligence applications such as time series analysis, consistency For artificial intelligence applications such as classification, link prediction and recommendation systems and elimination or reduction of the human factor for artificial intelligence applications such as autonomous vehicles and cybersecurity, in terms of methodological, epistemological and logical philosophy. On the other hand, the fact that machine learning constitutes "black boxes" makes this examination and its real-life application much more challenging and difficult to understand. The increase in studies focusing on artificial intelligence in the psychological, organizational, philosophical and logical fields and the spread of an interdisciplinary understanding including computer science and mathematics are of great importance at this point.

References

Aggarwal, C., & Subbian, K. (2014). Evolutionary network analysis. *ACM Computing Surveys*, *47*(1), 1–36. https://doi.org/10.1145/2601412

Albaity, M., & Mahmood, T. (2022). Medical diagnosis and pattern recognition based on generalized dice similarity measures for managing intuitionistic hesitant fuzzy information. *Mathematics*, *10*(15), 2815. https://doi.org/10.3390/math10152815

Ali, L. A., & Mustafa, F. A. (2023). Mosque typo-morphological classification for pattern recognition using shape grammar theory and graph-based techniques. *Buildings*, *13*(3), 741. https://doi.org/10.3390/buildings13030741

Araujo, T., Helberger, N., Kruikemeier, S., & de Vreese, C. H. (2020). In AI we trust? Perceptions about automated decision-making by Artificial Intelligence. *AI & Society*, *35*(3), 611–623. https://doi.org/10.1007/s00146-019-00931-w

Asan, O., Bayrak, A. E., & Choudhury, A. (2020). Artificial Intelligence and human trust in healthcare: Focus on clinicians. *Journal of Medical Internet Research*, *22*(6), e15154. https://doi.org/10.2196/15154

Asjad, M., Alam, A., & Hasan, F. (2018). A comparative study of classifier techniques for lift index data analysis. *Benchmarking: An International Journal*, *25*(2), 632–641. https://doi.org/10.1108/bij-09-2016-0137

Barron-Estrada, M. L., Zatarain-Cabada, R., Romero-Polo, J. A., & Monroy, J. N. (2021). Patrony: A mobile application for pattern recognition learning. *Education and Information Technologies*, *27*(1), 1237–1260. https://doi.org/10.1007/s10639-021-10636-7

Carabantes, M. (2019). Black-Box Artificial Intelligence: An epistemological and critical analysis. *AI & Society*, *35*(2), 309–317. https://doi.org/10.1007/s00146-019-00888-w

Clauset, A., Moore, C., & Newman, M. E. (2008). Hierarchical structure and the prediction of missing links in networks. *Nature*, *453*(7191), 98–101. https://doi.org/10.1038/nature06830

Dunlavy, D. M., Kolda, T. G., & Acar, E. (2011). Temporal link prediction using matrix and tensor factorizations. *ACM Transactions on Knowledge Discovery from Data*, *5*(2), 1–27. https://doi.org/10.1145/1921632.1921636

Dziczkowski, G., Głowania, S., & Zacny, B. (2022). Trust in machine-learning systems. In J. Paliszkiewicz & K. Chen (Eds.), *Trust, Organizations and the Digital Economy: Theory and Practice* (pp. 108–121). Routledge.

Felzmann, H., Villaronga, E. F., Lutz, C., & Tamo-Larrieux, A. (2019). Transparency you can trust: Transparency requirements for artificial intelligence between legal norms and contextual concerns. *Big Data & Society*, *6*(1), 205395171986054. https://doi.org/10.1177/2053951719860542

Guerrero-Cusumano, J. L. (2023). Foreword: The evolution of trust. In J. Paliszkiewicz, K. Chen & M. Launer Trust (Eds.), *Trust and Digital Business: Theory and Practice* (pp. xxv–xxviii). Routledge.

Gillath, O., Ai, T., Branicky, M. S., Keshmiri, S., Davison, R. B., & Spaulding, R. (2021). Attachment and trust in Artificial Intelligence. *Computers in Human Behavior*, *115*, 106607. https://doi.org/10.1016/j.chb.2020.106607

Haghani, S., & Keyvanpour, M. R. (2017). A systemic analysis of link prediction in social network. *Artificial Intelligence Review*, *52*(3), 1961–1995. https://doi.org/10.1007/s10462-017-9590-2

Kozuch, B. (2022). The dimensions of trust in the digital era. In J. Paliszkiewicz and K. Chen (Eds.), *Trust, Organizations and the Digital Economy: Theory and Practice* (pp. 16–26). Routledge.

Liben-Nowell, D., & Kleinberg, J. (2007). The link-prediction problem for social networks. *Journal of the American Society for Information Science and Technology, 58*(7), 1019–1031. https://doi.org/10.1002/asi.20591

Lü, L., & Zhou, T. (2011). Link prediction in complex networks: A survey. *Physica A: Statistical Mechanics and Its Applications, 390*(6), 1150–1170. https://doi.org/10.1016/j.physa.2010.11.027

Mayer, R. C., Davis, J. H., & Schoorman, F. D. (1995). An integrative model of organizational trust. *The Academy of Management Review, 20*(3), 709–734. https://doi.org/10.2307/258792

McCarthy, J. (1981). Epistemological problems of artificial intelligence. *Readings in Artificial Intelligence*, 459–465. https://doi.org/10.1016/b978-0-934613-03-3.50035-0

Paliszkiewicz, J., & Chen, K. (2023). Building digital trust in business. In J. Paliszkiewicz, K. Chen & M. Launer Trust (Eds.), *Trust and Digital Business: Theory and Practice* (pp. 3–12). Routledge.

Paliszkiewicz, J., Goluchowski, J., & Chen, K. (2022). The meaning and interdisciplinary nature of trust in the digital economy – Future directions. In J. Paliszkiewicz & K. Chen (Eds.), *Trust, Organizations and the Digital Economy: Theory and Practice* (pp. 3–15). Routledge.

Quine, W. V. (1986). *Philosophy of Logic* (2nd Edition). Harvard University Press.

Rousseau, D. M., Sitkin, S. B., Burt, R. S., & Camerer, C. (1998). Not so different after all: A cross-discipline view of trust. *Academy of Management Review, 23*(3), 393–404. https://doi.org/10.5465/amr.1998.926617

Ryan, M. (2020). In AI we trust: Ethics, artificial intelligence, and reliability. *Science and Engineering Ethics, 26*(5), 2749–2767. https://doi.org/10.1007/s11948-020-00228-y

Wheeler, G. R., & Pereira, L. M. (2004). Epistemology and artificial intelligence. *Journal of Applied Logic, 2*(4), 469–493. https://doi.org/10.1016/j.jal.2004.07.007

Winfield, A. F., & Jirotka, M. (2018). Ethical governance is essential to building trust in robotics and artificial intelligence systems. *Philosophical Transactions of the Royal Society A: Mathematical, Physical and Engineering Sciences, 376*(2133), 20180085. https://doi.org/10.1098/rsta.2018.0085

Index